Olive Oil Sensory Science

Edited by

Erminio Monteleone

Dipartimento di Gestione dei Sistemi Agrari, Alimentari e Forestali (GESAAF), University of Florence (Università degli Studi di Firenze), Florence, Italy

Susan Langstaff

Applied Sensory, LLC, Fairfield, California, USA

WILEY Blackwell

This edition first published 2014 © 2014 by John Wiley & Sons, Ltd

Registered office: John Wiley & Sons, Ltd, The Atrium, Southern Gate, Chichester, West Sussex, PO19 8SQ, UK

Editorial offices: 9600 Garsington Road, Oxford, OX4 2DQ, UK
The Atrium, Southern Gate, Chichester, West Sussex, PO19 8SQ, UK
111 River Street, Hoboken, NJ 07030-5774, USA

For details of our global editorial offices, for customer services and for information about how to apply for permission to reuse the copyright material in this book please see our website at www.wiley.com/wiley-blackwell.

Library of Congress Cataloging-in-Publication Data

Monteleone, Erminio.
Olive oil sensory science / Erminio Monteleone and Susan Langstaff.
pages cm
Includes index.
ISBN 978-1-118-33252-8 (cloth)
1. Olive oil. 2. Olive oil industry. 3. Olive oil–Analysis. 4. Food–Sensory analysis.
I. Langstaff, Susan. II. Title.
TP683.M577 2013
664′.362–dc23

2013024199

A catalogue record for this book is available from the British Library.

Wiley also publishes its books in a variety of electronic formats. Some content that appears in print may not be available in electronic books.

Cover illustration by www.cargocollective.com/hannaherrera
Cover design by www.hisandhersdesign.co.uk

Set in 10.5/12.5pt Times Ten by Aptara Inc., New Delhi, India

1 2014

Contents

Part II

List of Contributors

Efi Christopoulou, Chemist, Taster of VOO, external expert of EU, IOC, and former Ministry of Development expert, Athens, Greece

Anna Claret, Researcher, IRTA, Food Technology Centre, Monells, Spain

Claudia Delgado, Department of Food Science and Technology, University of California, Davis, CA, USA. Current: Science Leader – Sensory and Statistics, Gustatec Division, Chromocell Corporation, North Brunswick, NJ, USA

Caterina Dinnella, Senior Researcher, Dipartimento di Gestione dei Sistemi Agrari, Alimentari e Forestali – GESAAF, University of Florence, Florence, Italy

Margaret Edwards, Owner, Matiatia Grove/The Waiheke Olive Oil Company Ltd, Oneroa, Waiheke Island, New Zealand

Sonia Esposto, Researcher, Dipartimento di Scienze Economiche Estimative e degli Alimenti (DSEEA), University of Perugia, Perugia, Italy

Aurora Gómez-Rico, Department of Food Science and Technology, University of California, Davis, CA, USA. Current: AINIA, Parque Tecnológico de Valencia, Paterna Valencia, Spain

Luis Guerrero, Researcher, IRTA, Food Technology Centre, Monells, Spain

Jean-Xavier Guinard, Professor, Department of Food Science and Technology, University of California, Davis, CA, USA

Alexandra Kicenik Devarenne, Owner, CalAthena, Petaluma, CA, USA

Kostas Kontothanasis, Journalist, Olive and Olive Oil Magazine, correspondent in Kalamata, Greece

Susan Langstaff, Owner, Applied Sensory LLC, Fairfield, CA, USA

Susana Mattar, Researcher, CRESA: Laboratorio de Análisis Sensorial de Aceite de Oliva y Alimentos, Facultad de Ciencias de la Alimentación, Universidad Católica de Cuyo, Rivadavia, San Juan, Argentina

Marzia Migliorini, Head of Research Division, Metropoli, Azienda Speciale della Camera di Commercio di Firenze – Divisione Laboratorio Chimico Merceologico, Florence, Italy

Erminio Monteleone, Associate Professor, Dipartimento di Gestione dei Sistemi Agrari, Alimentari e Forestali (GESAAF), University of Florence (Università degli Studi di Firenze), Florence, Italy

Claudio Peri, Professor Emeritus, University of Milan, Milan, Italy

Leandro Ravetti, Technical Director, Modern Olives/Boundary Bend Ltd, Lara, Victoria, Australia

Annamaria Recchia, Postdoctoral Fellow, Dipartimento di Gestione dei Sistemi Agrari, Alimentari e Forestali (GESAAF), Università degli Studi di Firenze, Florence, Italy

Agustí Romero, Researcher, IRTA, Mas de Bover, Olive Production, Oil Technology and Nuts, Constantí, Spain

Metta Santosa, Department of Food Science and Technology, University of California, Davis, CA, USA. Current: Sensory Scientist and Consultant, Hayward, CA, USA

Maurizio Servili, Professor, Dipartimento di Scienze Economiche Estimative e degli Alimenti (DSEEA), University of Perugia, Perugia, Italy

Sara Spinelli, Consultant in Communication, Sensory, and Consumer Research, SemioSensory, Carmignano – Prato, Italy

Agnese Taticchi, Researcher, Dipartimento di Scienze Economiche Estimative e degli Alimenti (DSEEA), University of Perugia, Perugia, Italy

Hely Tuorila, Professor, Department of Food and Environmental Sciences, University of Helsinki, Helsinki, Finland

Adriana Turcato, Directora Área Extensión, CRESA: Laboratorio de Análisis Sensorial de Aceite de Oliva y Alimentos, Facultad de Ciencias de la Alimentación, Universidad Católica de Cuyo, Rivadavia, San Juan, Argentina

Vassilis Zampounis, Agroeconomist, Axion Agro Ltd, Athens, Greece

Olive Oil Sensory Science: an Overview

Erminio Monteleone[1] and Susan Langstaff[2]
[1]*Dipartimento di Gestione dei Sistemi Agrari, Alimentari e Forestali (GESAAF), University of Florence (Università degli Studi di Firenze), Florence, Italy*
[2]*Applied Sensory LLC, Fairfield, CA, USA*

According to data from the International Olive Council, the olive oil market is increasing in a global dimension. Consumption and production are particularly growing in "*new*" areas outside the Mediterranean region and new needs in product optimization and development as well as in marketing strategies are emerging (Delgado and Guinard, 2011). For instance, in the United States, interest in and consumption of olive oil have been growing exponentially during the past 20 years (an increase of 228%). Both traditional and new olive oil consumers are interested in olive oil for two main reasons: health benefits and flavor.

The sensory properties of extra virgin olive oil (EVOO) vary widely depending on a number of factors. Differences in genetic resources, environmental conditions, process specifications, and local know-how induce sensory differences among oils (Caporale *et al.*, 2006). The success of olive oils in the market can be built on these differences and on the skill of producers in controlling the critical factors of sensory quality and in interpreting consumer hedonic and sensory expectations.

Currently, most of the attention on sensory properties of olive oil is focused on how to evaluate whether a given oil is free of defects and how EVOO is qualified. It is known that International Olive Council standards for the sensory evaluation of oils represent an effective method to qualify oils in categories such as Extra Virgin or Virgin. These standards consist in evaluating both "positive" and "negative" attributes. The latter are the category of defects that have to be absent in an EVOO. Positive sensory attributes are Bitterness, Pungency, and Fruity notes. It should be underlined that this necessary evaluation is not sufficient to describe the sensory diversity among EVOOs.

Modern product development and hard competition within the oil–food industry require a clear understanding of sensory aspects of EVOOs, and adequate sensory and consumer research techniques. The use of sound sensory methods is rarely applied in olive oil production and in olive oil consumer studies. There is a clear need for information on when and how to apply them. From an operative point of view, sensory evaluation is essential to understanding the marketplace since it provides information on how to manipulate the sensory properties or the culinary use of products in relation to consumers' hedonic responses.

Analytical sensory evaluation serves to identify perceived product attributes and differences. In this capacity, its function is similar to the chemical, physical, and microbiological characterization of products. However, being closer to consumer perceptions than physicochemical and biological assays, sensory analysis helps in interpreting the consequences of process modifications and the culinary use or the sensory functionality of EVOOs. Also, by using methodologies applicable to consumers, sensory research can directly acquire information about consumer response to products (Tuorila and Monteleone 2009). The need to combine the information on product characteristics and consumer behavior in product development has led sensory scientists to adopt methods from other scientific disciplines that can be successfully applied to investigating consumer appreciation of EVOOs.

Sensory evaluation of olive oils requires many skills, but this is not always appreciated. In fact, sensory data may appear easy to collect because human subjects respond to questions and perform tasks, even when a task is obscure or inappropriate. This creates an insidious potential for collecting data of poor validity (Tuorila and Monteleone 2009).

This book, *Olive Oil Sensory Science*, is aimed at covering the lack of information on "why," "how," and "when" to apply conventional and new sensory methods in relation to the specific properties of olive oils. Thus its topic is the potential contribution of sensory science to the success of EVOOs in the market. Appropriate sensory methods for product optimization/development and consumer testing purposes are presented. For each method, procedures and tools necessary to guarantee the quality of data are reported. Furthermore, the book considers methods suitable to investigate the sensory functionality of olive oils and multidisciplinary approaches to communicate the culinary use of this product.

Three introductory chapters precede the description of the sensory methods and methodologies. In the first, the quality dimensions of EVOO are critically reviewed. The concept of excellence in olive oil production, including chain control and certification, is presented and in this vision, sensory profile and health components are proposed as characterizing elements of brands based on origin, cultivar, and processing conditions. The second chapter is dedicated to agronomic and technological aspects of production that affect sensory properties and their occurrence in olive oil. Finally, the third chapter introduces the reader to the importance of sensory perception in determining consumer choice of olive oil. The authors clearly and specifically explain why robust sensory and consumer studies are needed to guarantee the success of EVOOs in the market. They highlight the importance of identifying the existing and forthcoming segments of consumers who are inclined to like and use EVOOs, and of providing information (on sensory properties and culinary uses) to support specific interest in the product. Given the special nature and prestige of EVOOs among consumers, segmentation based on involvement, familiarity with and knowledge of EVOOs is expected, and instruments measuring these aspects are described.

The core of the book is represented by four chapters dedicated to the application of sensory methods and approaches in quality control, product development, and consumer studies. The chapter on sensory quality control reviews the background and development of the official International Olive Council methods for olive oil sensory quality evaluation and categorization. Chapter 5 presents sensory methods used in product optimization. A large part of this contribution is dedicated to definitions and reference standards of sensory descriptors of olive oils. Statistical tools used to validate both panel and assessor performance and to analyze the data are described in detail, offering the olive oil world the opportunity to know more about how to interpret sensory data. Particular relevance is given to the appropriate methods used to obtain the so-called perceptual maps, pictorial representations describing differences and similarities among a set of EVOOs.

In addition to the conventional Descriptive Analysis (DA) and the Free Choice Profiling (FCP), other recent sensory methods, such as sorting and napping, are presented, considering their specific applications to olive oils. Perceptual mapping based on descriptive data is proposed as an effective tool when the description of the sensory style of an oil is of interest. The term "style" is referred to the sensory profile that describes an oil (or a group of oils) as different from others. The description of the sensory profile of oils is suggested as a necessary step to link the world of production with the world of use and consumption. In relation to the production system, the sensory profile of an oil represents a product specification describing characteristics due to cultivar type, climatic conditions, and operational process conditions; it also represents the product specifications necessary to verify the producer's capability to control the critical factors affecting the sensory characteristics and to guarantee the consistency of the sensory profile of an oil over time. In relation to the use and consumption of oils, the sensory profile is necessary to the communication of the sensory style and the culinary use of a product; it is also necessary to study the sensory functionality of an oil in dish preparations and oil–food pairings and to understand and interpret consumer likes and dislikes. Furthermore, methods that analyze temporal aspects of sensory characteristics of olive oils such as Time Intensity (TI) and Temporal Dominance of Sensations (TDS) are presented. The latter method is presented in Chapter 7 as a reference method to study the sensory functionality of EVOOs. Current methods used to classify EVOOs into sensory quality categories and to test consumer preferences involve evaluations of oils on their own and thus do not consider the conditions that consumers experience when consuming olive oil in real life (i.e., mixed with other food) (Dinnella *et al.* 2012). EVOOs are never consumed in isolation – they are used as an ingredient in the preparation of dishes or paired with other food, hence they should not be judged for their sensory attributes *per se*, but for their sensory functionality in combinations. In this book, the sensory functionality is defined as the ability of an EVOO to modify the sensory profile and acceptability of a given oil–food combination. Hence the study of sensory functionality of EVOOs requires the collection of both analytical (DA and TDS) and affective responses. Possible approaches to investigate the sensory performance of EVOOs are proposed and the many factors involved are identified and described.

Chapter 6 reports the methods appropriate to understanding the perceptions of and preferences for olive oil, and the knowledge and attitudes of consumers towards olive oil. It describes a possible approach to the study of consumer behavior as it pertains to olive oil. A three-factor consumer research model is proposed. The model relates consumer behavior to three types of variables – product variables, consumer variables, and context variables. Product variables are all the characteristics of the product, such as price, brand image, package, label information, and sensory properties, among others. Consumer variables are mostly demographics and psychographics, but also include anatomy, physiology, and genetics. Context variables include the physical environment of consumption but also the nonphysical context (e.g., the location of consumption, the presence or absence of others, the political, economic, and social climate, and societal trends and pressures). The consumption behavior measures that can be modeled with this model are variables such as preferences, liking, purchase intent, repeat purchase, and satisfaction, among others. Both qualitative and quantitative consumer research methods are presented, such as focus group, means–end chain analysis, sorting, and preference mapping, and a range of multivariate statistical techniques necessary to interpret consumer responses are described.

Current ways to communicate EVOO sensory quality are unable to differentiate among products. Information on the sensory properties of oils is confusing because general terms (such as "robust," "fragrant," "delicate," and "gentle") are often ambiguous. On the other hand, even when sensory information is correctly given (e.g., green or ripe fruity), it can be very difficult for consumers to understand what olive oil producers would hope to communicate. This happens because brands usually resort to the experts' descriptive language. Although technically correct, this language can be difficult or meaningless for consumers without any previous knowledge of olive oil. In this kind of communication, consumers are invited to play the role of experts, namely of skilled tasters who have the pleasure of experimenting with different sensory properties. Very often, particularly in new markets, this approach does not work. A possible alternative is to focus the communication on the culinary use of varied sensory styles of EVOOs. Studies are required on the role that oils with different sensory profiles could have in culinary preparations, showing which "oil sensory styles" best highlight which food flavor. Chapter 8 presents possible methodological approaches in investigating the culinary use of oils by chefs in order to gather information that can be used effectively for marketing purposes. In particular, the contribution highlights the need for multidisciplinary methodologies that involve sensory and human sciences such as semiotics.

Olive oil is on an upward trajectory, globally, in terms of production and consumption. Six chapters in this book are dedicated to the sensory diversity among olive oils from around the world, considering traditional and new production countries. Chapters 9, 10, and 11 examine the major European olive-growing countries: Spain, Italy, and Greece, respectively. These "Old World" countries are also the leading consumers of olive oil and they have developed their own geographical guarantee for many of their oils based on the Appellation Contrôllée system begun in France for wine. This system attempts to ensure consistency of quality, determines which olive varieties can be grown where, and what the essential qualities of their oils should be. This approach assumes that the composition of virgin olive oils is related to the geographical area where they are produced. The belief is that the sensory and chemical characterization of the geographical origin of virgin olive oils is needed not only to protect denominations of controlled origin, but also to identify the "best" olive oil for certain consumers.

Olive growing and the making of olive oil were all but unknown outside Europe and the Mediterranean basin until the olive tree was exported by Europeans who settled in new lands over the past 500 years. Now, olive oil is a part of these "New World" countries where the climate allows the trees to prosper. The United States – California in particular – Australia, New Zealand, and many countries in South America all have healthy olive oil industries that strive to stand on their own merits. Chapters 12, 13, and 14 examine these areas as producers of olive oil as they continue to experiment and challenge accepted norms from the Old World.

In the end, what has made olive oil fascinating has been the array of different oils with different sensory properties, from different olive types, grown by different growers, in different conditions, in different corners of the planet. This book is aimed at all those working in the olive oil field: research and development scientists and food technologists with olive oil companies, olive oil producers and marketers, sensory and consumer food researchers, students of food science and technology, and culinary students. Also, those in the health and wellness communities will be interested because of the health aspects of olive oil.

The twenty-first century consumer in affluent countries is sophisticated and savvy and is ready to spend money on products that meet his or her needs. Olive oil companies need to understand their products better and be able to explain their products to these consumers using the sensory techniques presented in this long overdue book.

References

Caporale, G., Policastro, S., Carlucci, A., and Monteleone, E. (2006). Consumer's expectations for sensory properties in virgin olive oils. *Food Quality and Preference*, **17**, 116–125.

Delgado, C., and Guinard, J. (2011). How do consumer hedonic ratings for extra-virgin olive oil relate to quality ratings by experts and descriptive analysis ratings? *Food Quality and Preference*, **22**, 213–225.

Dinnella, C., Masi, C., Zoboli, G., Monteleone, E. (2012). Sensory functionality of extra-virgin olive oil in vegetable foods assessed by Temporal Dominance of Sensations and Descriptive Analysis. *Food Quality and Preference*, **26**, 141–150.

Tuorila, H. and Monteleone, E. (2009). Sensory food science in the changing society: opportunities, needs, and challenges. *Trends in Food Science and Technology*, **20**, 54–62.

Part I

1
Quality Excellence in Extra Virgin Olive Oils

Claudio Peri
University of Milan, Milan, Italy

1.1 Introduction

The aim of this chapter is to present the standards of excellence for extra virgin olive oil and the critical factors in the production and marketing process. The topics are presented in three parts (Table 1.1):

Table 1.1 Topics covered in this chapter.

Part 1 – The standards of excellent olive oil	1.1. Genuineness as the prerequisite to excellence 1.2. Product traceability from field to table 1.3. The chemical standards of excellent olive oils 1.4. Sensory standards I: absence of sensory defects 1.5. Sensory standards II: sensory characteristics and performances 1.6. Nutritional and health properties 1.7. Conclusion: the standards of excellent olive oil
Part 2 – The control of critical processing parameters	2.1. Coordination of the harvesting and milling operations 2.2. Control of time–temperature conditions in malaxation 2.3. The problems of "hygienic design" and "residence time distribution" 2.4. Storage conditions of excellent oils
Part 3 – The marketing of excellent olive oils	3.1. Conditions and opportunities for successfully marketing excellent olive oils

Olive Oil Sensory Science, First Edition. Edited by Erminio Monteleone and Susan Langstaff.

- Part 1 defines the standards of excellence for extra virgin olive oil.
- Part 2 presents the problems of process control as critical points for achieving the standards of excellence.
- Part 3 presents some conditions and opportunities for successfully marketing excellent olive oils.

1.2 Part 1. The standards of excellent olive oil

The first part is an itinerary into the main aspects of extra virgin olive oil quality, aimed at defining coherent, selective, measurable, and controllable standards of excellence.

1.2.1 Point 1. Genuineness as the prerequisite of excellence

We define genuineness as conformity with legal prescriptions or, in general, as the agreement – validated by documents and data – between what is claimed about the product and what is actually offered to the consumer. In the idiom of quality experts, an expression representing such a commitment of producers is:

- We declare what we want to do;
- We do what we declare; and
- We document what we do.

This is the foundation of producers' trustworthiness and hence the prerequisite of excellence in olive oil.

1.2.1.1 The fundamental criterion of genuineness The law that defines extra virgin olive oil is motivated most of all by the concern to prevent fraud. It is sufficient to scan the articles and annexes of the European laws on the definitions of olive oil (European Commission, 2007) to see that the aim of a large number of parameters is to expose the fraudulent practice of mixing of virgin olive oils with refined oil or with other vegetable oils. These mixings are a violation of the fundamental criterion of genuineness that defines *virgin olive oils as oils obtained "exclusively from olive" with "purely mechanical systems."* The "purely mechanical systems" include cleaning and washing of the olives, crushing, malaxation, extraction with a decanter or by pressure or percolation, centrifugal separation, and filtration. Except for potable water, considered as an adjuvant, no additive can be used.

1.2.1.2 False claims of identity Problems arise from the fact that the marketing of extra virgin oils is based primarily on the declaration of an origin or a particular method of production, which are impossible to prove by analytical means. It is easy to understand that for an oil sold at a high price, such as Tuscan, if it is not possible to recognize the presence of an oil from Apulia or perhaps from Spain or Australia, that costs much less, the temptation to commit fraud will be very great and such frauds will consequently be very frequent.

We can list the easiest and most frequent types of false declarations:

- Declare the oil as "our own production," but sell an extra virgin oil of different origin with your own label.
- Declare an origin, but offer a product with a different origin.
- Declare as "new oil" an oil mixed with oil from the previous year.
- Declare as "organic oil" an oil mixed with nonorganic oil.
- Declare as monocultivar an oil produced from various cultivars.

These deceptions are in no way identifiable by means of analysis. The promise of experts that they are able to demonstrate analytically the authenticity of an oil should also be strongly disapproved of because it is misleading information to consumers. In fact, an analytical system can demonstrate the similarity or the difference between two samples, but can never be a crucial piece of evidence regarding an origin or a processing method if documented traceability of the oil is not available.

1.2.2 Point 2. Product traceability from field to table

The problem of genuineness raises the topic of traceability (Peri and Di Martino, 2004), which is the only effective means to demonstrate the reliability of claims concerning the origin of an oil or the application of a particular technology.

Traceability should allow a response to questions such as "what is the origin of this oil?" or "were appropriate techniques used for the denomination?" or "who or what company is responsible for producing the olives?, who for the milling?, who for the analysis?, who for the storage and packaging?" The final and most important question, which becomes essential in the case that fraud is detected, is "in case there is a defect in the product or a deception, who is accountable?"

The need to be able to respond to this last question has prompted lawmakers to propose traceability as a mandatory requirement for all consumer goods, including food.

The conclusion to this argument is that, first of all, traceability must be a traceability of responsibility.

1.2.2.1 Traceability of the chain If traceability is understood as a method to protect the consumer from fraud and risks, the only traceability that makes sense is that which is applied to the entire chain "from field to table." Fraud and loss of identity can occur, in fact, at each point in the chain and at any time during the commercial life of an oil.

There are two fundamental tools that can be used to identify the responsibility at each step of the chain:

1. Documented evidence of product identity based on uninterrupted monitoring of material flow of oil lots: their origin, identity, quantity, location, and destination (a lot is a homogeneous quantity of product; for a liquid such as olive oil, a lot is coincident with the contents of one container or the portions of the product deriving from the same container).
2. Analytical evidence of product identity based on an analytical certificate of the composition of the oil.

Of these two tools, the first is the most important: if perfect monitoring of material flow can be guaranteed, responsibilities would be perfectly defined. The analytical tool can only give supporting evidence of product identity as substantiated by the documented material flow (Peri, 2010).

A suitable combination of the two tools can give credibility to producer's and seller's claims.

1.2.3 Point 3. The chemical standards of excellent olive oils

The standards prescribed by law (European Commission, 2007) to attribute various levels of quality to virgin olive oils (extra virgin, virgin, and lamp) have a modest discriminatory ability.

The definition of extra virgin oil is based on the following limit values:

- free acidity $<0.8\%$;
- peroxide number <20 meq. of oxygen per kilogram; and
- a UV absorption index (K_{232}) <2.50.

These limits are compatible with a mediocre oil and the fact that the law suggests, actually imposes, that an oil with these standards should be defined on the label as an oil "of superior quality" seems to be a source of misunderstanding for the consumer. In fact, within these wide limits, both mediocre oils and excellent oils can qualify as extra virgin. Just the perception of this uncertainty breaks down consumer confidence, because they have no other way to orient themselves, if not by the definition of extra virgin.

An extra virgin oil that aspires to be qualified as excellent should have standards that are much more restrictive that those prescribed by law. A model of excellence was set up in the past by the 3E Association (Ethics, Excellence, Economics), a not-for-profit association founded by a group of

Italian producers with the aim of promoting the excellence of extra virgin olive oil. The present author was President from its foundation in August 2004 until December 2011. The 3E model proposes the following standards (Peri, Kicenic Devarenne, and Pinton, 2010):

- free acidity ≤ 0.3 (± 0.02);
- peroxide number ≤ 7.5 (± 0.2); and
- a UV absorption index (K_{232}) ≤ 1.85 (± 0.02).

1.2.4 Point 4. Sensory standards I: absence of sensory defects

The law requires the absence of sensory defects in extra virgin olive oil, so let us examine the effects of this requirement.

If an oil that complies with the chemical standards of extra virgin has a sensory defect, it is declassified to virgin with serious consequences for the producer because no consumer would consciously buy virgin oil if they were simultaneously offered an extra virgin oil at an acceptable price. In other words, declassification of extra virgin causes the exclusion of the oil from suitable commercial exploitation. This outcome is difficult for the producer to accept and therefore the remedy (which is worse than the defect) frequently consists in marketing as extra virgin, oils that have sensory defects and therefore should be categorized as virgin. Tests carried out in the market indicate that, in reality, many oils marketed as extra virgin, or served as extra virgin in restaurants, have sensory defects (Mueller, 2012).

Since the sensory response is the result of a statistical elaboration of individual perceptions, it is sufficient that the judges of a panel "pretend to not perceive" the defect in order to declare the oil to be extra virgin. The risk of fraud (that is, legally incorrect classification) is very high. On the other hand, it can happen that an odor or a particular taste, however natural, can be perceived as "strange," suggesting to the taster that is should be qualified as a defect, causing unjust damage to the producer. It is not surprising, therefore, that this rule generates many misunderstandings and disputes.

1.2.4.1 The causes of the problem The legislators and technicians who promoted this rule probably missed the fact that there are two types of possible causes of sensory defects:

1. An oil has a sensory defect because it comes from altered, unhealthy, poorly stored olives. Generally, when this happens, the oil not only has sensory defects, but is also outside the chemical parameters of extra virgin. In this case, the sensory analysis only confirms the data from the analytical profile.

2. An oil has a sensory defect, although coming from unaltered healthy olives and hence has good or very good chemical parameters. In this case, the defect is due to "contamination" of a good oil with a small or very small amount of a defective oil. It can happen that the oil was poured into a container that had not been adequately cleaned, or that in the mill there were dead spots in which the olive paste or oil stood for hours. In these cases, which are far from rare, the oil acquires a sensory defect by contamination of the mass of good oil with a small amount of greatly altered oil.

Whereas analytical defects can only derive from massive alterations of the oil, sensory defects can arise from point sources of contamination, which are difficult to detect.

The contradiction is obvious: on the one hand, the chemical parameters for extra virgin oil are very permissive, whereas on the other, for the sensory defects, a very rigorous standard has been adopted, to which only truly excellent oils correspond. The tensions that are generated in this contrast and the swindling that is a consequence were predictable and foreseen long ago (Peri, Garrido, and Lopez, 1993).

The solution to the problem of sensory defects involves interventions in the process to avoid contamination of excellent oils with small amounts of defective oil. This topic is developed further in point 10 (Section 1.3.3).

1.2.5 Point 5. Sensory standards II: sensory characteristics and performances

We define "sensory characteristics" as the sensations recorded by a panel of sensory judges when oil is tasted as such. These characteristics are conventionally described with general expressions, such as sensations of "bitterness" or sensations of "astringency," or for similarity to other known products, for example, the flavor of "almond" or the odor of tomato leaf. The sum of sensory characteristics is what we call a "sensory profile."

We define "sensory performance" or "sensory properties" as the sensations perceived in the dishes in which the oil is used as a condiment. They are a combination of the sensory characteristics of the oil with that of the dish itself. The sensations perceived in the oil as such can be reduced or enhanced, or combined, giving rise to new sensations, not detectable as such, either in the base food of the dish or in the oil as such.

Owing to a widespread conceptual error, sensory characteristics are confused with sensory performance. Oil, however, is a condiment and its sensory role should not be evaluated outside its combination with a food. All contests and awards that follow tasting competitions are based on a great misunderstanding. The oils that receive prizes are those that are the most "harmonic" or "balanced," neglecting the fact that the sensory descriptors of oil are completely changed on combining the oil with different foods and that the sensory characteristic of a condiment must be unbalanced and somehow excessive by definition. In fact, a very small quantity of a condiment must impart a sensory change to a dish. Would you consider salt as too salty or sugar as too sweet or hot pepper as too pungent? Or would you consider that these are precisely the characteristics that give them an interesting sensory role in a dish? The different meaning and role of sensory characteristics and sensory performances of extra virgin olive oil are the object of interesting comparisons and innovative research by Monteleone's group (Monteleone, 2010).

> ***Box 1.1***
>
> With the various competitions and awards, based on tasting the oils as such, the topic of the diversity of sensory profiles has been dramatically distorted. Rather than considering the diversity of sensory profiles as an opportunity to play with in the various combinations possible in culinary preparations, there is a fossilized idea that there is an ideal oil profile and that an oil is much better the closer it gets to that profile. Famous tasters judge olive oils as "harmonic" or "balanced" or "excellent" or "disharmonic" or "poor" as though an ideal sensory profile exists, rather than a sensory performance appropriate to the dishes. This attitude destroys the merits of biodiversity and leads to an unfortunate end – that in which all the oils of the world end up with the same sensory profile. It is another classic case of conflict of business interests with those of biodiversity.

1.2.5.1 The prejudice "my oil is better than yours" and its effects on the credibility of the sector In each region, it is said that the oil in their own region is better than oil from other regions. In regions in which there are PDO (Protected Designation of Origin) oils, it is said that one's own PDO is better than other PDOs, and obviously in Italy it is said that Italian oils are the best in the world. Not much insight is needed to understand that these expressions are wrong and that the quality of an oil depends on the ability and honesty of the producer. Poor oils are also produced in regions with great traditions and fame and excellent oils can be produced in any part of the world where it is possible to grow olives.

Above all, the effect of this prejudice is disastrous because it tends to mislead consumers. In countries that are new consumers of olive oil, primarily those in which the marketing perspectives are more interesting, such as North America, Northern Europe, Japan, and the elite markets of Asia, statements such as these end up generating suspicion and mistrust: how can one believe in a market in which everyone claims to offer the best product?

The explanation of this problem, however, cannot be attributed only to parochialism of the producers or to exaggerations in advertising. The roots of this prejudice are, in fact, inherent in the very diversity of the oils that are produced in various parts of the world. The biodiversity of the olive produces diversity of the sensory profiles and, as is well known to food habit experts, people's preferences are influenced by personal experiences of food, above all by those experienced in childhood. Hence it is perfectly justifiable that for someone born and raised throughout their entire childhood in a given region, the taste of oil recognized and perceived as excellent is the oil from their own region.

It can also be understood that this problem arises in a completely different way in countries in which olive oil is not a familiar product. In these countries, which are of increasing importance for the marketing of excellent oils, the topic of the differences in sensory profiles must be presented as an opportunity, not as a limit.

Box 1.2

The variety of sensory profiles is an extraordinary richness connected to the biodiversity of the olive and to the ability of the producers. From the prejudice "my oil is better than yours," which is false, what should be claimed, substantiated with sensory profile data, is "my oil is different from yours," in order to reach a deepening of its uses that allows one to say to the consumer, "my oil matches well with ... " This point was masterly highlighted by Greg Drescher at the BEV IV Conference with these words (Drescher, 2010): *"(...) the consumer mind-set at the moment is "how do I find a bottle of good olive oil." Of course we want to shift their attention from "good" to "excellent" but we also need to transform consumer thinking to a concept that seeks out a variety of excellent olive oils for various uses and pairing opportunities. Most consumers don't just look for one good or excellent type of wine, but rather seek several types of wine according to their meal plans (...). And flavour diversity is central to our mission to both preserve biodiversity in the olive oil world, but also to create additional complexity, intrigue, and excitement that has worked so well for the premium and super-premium worlds of wine, cheese, coffee, and more."*

An extraordinary contribution in defense of the excellence of extra virgin olive oils and of the variety of sensory profiles is a declaration of intent that was signed by prestigious organizations in a "Leadership Retreat" organized in 2008 in Florence by the Georgofili Academy and the 3E Association (Anon., 2008).

In conclusion:

1. Sensory characteristics are an extremely important element in characterizing an oil, its origin, the cultivar, and the processing conditions. This implies that a producer seeking excellence should:
   - identify the descriptors of the olfactory, gustatory, visual, tactile, and chemesthetic sensations that characterize their oil brand(s) (sensory profile) (chemesthetic sensations arise when chemical compounds activate receptors associated with other senses that mediate pain, touch, and thermal perception; chemesthetic sensations include piquancy of extra virgin olive oil, and they may partly overlap with the tactile sensation of astringency);
   - select a panel of well-trained tasters able to evaluate the sensory profile in a reliable and reproducible manner;
   - document the profile of the sensory characteristics of the oil produced each year in order to characterize the "sensory style" of their brand(s) and replicate it in subsequent years.
2. Sensory performance is the most important element of communication and guidance to the consumer, and, therefore, of marketing. This implies that producers involve chefs and gastronomic experts to search for dishes that best enhance the performance of their oil. The innovative and creative element in this area is a powerful tool for communication and marketing.

1.2.6 Point 6. Nutritional and health properties

Extra virgin olive oil is a product with extraordinary health properties connected to protecting the body from cardiovascular diseases, from some forms of tumors, including prostate, breast, and colon, and numerous other degenerative, chronic diseases. It is beyond the scope of this chapter to survey the vast scientific literature dedicated to this topic. Mention will only be made of articles in which a wealth of information and citations or critical information and comments are reported (Servili and Montedoro, 2002; Cicerale *et al.*, 2009; Pelucchi *et al.*, 2011; Visioli, 2012).

Table 1.2 summarizes the composition of extra virgin olive oil.

This framework is unusual for edible oils, especially due to the presence of a fraction in which polar and nonpolar substances coexist. Substances such as chlorophyll, esters, and other volatiles, which play an essential role in the sensory profile of the oil, are also listed in Table 1.2.

In greater detail, Figure 1.1 (Vera Lavelli, personal communication, 2002) summarizes the most important information about the main groups of healthy compounds. The arrows visually represent the stage at which the studies on the health properties of the compounds in olive oil are: from the initial

Table 1.2 Composition of extra virgin olive oil (compounds having a major health importance are in *italics*).

Component	Content (wt%)
Triglycerides (*oleic acid, linoleic acid, α-linolenic acid*)	97–99
Polyphenols *α-Tocopherol* *Squalene* *Carotenoids* *Phytosterols* *Terpenic acids* Chlorophyll Mono- and diglycerides Free fatty acids Esters and other volatile substances Water	1–3

Category of compounds	Specific features	*In vitro* and animal experiments	Epidemiological studies	Clinical evidence
Oleic acid	Common in many oils and fats, but in none at such a high concentration as in virgin or refined olive oil			
Optimum relationship ω3/ω6/ω9	In both virgin and refined olive oils			
Squalene	In fish oil, in both virgin and refined olive oils			
α-Tocopherol	In many plant products and in virgin olive oil			
Polyphenols derived from oleuropein	Exclusively in virgin olive oil			
Pool of protective antioxidant substances	Specific features of the pool of health substances in virgin olive oil			

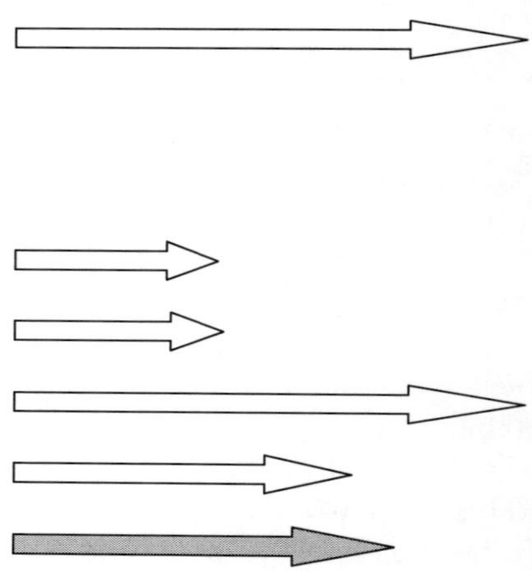

Figure 1.1 Some olive oil components with biological and health activity. *Source:* V. Lavelli, University of Milan, 2002a. Reproduced with permission of V. Lavelli.

level of *in vitro* experiments to clinical evidence obtained from experiments on humans.

As can be seen, there is clinical evidence that oleic acid and α-tocopherol are associated with health benefits, and polyphenols are also close to that level. However, the most important observation is about the benefits of consuming oil as such, due to the synergy of all the biologically active compounds. If there is a shortcoming in the attitude of specialists and in the various promotional campaigns, from which would be good to be freed, it is the tendency to emphasize some aspects rather than others or to link the health value to only certain biologically active compounds in the oil. Attempts to isolate some minor compounds from the oil in order to manufacture health pills are generally inappropriate and sometimes misleading to the consumer.

1.2.6.1 Two shining stars The aim of this final consideration is to emphasize two aspects that render olive oil a unique product and hence interesting in terms of health: the excellent framework of fatty acids that make up the triglycerides and the exclusive presence of phenolic compounds.

1.2.6.1.1 Oleic acid By far the most abundant component in olive oil is oleic acid, a monounsaturated fatty acid with 18 carbon atoms. It has a simple linear structure folded in the center due to a double bond. This is the basis of its extraordinary properties.

Table 1.3 Physical and chemical properties of fatty acids.

Fatty acid	Molecular weight	Melting point (°C)	Relative rate of oxidation
Stearic acid	284	69	–
Oleic acid	282	13	1
Linoleic acid	280	−5	64

Table 1.3 gives some properties of three fatty acids, all of which have 18 carbons atoms, but are characterized by different degrees of unsaturation: stearic acid is saturated, oleic acid is monounsaturated and linoleic acid is diunsaturated. The differences in molecular weight are very slight, but the differences in physical and chemical properties are exceptional:

- Whereas stearic acid has a melting point of 69 °C, very much higher than the normal human body temperature, which is about 37 °C, oleic acid has a melting point of 13 °C, very much lower that the human body temperature. This means that in the body, stearic acid is solid and it can easily be understood how it can increase the tendency to form deposits in arteries. Oleic acid, in contrast, is liquid and its fluidity not only reduces the tendency to form deposits in arteries, but also favors the mobility and removal of deposits of solid fatty acids. Linoleic acid has an even lower melting point and is also a liquid at body temperature.

- Linoleic acid oxidizes much more easily and rapidly than oleic acid. Its rate of oxidation is 64 times higher than that of oleic acid, which, therefore, appears much more stable.

The situation can be summarized as follows: stearic acid is very stable, but unfortunately is solid at body temperature; linoleic acid is very fluid, but unfortunately is very susceptible to oxidation with all of the associated damage. Oleic acid is a true miracle: fluid and stable, and ideal for our nutrition and for taking part in the formation of vital components in our cells.

Another point, however, should be made clear: linoleic acid is a valuable essential fatty acid, necessary for our bodies. Modern nutritional science, which has studied the significance and the role of fatty acids, has discovered that a series of polyunsaturated fatty acids with a double bond in the omega–3 and omega–6 positions carry out very important functions in the prevention of cardiovascular diseases.

Table 1.4 Distribution of polyunsaturated fatty acids.

Type	Content considered optimal by nutritionists (%)	Content in olive oil (%)
Saturated	6–8	8–14
Monounsaturated (oleic)	12–14	65–83
Polyunsaturated (omega-6, linoleic)	6–7	6–15
Polyunsaturated (omega-3, α-linolenic)	0.5–1.5	0.2–1.5

Source: Viola and Viola (2009).

Table 1.4 (Viola and Viola, 2009) shows the distribution of polyunsaturated fatty acids in olive oil. Except for the high content of oleic acid, of which we have indicated the benefits, the amount and ratio of omega–6 to omega–3 in olive oil are remarkably close to what nutritionists have indicated as optimal.

In conclusion, olive oil seems to be the lipid product best adapted to human nutritional needs, and this applies mainly to extra virgin olive oil but also to virgin and refined oil.

1.2.6.1.2 Phenolic compounds In virgin olive oil there are numerous phenolic compounds that carry out an important antioxidant activity, ensuring the preservation of the most important health benefits and sensory characteristics (Servili and Montedoro, 2002; Cicerale *et al.*, 2009; Visioli, 2012).

By way of example, Figure 1.2 shows the transformations of oleuropein, the most abundant phenolic compound in the olive. It is a bitter compound, with low solubility in water and even lower in oil. It is found in olive oil at a very low concentration (a few milligrams per gram). If sugar (glucopyranose)

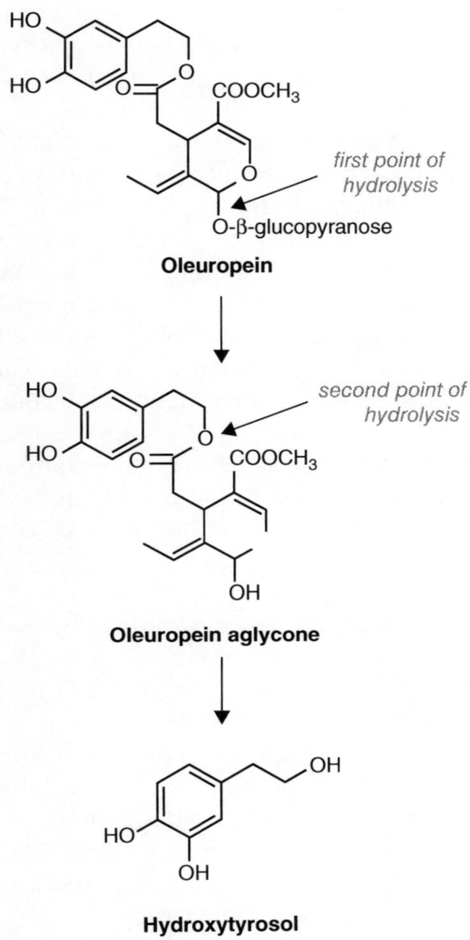

Figure 1.2 The transformation of oleuropein to hydroxytyrosol.

is removed from the oleuropein molecule by hydrolysis, oleuropein aglycone is obtained, which is less polar than oleuropein owing to the loss of the sugar moiety. As a consequence, the aglycones tend to pass more easily into the oil. The aglycone concentration can be in the range of hundreds of milligrams per kilogram, thus contributing to the bitterness and antioxidant potential of the oil. In a completely analogous way, another phenolic compound, ligstroside, which is very similar to oleuropein, is transformed into the aglycone and reaches high concentrations in the oil, also contributing to bitterness and to the antioxidant potential. These transformations are due to hydrolyzing enzymes (β-glucosidases) and they take place mainly during malaxation.

A further transformation consists in a second hydrolysis and the formation of hydroxytyrosol, a polar molecule with strong antioxidant power. This confirms the great importance of the fraction of polar compounds, which, although present in low concentrations in oil, plays a very important biological role.

1.2.6.2 Analytical evaluation of the health properties of extra virgin olive oil The information on the health properties of extra virgin olive oil is complex and not precisely definable on the basis of analytical data. It cannot be determined with certainty, as it depends on the concentration of the various health components, on the synergy that is established among them, on the reciprocal protective effect from the oxidative alterations, on the basic lipidic structure of the oil, and in particular on the ratio of the concentrations of the saturated and mono- and polyunsaturated fatty acids. Hence it is not advisable to connect the overall health benefit to the concentration of some specific constituent.

On the other hand, determination of antioxidant compounds and in particular polyphenols may be very useful for optimizing the production process.

The following points may be considered for characterizing oils in terms of health potential:

- *Oleic acid*. It is not particularly important to determine the percentage of oleic acid in extra virgin olive oil. In fact, all extra virgin olive oils have high concentrations and the possible differences due to cultivar are not discriminant in terms of quality.
- *Phenolic substances*. The evaluation of phenolic substances is certainly a characterizing element of the oil, very useful for monitoring the milling process and also as an indicator of storage stability. The simplest approach is determination of total polyphenols with a colorimetric method. A more significant framework can be obtained by a high-performance liquid chromatographic (HPLC) evaluation of the various components of the phenolic fraction and in particular of the aglycones of oleuropein and ligstroside.
- *α-Tocopherol*. α-Tocopherol is also interesting, not only for its high health value, but also as an indicator of the storage conditions of the oil. Under incorrect storage conditions and with a low concentration of polyphenolic compounds, the α-tocopherol concentration rapidly decreases. On the other hand, α-tocopherol is associated with an interesting antioxidant activity without contributing, as polyphenols do, to a bitter taste.
- Further useful information can be obtained by evaluating the total antioxidant potential with determinations based on enzymatic or chemical models (Lavelli, 2002).
- Other evaluations (sterols, carotenoids, squalene, terpenes, intermediate oxidation products of unsaturated fatty acids, etc.) suggested in research papers are not necessary in the context of process control and in characterizing the product for commercial purposes. Methods that allow the total amount of antioxidant substances and their actual state of oxidation to be compared could be interesting (Rovellini *et al.*, 2010).

Knowledge of the wealth of antioxidants that characterize one's own oil is, together with the definition of the sensory profile, the most useful information both for optimizing the production process and for the consumer.

1.2.7 Point 7. Conclusion: the suggested standards of excellent extra virgin olive oil

Table 1.5, Table 1.6, and Table 1.7 present the standards of excellent extra virgin olive oil, which derive from the previous six points. They are the fruit of the experience of the 3E Association and of successive experiments in various production and marketing conditions. They have been the object of discussion and approval by important scientific and professional organizations in particular during the five international conferences "Beyond Extra Virgin," devoted to excellence in olive oil (Anon., 2007–2011).

1.2.7.1 A worldwide network The only way to create a marketing segment of excellence in olive oil is to establish a network of companies adopting the same definitions and scheme of excellence on a worldwide basis. Table 1.5 and Table 1.6 suggest the six parameters that could be adopted as the common concept and scheme of excellence in olive oil. It is interesting that the five parameters in Table 1.5 (traceability, free acidity, peroxide number, UV absorption index, and sensory defects) require conformity to very selective standards, whereas the sixth parameter in Table 1.6 (the sensory profile based on sensorial characteristics) has only a documenting and not a selecting purpose. In other words, we do not propose a standard or a model of a sensory profile: we simply recommend that the sensory profile be available as an

Table 1.5 Base requirements, common to all excellent olive oils independent of origin, cultivar, and production techniques.

Chain traceability	Documentation of material balances from field to table
Chain traceability is documented evidence of the producer's transparency, a prerequisite of excellence	
Free acidity	$\leq$0.3 ($\pm$0.02)
Peroxide number	$\leq$7.5 ($\pm$0.2)
K_{232}	$\leq$1.85 ($\pm$0.02)
Chemical data document the good quality of olives and effective control of processing conditions	
Sensory defects	Absent according to the official method of analysis
Absence of sensory defects documents the good quality of olives and perfect control of processing conditions	

Table 1.6 Requirements depending on origin, cultivar, and the producer's know-how (the brand "style").

Sensory characteristics	An "analytical" sensory profile of olfactory, gustatory, visual, tactile, and chemesthetic sensations
Consistency and conformity to a characteristic sensory style are an essential feature of all excellent extra virgin olive oils	

essential piece of information about the oil and as a reference of its particular sensory style.

The two parameters listed in Table 1.7 (sensory performance and nutritional and health properties) are not a mandatory requirement of excellence. They can be adopted by companies to characterize further the particular composition and performances of a brand in connection with specific consumer and marketing targets.

We have tried to spread this definition of excellence in the olive oil world through several meetings of the International Conference "Beyond Extra Virgin," successfully involving scientific and educational organizations such as several universities in different countries, the Georgofili Academy in Florence, and the Culinary Institute of America in Napa Valley (California) (Anon., 2007–2011). Unfortunately, the business world has not followed suit. So far, a deep-rooted tendency towards individual approaches and regional or local patriotism has prevented the establishment of a common network of people and companies to pursue excellence in the olive oil sector. The consequence is confusion of the levels based only on the current definition of extra virgin olive oil and on a tendency for price competition, which obviously damages the best producers and the best oils. This subject is further developed in point 12 (Section 1.4.1).

Table 1.7 Other optional requirements depending on origin, cultivar, and the producer's know-how (the brand "style").

Sensory performance	An archive of culinary recipes
Indication of culinary uses and pairing opportunities is a most effective communication tool with consumers	
Nutritional and health properties	Total polyphenols (mg/kg) HPLC evaluation of various components of the phenolic fraction (mg/kg) α-Tocopherol (mg/kg)
Whereas the sensory profile of sensory characteristics should be considered as a necessary requirement of excellent olive oils, the data on sensory performance and on nutritional and health properties can be considered as optional requirements, depending on the marketing and consumer target	

1.3 Part 2. The control of critical processing parameters

This second part presents the critical points of olive processing and oil handling. We define as "critical" a point in which there is a significant risk that the process may not achieve the planned results. In other words, a critical point is one in which the risk of compromising the achievement of excellence in the final product is high. Identification of critical points is essential for effectively controlling the process. Such identification is based on a risk analysis that takes into account the gravity and the probability that negative consequences may follow a mistake or nonconformity in the point under consideration.

Risk analysis is beyond the scope of this chapter and therefore the following points (8–11) have been selected as being considered the most critical, based on the author's experience and knowledge, as summarized in Table 1.8.

Table 1.8 Gravity of risk[a] at critical process points.

Critical point	Gravity of the consequences in terms of product quality	Probability of occurrence of negative consequences	Gravity of risk at the point under consideration, from 1 (low) to 3 (high)
Lack of coordination of the harvesting and milling operations – improper olive handling	Medium	Medium	2.0
Inaccurate control of time–temperature relationship in malaxation	Medium	Low	1.5
Deficiencies in the hygienic design of plants – lack of control over residence times	High	High	2.5
Lack of control over the oil storage and distribution conditions	High	High	3.0

[a]Risk is the probability of not meeting the excellence standards as defined in Table 1.5.

1.3.1 Point 8. Coordination of the harvesting and milling operations

It is said that milling of the olive should follow harvesting "in the shortest time possible." However, this rule, valid in principle, frequently leads to irrational decisions and procedures when applied in a rigid or too restrictive way.

It is true that olive storage is a critical point for the quality of the oil and that storage in improper conditions causes oil degradation, but it is important to make a distinction. First, olives should be harvested, transported, and stored in suitable conditions, that is, avoiding mechanical damage and exposure to the elements, be put in a well-ventilated environment, with relative humidity not greater than 80%, in rigid containers, in layers not higher than 30 cm, at appropriate temperatures, and for appropriate times. Furthermore, it is necessary that the olives are healthy; if they have been attacked by parasites, they are not only rapidly altered after harvesting, but have already been altered at the time of harvesting.

If these rules are observed, storage for some or even many hours does not damage the quality of the oil.

1.3.1.1 Rationalization of the harvesting–milling process The problem that arises is that of rationalization of the process. If one tries to mill the olives immediately after harvesting, this creates a rigid connection between the harvesting operation and that of processing, which produces organizational inconsistencies and diseconomies of scale. Micro-mills with a very low hourly working capacity are becoming common. They are generally manually run and lack suitable automatic control. They operate in direct connection with the harvesting, adapting the rate of milling to the rate of harvesting. The labor costs are high and it becomes impossible to carry out the milling process under constant conditions.

Instead, if the olives are stored under proper conditions until suitably sized lots for the mill are obtained, they can be milled in a short time in larger sized mills, in standardized conditions, and with more reliable qualitative and quantitative results. Optimizing this step from harvesting to the milling of olives to producing oil is critical for optimal sizing of the mills.

1.3.1.2 Olive storage and handling When we say that after harvesting the olives should be processed "in the shortest time possible," we are making an incomplete and inaccurate statement. In fact, the outcome of any transformation involving biological and chemical changes depends on the relationship of both time and temperature. This point seems to be not well understood in olive oil technology.

In general, an exponential relationship exists between the time and temperature of any chemical–physical transformation. Hence a very frequent way to represent time–temperature relationships of technological operations is in a semilogarithmic plot, with the temperature on the abscissa with a linear scale and the time on the ordinate with a logarithmic scale.

Figure 1.3 shows the time–temperature relationship for the storage of healthy olives. The abscissa gives the temperature scale between 0 and 30 °C and the ordinate gives the time scale between 1 and 100 h. The two oblique lines divide the diagram into two parts: the lower one is compatible and the

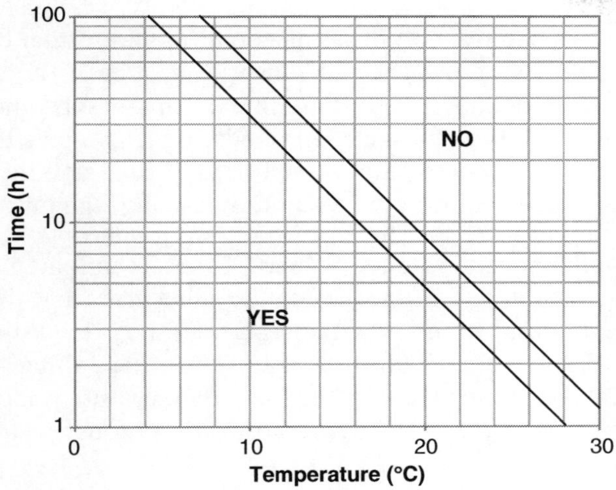

Figure 1.3 Time–temperature relationship for olive storage during the period from harvesting to milling.

upper one incompatible with good olive storage. Between the two lines, conditions can vary in relation to the cultivar and the degree of maturity of the olives. It can be seen, for example, that it is advisable not to exceed a holding time of 1 h at temperatures of 29–30 °C, whereas at 5–7 °C storability can be extended to up to 4 days. At 15 °C the olives can be safely stored for up to about 20 h, and at 10 °C for up to 50 h.

1.3.2 Point 9. Control of time–temperature conditions in malaxation

The condition of maximum vulnerability for oil quality is in the processing phases, where the olive paste is present. The tissues and cells of the olives are broken by crushing: the oil is in extensive contact with the aqueous phase in which various lipases and oxidases are active. The state of vulnerability ceases with the removal of the water and the solid substances by centrifugation and the finishing filtration.

Malaxation, which takes place in the paste state, is very important both for the extraction yield and the quality of the oil. It produces two types of effects:

- Effects of a physical–chemical nature, consisting in separation by differences in polarity, solubility, and density of the oil from the aqueous phase. The phenomenon, called "coalescence," is caused by the combination of small drops of oil dispersed in the paste from a continuous mass of oil; this renders an easier and more complete separation of the oil and thus influences the yield.

- Effects of a biochemical and physical–chemical nature that cause the formation of a specific chemical and sensorial profile of the oil. Complex enzymatic reactions cause the formation of aromatic substances and modification of the polarity of some compounds, while complex diffusion phenomena allow compounds important for the taste, stability, and health value to be transferred from the aqueous phase into the oil.

Many factors influence the rate and extent of these phenomena, such as the degree of olive ripening, the fineness of the olive crushing, the amount of water that determines the fluidity of the paste, and the presence of stone fragments that determines the drainage capacity. All these things being equal, the main critical controlling factor is the time–temperature relationship. Various effects of temperature on the desired transformations must be taken into account:

- Coalescence of the oil is facilitated by an increase in temperature and hence a decrease in viscosity.
- Transfer of phenolic substances from the paste to the oil is also facilitated by an increase in temperature.
- The formation of aromas through complex enzymatic pathways occurs at a maximum "intermediate temperature," and it is depressed by both decreases and increases in temperature.
- The loss of aromas increases with increase in the vapor pressure of volatile molecules and hence with increase in temperature.

It is necessary to strike a balance among the various needs. Figure 1.4 shows the time–temperature relationships within which the process of

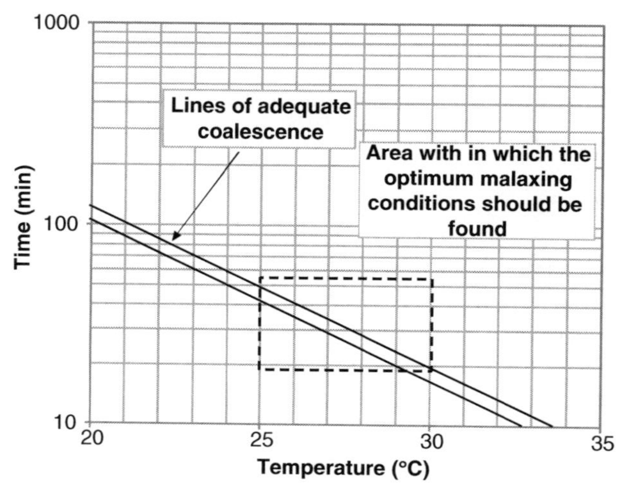

Figure 1.4 Time–temperature relationship for malaxation.

malaxation should take place. The temperature interval on the abscissa is between 20 and 35 °C and the time on the ordinate on a logarithmic scale varies from 10 to 1000 min. The rectangular area traced in the center of the diagram, between 25 and 30 °C and between 20 and 50 min, is the area in which each miller should seek the best compromise between yield and quality. The two oblique lines delimit the conditions within which suitable coalescence phenomena take place.

Box 1.3

The discussion on optimizing the time–temperature relationship of malaxation allows us to clarify one of the most important points for a producer of excellent olive oil, which is that, above all, he/she must be an experimenter. The conditions that maximize quality without reducing the extraction yield cannot be fixed one time for everyone, nor can they be taken from a handbook. They can only derive from critical and creative experience.

We also consider that the definition of "cold extraction" of a process in which the temperature of malaxation does not exceed 27 °C, is incorrect information: first, because there is nothing that says that 27 °C is a temperature tied to some particular phenomenon, and second, because indicating a temperature without indicating a time is incomplete and misleading.

Regarding the definition and control of the time–temperature relationship, there is still a further critical point to clarify, widely neglected in practice, namely control of residence times of the paste at the indicated temperature (see point 10, Section 1.3.3).

1.3.3 Point 10. The problems of "hygienic design" and "residence time distribution"

The hygienic design of installations – a concept well known to plant designers in all food sectors – is a topic that has been underestimated for a long time by olive mill builders. The fact that oil is normally not subject to microbial alteration has contributed to making the topic of hygienic design be substantially ignored; only recently have wiser builders considered it an important concept in plant design.

In a few words, "hygienic design" means compatibility of materials, absence of dead spots in the product circuit, and ease of emptying, washing and cleaning.

Dead spots are those points in which some part of the product may stay for a long time, giving rise to possible alterations, which successively end up contaminating the entire mass. If there is a dead spot in a conduit or in an apparatus in which the olive paste or oil remains for long periods, alterations can occur with production of off-odors and off-flavors that can be passed on to the paste or oil, triggering or accelerating degradative reactions in the entire mass.

The problem of residence time distribution is tied to analogous problems. Let us suppose, for example, that loading of a malaxer takes 15 min, the actual malaxation lasts 30 min, and the time for emptying the malaxer is a further 15 min. In this situation, given that the contents of the malaxer are continuously kept homogeneous by the rotating arms, there will be a part of the paste that enters the malaxer at the beginning of filling and that leaves at the end of emptying, and for this fraction of paste, the residence time will not be 30 min, but closer to 60 min. Suppose that emptying of the malaxer at the end of malaxing is not complete, but a part of the paste already malaxed remains on the bottom or clings to the walls of the malaxer. This residue will undergo a second malaxation and the time of exposure of a small fraction of the paste to the temperature of the malaxer could become about 120 min. A fraction of this fraction could well remain for a third period of malaxing, and so on: the residence time is distributed in a very flattened Gaussian distribution with small amounts that remain at the temperature of the malaxer for very long times. Considering that enzymes operate very efficiently at malaxation temperatures, it can be said that in these cases the malaxer becomes a primer reactor of the degradation reactions, in such a way that after some hours of working, the malaxer becomes a propagator of sensory defects in the oil produced.

These risks become very serious when lots of unhealthy olives are inserted into the working cycle between lots of good, healthy olives. In this case, there is a risk of all the oil being contaminated by off-odors and off-flavors.

The problems due to poor sanitary design or to excessive dispersion of residence times can be resolved by acquiring correctly designed plants, with:

- conduits, equipment, and containers without dead points;
- complete emptying between one working batch and another;
- complete access of water and of washing solutions or, in the case of difficult-to-reach areas, ease of dismantling and reassembly for manual washing;
- malaxers with complete emptying and automatic washing after each batch;
- malaxers assembled in parallel and never in series.

As an alternative, if the plant does not have these characteristics, it is necessary to alternate working periods of 5–6 h with complete emptying and washing cycles.

1.3.4 Point 11. Storage conditions of excellent oils

Storage (and also transport and distribution within the commercial network and home storage by the consumer) is one of the most critical phases for the quality of extra virgin olive oil. The reason for the frequent alteration of oil during storage and distribution is, first, the consequence of widespread underestimation of the perishability of excellent olive oils. The fact that oil is not normally subject to microbiological alteration has contributed to the spread of the erroneous perception that oil is a stable product, that it can be put under mechanical and thermal stress without damaging it. The normal way of using olive oil in restaurants or by families is often entirely inappropriate: half empty bottles open for days, stored in places with inappropriate temperatures.

This is even more of a mistake the higher is the level of quality of the oil. The value of an excellent oil, in fact, is in the complexity of the sensory and health profiles, which are due to very vulnerable compounds.

Storage of oil under appropriate conditions allows not only compliance with the requirements for extra virgin oil (that are relatively easy to guarantee) to be maintained, but also the sensorial requirements and flavor nuances, which characterize excellence.

The problem can be presented in terms of time–temperature relationships, if all of the other conditions are satisfactory, in particular:

- containers are of inert materials, impermeable to light and closed tightly;
- inert gas and no air is present in the headspace of storage containers;
- the oil is perfectly clear and deposits or suspended material are absent;
- good hygienic practices are applied to the premises, equipment, and personnel involved in storage and handling of the oil.

Figure 1.5 reports, in semilogarithmic coordinates, the time–temperature relationship that allows the best storage of oil. The temperatures on the abscissa are between 10 and 25 °C and the times on the ordinate, on a

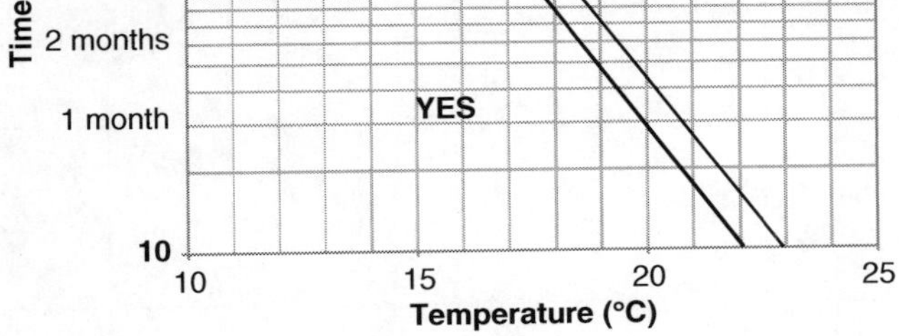

Figure 1.5 Time–temperature relationship for olive oil storage.

logarithmic scale, are between 10 and 1000 days. The two oblique lines divide the diagram into two areas: the area below the oblique lines is compatible, whereas that above is incompatible with good storage of excellent oil. The space between the two oblique lines is an area of variability depending on the chemical stability of the oil and especially on its supply of antioxidants. The area below 12 °C should also be avoided because in these temperature conditions, crystallization and solidification of the oil begin. They do not have negative effects on the quality, but cause changes in appearance not appreciated by consumers and handling difficulties in pouring, bottling, and serving.

1.3.4.1 The quality–proximity matrix This is a fundamental consideration: the systems implemented thus far in certifying oil quality are insufficient to guarantee that the excellence obtained at the production stage reaches the consumer. It is necessary that the quality control systems continuously monitor the conditions of storage, transport, and distribution. This consideration, which implies a radical operating change in the entire chain of excellent extra virgin oil, led us to set up a new system of guarantees and certification according to the quality/proximity matrix shown in Figure 1.6 (Peri, 2012).

In practice, it suggests not only that the consumer is offered oil that has been checked and certified at the production phase, as for the various brands and denominations today, but also that the optimal conditions have been checked and guaranteed both in the marketing phase and in the phases of storage and use in restaurants.

QUALITY LEVEL	PROXIMITY LEVEL		
	FIRST LEVEL: Control of storage conditions from field to bottle	*SECOND LEVEL:* Control of storage conditions from field to retail shelves	*THIRD LEVEL:* Control of storage conditions from field to table
EXCELLENT (e.g., Table 1.5, Table 1.6, and Table 1.7)		*AREA OF EXCELLENCE*	
GOOD (e.g., Protected Designation of Origin)			
COMMON (e.g., legal standards of extra virgin olive oil)			

Figure 1.6 The quality–proximity matrix. *Source:* Peri (2012).

As can be seen in Figure 1.6, different quality levels can be achieved at the production stage: a level of common quality conforming to the legal definition of extra virgin olive oil; a higher level as, for example, in product certification according to the standards of local and regional designations of origin; and a further and higher level of excellence according to the standards presented in Part 1 of this chapter.

For each of these levels, monitoring of the handling and storage conditions can be carried out until the bottling and shipping of bottles for marketing. This is the most frequent case in practice, but also the least reliable in terms of guaranteeing quality at the consumer's table. Controlling and monitoring of the handling and storage conditions can continue during the commercial life of the product. This possibility would be a great step towards proximity and guarantee until the consumer's table. Finally, control of handling and storage in restaurants would be the most effective way of providing the consumer with the quality level that was achieved at the production stage.

Control systems applied without interruption until the last stage of the quality–proximity matrix are the true challenge in marketing excellent extra virgin olive oil.

1.4 Part 3. The marketing of excellent olive oils

This part provides an overall view of the marketing problems of excellent olive oils. It presents some ideas that may help in defining a marketing policy, which is consistent with the aim of valuing excellence in olive oil.

1.4.1 Point 12. Conditions and opportunities for successfully marketing excellent olive oils

The olive oil market is dominated by price competition that discourages the production of excellent oil and tends to homogenize the quality, destroying the variety of sensory profiles. The goal that oriented the action of the 3E Association was to reverse this trend: create a market segment that rewards the efforts of producers of excellent oil; stimulate the transfer of the culture of excellent olive oil to various gastronomic cultures and food habits; and update the system of excellent oil production, marketing, and use.

To understand what needs to be done to achieve such ambitious goals, it is first necessary to become aware of the problems and limitations of the current situation in olive oil marketing.

1.4.1.1 Asymmetric information The difficulty of commercially enhancing excellent olive oil is well explained by the "Theory of Asymmetric Information." Professor George Akerlof of the University of California, Berkeley, won the Nobel Prize in Economic Sciences in 2001 for a fundamental publication in 1970 (Akerlof, 1970; Peri, 2007). This theory is very simple to understand. The basic argument is that in many markets, purchasers have an idea of an average value of certain goods, whereas the sellers have more precise knowledge of actual values. This asymmetry is an incentive for sellers to sell goods that have a market quality lower than the average. Akerlof gives the example of the used car market, in which new cars and old cars, both good and bad, are sold. Because buyers are willing to spend a certain amount for a used car, which seems appropriate to them, the owner of a used car in good or very good condition is in trouble: the seller cannot find buyers willing to pay a higher price than the going price on the market. In conclusion, Akerlof says that the used car market ends up being dominated by low-quality cars, which are those for which the seller earns the most. Akerlof defines this as "adverse selection."

The Akerlof theory applies perfectly to the case of olive oil, which bears an obvious asymmetry of information between the seller (who knows well the origin of the oil, its quality, and the price that they paid to their supplier) and the consumer (who, instead, knows nothing and is not able to judge the quality of the oil). The signs of "adverse selection" are very obvious and have been the object of numerous and sometimes sensational reports (Mueller, 2012).

1.4.1.2 Conditions and opportunities for successful marketing Two main indications have emerged in the long debate about excellence in olive oil that has taken place worldwide in recent years, stimulated by the 3E Association (Anon., 2007–2011):

1. Commercial operators and consumers need to be aware of the fact that great oils are delicate and vulnerable products that require storage conditions and care that is normally given only to fresh products. New systems of selection, control, and chain certification need to be developed in order to guarantee excellence until the product reaches the consumer's table. At the same time, differences according to the concepts of the quality–proximity matrix should become evident in product presentation on the retail shelves.
2. A new culture of excellence in olive oil should spread among restaurants because, as Greg Drescher puts it (Drescher, 2010): "*(…) chefs and influential restaurants have become the trend setters that have driven new trends, and new food cultures, through society. Chefs are in a position to 'own' new flavours and standards of culinary excellence in a way that the retail sector never will. They are in a position to carry a new narrative directly to influential customers, and to the media. Also, through their creativity and insight, we will learn more as a community about how to best advance knowledge and appreciation of excellent olive oil. Many, if not most, of these chef-driven insights will transfer, perhaps with some adaptation, to ideas for retailers and home cooks.*"

Obviously, the motor behind the success of these initiatives will be to convince retailers and chefs that excellence can turn olive oil from a commodity to a specialty and, in restaurants, from a cost to a profit center.

1.4.1.3 Attractive marketing arguments We summarize in the following points a series of arguments that can be attractive to consumers. Excellent extra virgin olive oils can offer a unique combination of pleasure and health, biodiversity and culinary traditions, the beauty of the olive landscapes, and historical and mythical connections with the greatest civilizations.

The oil comes from fruit and not from seed Extra virgin olive oil, different from all other edible oils, comes from fruit and not from seed.

⇩

Naturalness Because of the high water content of the olive, it can be extracted by purely mechanical means. Seed oils, instead, must be extracted with solvents, which involves a series of technological operations to desolventize, neutralize, decolor, and deodorize them, and so on. This makes extra virgin olive oil a very natural product, whereas other edible oils are products of technology rather than Nature.

⇩

Richness and complexity of the minor components For the same reason as coming from a fruit which is a living organism, extra virgin olive oil contains a valuable minor fraction of hydrophilic components (in particular polyphenols) that confer extraordinary and unique sensorial properties, health benefits, and storability.

⇩

A healthy product It contains a unique pool of protective substances with respect to degenerative diseases, including cardiovascular diseases and tumors. Its consumption can be recommended as a good dietary practice for consumers of all ages, from weaning in early childhood to older ages.

⇩

An extraordinary variety of sensorial profiles These derive from the numerous varieties of origin, olive genotypes, and the ability and art of the producer.

⇩

An extraordinary culinary versatility Variety in sensory profile means great culinary versatility, as demonstrated by the innumerable expressions of the Mediterranean cuisine. There is an unexplored potential of new dishes, which chefs, with their creativity, can develop to enhance the use of excellent olive oils

And further

A valuable role of olive in maintaining biodiversity in the natural and man-made areas of the different territories of the Mediterranean and also in other areas where olive growing has begun in the two hemispheres of the globe (Loumau and Giourga, 2003).

And finally

A powerful reminder of myth, symbolism, and religious traditions that are part of the culture that started around the Mediterranean. The olive tree is a light-loving plant (heliophilic), a symbol of peace, frugal and strong, able to survive in rocky and poor soil, and to regrow from its own roots, up to millennia.

⇩

And a messenger of beauty in the hills populated with olive trees, sung by poets in all languages throughout the ages (Ciurnelli, Pascale, and Ponti, 2011).

References

Anon. (2007–2011) Five annual meetings of the International Conference "Beyond Extra Virgin" (BEV) have taken place as follows:

- BEV I at the University of California, Davis, CA, USA, 22–23 May 2007;
- BEV II at the Academy of Georgofili, Florence, Italy, 15 May 2008;
- BEV III at the University of California, Davis, and the Culinary Institute of America, CA, USA, 21–23 June 2009;
- BEV IV in Verona, Italy, 20–22 September 2010;
- BEV V in Cordoba, Spain, 8–10 June 2011.

Anon. (2008) Beyond Extra Virgin – Olive Oil Excellence and World Heritage Project: Fostering Innovation in Quality and Flavor Discovery at the Intersection of Agriculture, Science and Technology, and the Culinary Arts, a declaration of intent signed by the Academy of Georgofili (Centre for Quality Studies), University of California, Davis (UC Davis Olive Center, California Institute of Food and Agriculture Research, Food Science and Technology Department, Robert Mondavi Institute for Wine and Food Science), the Culinary Institute of America, and Association TRE-E, Florence, 7 June 2008 (text available on request from C. Peri at claudio.peri@fastwebnet.it).

Akerlof, G.A. (1970) The market for "lemons". Quality uncertainty and the market mechanism. *The Quarterly Journal of Economics*, **84** (3), 488–500.

Cicerale, S, Conlan, X.A., Sinclair, A.J., and Keast, R.S.J. (2009) Chemistry and health of olive oil phenolics. *Critical Reviews in Food Science and Nutrition*, **49**(3), 218–236.

Ciurnelli, O., Pascale, M., and Ponti, A.C. (2011) *OliveTolive una Antologia di Poesia dell'Olivo da Omero a Oggi*, Fabrizio Fabbri Editore, Perugia.

Drescher, G. (2010) The super-premium olive oil experience: developing a culture of flavor discovery. Presented at BEV IV, Verona, 22 September 2010.

European Commission (2007) Commission Regulation (EC) No. 702/2007 of 21 June 2007 Amending Commission Regulation (EEC) No. 2568/91 on the Characteristics of Olive Oil and Olive-Residue Oil and the Relevant Methods of Analysis. European Commission, Brussels.

Lavelli, V. (2002) Comparison of the antioxidant activities of extra virgin olive oils. *Journal of Agricultural and Food Chemistry*, **50** (26), 7704–7708.

Loumau, A., and Giourga, C. (2003) Olive groves: the life and identity of the Mediterranean. *Agriculture and Human Values*, **20**, 87–95.

Monteleone, E (2010) From sensory characteristics to sensory functionality of super-premium olive oil. Presented at BEV IV, Verona, 21 September 2010.

Mueller, T. (2012), *Extra Virginity – the Sublime and the Scandalous World of Olive Oil*. W.W. Norton, New York.

Pelucchi, C., Bosetti, C., Negri, E., Lipworth, L., and La Vecchia, C. (2011) Olive oil and cancer risk: an update of epidemiological findings, through 2010. *Current Pharmaceutical Design*, **17**, 805–812.

Peri, C. (2007) A vision, a name and a strategy for excellence in olive oil. Presented at BEV I, University of California, Davis, CA, 22 May 2007.

Peri, C. (2010) Strumenti gestionali e indicatori molecolari per la tracciabilità di filiera dei prodotti alimentari. Presented at the Conference "Frontiere della Tracciabilità Molecolare e Sicurezza dei Prodotti Alimentari," Academy of Georgofili, Florence, 18 March 2010.

Peri, C. (2012) La matrice qualità–prossimità: un nuovo modo di concepire la garanzia e la certificazione dei prodotti alimentari. Lecture delivered at the Academy of Georgofili, Florence, 16 February 2012.

Peri, C., and Di Martino, P. (2004) La rintracciabilità di filiera: aspetti tecnici e giuridici. *Alimenta*, **12** (1), 7–12.

Peri, C, Garrido, A., and Lopez, A. (eds) (1993) *The Sensory and Nutritional Quality of Virgin Olive Oil*, A FLAIR Research Project, Milan, 23–24 November 1993.

Peri, C., Kicenik Devarenne, A., and Pinton, S. (2010) 3E Super-Premium Selection for Extra Virgin Olive Oil, a Guideline of the 3E Association and CSQA. Presented at BEV IV, Verona, 20–22 September 2010.

Rovellini, P., Cortesi, N., Fusari, P., and Zaganelli, P. (2010) Nutritional health quality index evaluation of novel extra virgin olive oils, Stazione Sperimentale Oli e Grassi, Milan, 11 March 2010 (text available on request from P. Rovellini at rovellini@ssog.it).

Servili, M., and Montedoro, G. (2002) Contribution of phenolic compounds to virgin olive oil quality. *European Journal of Lipid Science and Technolnology*, **104** (9–10), 602–613.

Viola, P., and Viola, M. (2009) Virgin olive oil as a fundamental nutritional component and skin protector. *Clinics in Dermatology*, **27**, 159–165.

Visioli, F. (2012) Olive oil phenolics: where do we stand? Where should we go?, *Journal of the Science of Food and Agriculture*, **92** (10), 2017–2019.

2

The Basis of the Sensory Properties of Virgin Olive Oil

Agnese Taticchi, Sonia Esposto, and Maurizio Servili
Dipartimento di Scienze Economiche Estimative e degli Alimenti (DSEEA), University of Perugia, Perugia, Italy

2.1 Sensory attributes of virgin olive oil

The sensory attributes of virgin olive oil (VOO), due to visual, olfactory, gustatory, tactile, and chemesthetic sensations, are determined by the presence of several minor compounds, which contribute through different roles to the sensory quality of the product.

2.1.1 Visual attributes

The appearance of VOO is perceived as the combined result of turbidity/limpidity and color. The oil extracted from the olive paste is naturally turbid and opalescent and contains microdroplets of vegetation water and solid particles from the olive fruit. Owing to this physical state, the cloudy form is fairly stable, so that the veiled oils can maintain their turbidity for several weeks or even months before complete sedimentation of the residue. The fine emulsion is formed by a small amount of water (polar phase) in a large amount of the glycerides (lipophilic phase). During the mechanical extraction process, in fact, numerous amphiphilic compounds pass into the oil phase and form emulsions. In VOO, endogenous emulsifiers include a variety of minor components, such as free fatty acids (FFAs), partial glycerides, polyphenols, and phospholipids.

Olive Oil Sensory Science, First Edition. Edited by Erminio Monteleone and Susan Langstaff.

It is well known that cloudy olive oil also contains some material in suspension/dispersion, which, after storage for several months, can form a colored deposit containing proteins in admixture with polymerized phenols. Moreover, it has been established that oxidized polyunsaturated fatty acids (PUFAs) can form colored protein adducts (Boskou, 1996; Sotiroudis *et al.*, 2005).

During oil extraction, the pigment distribution of the olive paste between the pomace (solid phase) and the oil and wastewater (liquid phases) is regulated by mass partitioning phenomena. The affinity of those pigments for the oily phase is determined by the lipophilic nature, whereas the more hydrophilic anthocyanins are retained in the pomace and the vegetation waters. For that reason, the chloroplast pigments (chlorophylls and carotenoids) are mainly responsible for the color of VOO, ranging from yellow–green to greenish gold. The green components of the oils are chlorophyll pigments, whereas the yellow color is due to the presence of carotenoids.

Both chlorophyll *a* (bluish green) and chlorophyll *b* (yellowish green) can occur in VOOs, and the ratio between them has been reported to oscillate between approximately 6 and 8. Furthermore, pheophytins, chlorophyllides, pheophorbides, pyropheophytins, chlorins, rhodins, and purpurins are present. Their presence in olive oil depends on the genetic factors of the olive fruits (olive variety), the stage of fruit ripeness, environmental conditions, the extraction process, and storage conditions. During the extraction process, many physical and chemical modifications are responsible for the formation of chlorophyll derivatives. VOO abounds with pheophytins in particular: the pheophytins are formed as a consequence of the replacement of the magnesium ion in the chlorophyll molecules with two hydrogen atoms. This transformation of chlorophyll *a* into its pheophytin derivative causes the appearance of brownish nuances instead of green. The enzymatic hydrolysis of the phytol group leads to the formation of the chlorophyllides, which are transformed into pheophorbides (dephytilated pheophytins) following the loss of the magnesium cation. Heat exposure is, furthermore, at the basis of the transformation of the pheophytins into the corresponding pyropheophytins.

The yellow pigments in olive fruits are carotenoids, which include carotenes and xanthophylls, lutein being the main xanthophyll and β-carotene the principal carotene. However, there are also β-cryptoxanthin and epoxidized xanthophylls, such as neoxanthin, violaxanthin, antheraxanthin, and their furanoid isomers. Variable factors in the quantitative carotenoid composition of VOO also include the olive cultivar, stage of ripening, agroclimatic conditions, and processing and storing variables (Mínguez-Mosquera and Garrido-Fernández, 1989; Gandul-Rojas and Minguez-Mosquera, 1996; Ranalli *et al.*, 2005; Criado *et al.*, 2008; Moyano, Heredia, and Meléndez-Martinez, 2010).

Visually, the global appearance of an olive oil (in terms of luminosity, chromaticity, and color intensity) can be modified by the turbidity grade for the same pigment content. More transparent, filtered oils are, in fact, greener than those with higher turbidity, which are more yellowish, darker, and less saturated (Gordillo *et al.*, 2011).

One of the critical factors for the maintenance of the VOO color is the storage time, during which the pigment concentration, mainly of the chlorophylls, decreases gradually. The most unstable compound has been reported to be chlorophyll *a*, which is converted into the corresponding pheophytin, followed by chlorophyll *b*, which decreases more slowly. A nonenzymatic process, which involves the formation and subsequent accumulation of free radicals, is probably at the basis of the increase in pheophytin *a* and *b* during oil storage. The chlorophyllase activity leads to the formation of pheophorbides during the extraction process. However, their concentration decreases during oil storage. Among the carotenoids, α- and β-carotene disappear more rapidly than violaxanthin and lutein after several months of storage; the presence of oxygen and free radicals is a fundamental factor in their loss. The oxidation of carotenes and xanthophylls depends, in fact, on the simultaneous oxidation of unsaturated fats (Moyano, Heredia, and Meléndez-Martinez, 2010).

2.1.2 Olfactory attributes

The distinctive olfactory attributes of VOO are related to several compounds belonging to different chemical classes: aldehydes, alcohols, esters, hydrocarbons, ketones, furans, and probably other, as yet unidentified, volatile compounds. They are low molecular weight compounds (<300 Da), which vaporize readily at room temperature. Some volatile compounds reach the olfactory epithelium, dissolve in the mucus, and may bond with olfactory receptors to give an odor sensation (Kalua *et al.*, 2007; Servili *et al.*, 2009a). They can be perceived at very low concentrations [micrograms per kilogram or parts per billion (ppb)] (Angerosa, 2002; Kiritsakis, 1998; Vichi *et al.*, 2003). The mechanisms at the basis of the olfactory perception are very complex: it is generally assumed that the intensity of stimuli elicited by volatile substances is related to their concentration. Therefore, the simplest approach to searching for possible relationships between volatile compounds and sensory characteristics of VOOs is to relate the concentrations of volatiles to the intensities of sensory attributes. Thus, evidence has been found of some relationships between certain sensory notes and their corresponding volatile compounds, especially for the defects in VOOs. However, the concentration of each volatile compound may not always explain the aroma of VOOs. Indeed, it is not always true that higher concentrations of volatile compounds are the main contributors to the aroma, since many factors, such as antagonism and/or synergism among different molecules, volatility, hydrophobic character, size, shape, and conformational structure of the molecules, and type and position of functional groups, all appear to affect the odor intensity more than their concentration, due to their importance in establishing bonds with olfactory receptor proteins (Angerosa *et al.*, 2004). The minimum concentration of a compound able to elicit an olfactory response is the odor threshold value of that compound, assessed in model systems. Thus, the overall aroma is the result of the single contribution of each volatile substance, according to its own sensory threshold and its own odor properties. Several factors have been identified that contribute to the odor threshold value, and they can be summarized as intrinsic and extrinsic factors. The former include the following: stereochemistry (*cis–trans* isomers), chain length, degree of unsaturation, and occurrence of conjugated double bonds; the extrinsic factors described include variations of the sample matrix and the presence of carbohydrates, proteins, and other minor components (Kalua *et al.*, 2007).

The main odor descriptors associated with some of the most important VOO volatile compounds (Angerosa *et al.*, 2004) were discovered by means of sniffing tests, and are summarized in Table 2.1.

Interesting relationships have been highlighted by means of statistical elaborations applied to concentrations of volatile compounds and to intensities of different sensory notes, evaluated by the official methodology. A study using linear regression analysis showed some correlations between several volatile compounds and the descriptors of "green" olfactory attributes (Angerosa *et al.*, 2000). This study demonstrated that (*E*)-2-hexanal concentration affects the perception of "green" attributes, whereas other relationships occurring between the sensory notes and the concentrations of volatile compounds were discovered by Morales *et al.* (1995) by means of the multidimensional scaling (MDS) technique. Similar results were achieved by Servili *et al.* (2001), who used principal component analysis (PCA) and partial least-squares (PLS) to correlate concentrations of volatile compounds and sensory attributes. Those described as having a cut-grass odor were well correlated with the concentration of (*E*)-2-hexenal, hexanal, and (*Z*)-3-hexenal, whereas the floral note was linked with hexyl acetate. In the oils characterized by hay-like sensory notes, no impact compounds corresponding directly to hay odor were found. However, they were characterized by a low concentration of "herbaceous" aldehydes, but with higher amounts of some C_5 and C_6 alcohols.

Quantitatively, the C_6 and C_5 compounds, especially C_6 linear unsaturated and saturated aldehydes, represent the most important fraction of volatile compounds of high-quality VOOs. However, other neo-formation volatile compounds, such as C_7–C_{11} monounsaturated aldehydes, C_6–C_{10} dienals, C_5 branched aldehydes and alcohols, and some C_8 ketones, occur in high concentrations in VOOs affected by organoleptic defects (Angerosa *et al.*, 2004).

Table 2.1 Odor descriptors of some volatile compounds discovered by means of the sniffing technique.

Compound	Odor descriptor
Aldehydes	
Acetaldehyde	Pungent, sweet, floral
Propanal	Sweet, pungent, floral
2-Methylpropanal	Cooked, caramel
Hexanal	Green, apple, cut grass
Heptanal	Fatty
Octanol	Citrus-like, soapy
Nonanal	Soapy, citrus-like
Decanal	Soapy, citrus-like
2-Methylbutanal	Malty
3-Methylbutanal	Sweet, fruity, malty
2-Methyl-2-butenal	Apple
(*E*)-2-Pentenal	Green, apple, floral
(*Z*)-Pentanal	Green, pleasant
(*E*)-2-Hexenal	Bitter, almonds, green, green apple-like, fatty, bitter almond like, cut grass
(*Z*)-2-Hexenal	Green, fruity, sweet
(*E*)-3-Hexenal	Artichoke, green, floral
(*Z*)-3-Hexenal	Green leaves, grassy, green, apple-like, leaf-like, cut grass
2-Octenal	Fruity, soap, fatty
(*Z*)-2-Nonenal	Green, fatty
(*E*)-2-Nonenal	Paper-like, fatty, sharp, cut grass
2-Decenal	Fatty
2,4-Hexadienal	Cut grass
2,4-Heptadienal	Fatty, nutty
2,4-Nonadienal	Deep-fried
2,6-Nonadienal	Cucumber-like
2,4-Decadienal	Deep-fried
Benzaldehyde	Almond
Phenylacetaldehyde	Pungent, phenolic
Ketones	
Pentan-3-one	Sweet
1-Penten-3-one	Sweet, strawberry, sharp, pungent, green, metallic
1-Octen-3-one	Mushroom-like
Alcohols	
Ethanol	Alcoholic, ripe apple, floral
Pentan-1-ol	Pungent
Hexan-1-ol	Fruity, aromatic, soft, cut grass
2-Methylpropan-1-ol	Ethyl acetate-like
2-Methylbutan-1-ol	Fish oil
(*Z*)-2-Penten-1-ol	Banana
(*E*)-3-Hexen-1-ol	Fruity, fatty, pungent, cut grass
(*Z*)-3-Hexen-1-ol	Banana, leaf-like, green-fruity, pungent

(*continued*)

Table 2.1 (*Continued*)

Compound	Odor descriptor
(*E*)-2-Hexen-1-ol	Green, grassy, fruity, fatty, pungent
(*Z*)-2-Hexen-1-ol	Green fruit, green-fruity
1-Penten-3-ol	Wet earth
Esters	
Methyl acetate	Ester
Butyl acetate	Green, pungent, sweet
Ethyl acetate	Sweet, aromatic
Ethyl propanoate	Sweet, strawberry, apple
Ethyl butyrate	Cheesy, fruity
Ethyl isobutyrate	Fruity
Ethyl 2-methylbutyrate	Fruity
Ethyl 3-methylbutyrate	Fruity
(*Z*)-3-Hexenyl acetate	Green-banana, fruity, green, green leaves, floral, ester
Hexyl acetate	Sweet, fruity, floral
3-Methylbutyl acetate	Banana
Methyl 2-ethylbutyrate	Fruity
Methyl decanoate	Fresh
Methyl nonanoate	Fruity, sweet, floral
Acids	
Acetic acid	Pungent
Propanoic acid	Aromatic, pungent
Butanoic acid	Buttery, rancid
Pentanoic acid	Sweaty, pungent, putrid
Hexanoic acid	Sweaty, pungent
3-Methylbutyric acid	Sweaty
2-Methylbutyric acid	Sweaty
Others	
Methylbenzene	Glue, solvent-like
Ethylbenzene	Strong
Ethylfuran	Sweet, rancid
Dimethyl sulfide	Organic, wet earth
Dipropyl disulfide	Cooked meat
Cyclopropane	Musk

Source: Reprinted from Angerosa, F., Servili, M., Selvaggini, R., Taticchi, A., Esposto, S., and Montedoro, G.F. (2004) with kind permission from Elsevier.

The biogenesis of those compounds has been widely investigated. The lipoxygenase enzymatic pathway, which produces C_6 and C_5 compounds from PUFAs after being activated during the olive crushing operation, has been identified as the first pathway. The pathway begins with the production of 9- and 13-hydroperoxides of linoleic acid (LA) and linolenic acid (LnA) mediated by lipoxygenase (LOX). Very specific hydroperoxide lyases (HPLs) catalyze the subsequent cleavage of 13-hydroperoxides, which leads to C_6

aldehydes, of which the unsaturated aldehyde can isomerize from the *Z*-3 to the more stable *E*-2 form. The mediation of alcohol dehydrogenase (ADH) reduces the C_6 aldehydes to their corresponding alcohols, which can produce esters as a result of the catalytic activity of alcohol acetyl transferases (AAT). An additional LOX pathway is active when the substrate is LnA. In addition to the formation of hydroperoxide, LOX would also catalyze its cleavage via an alkoxy radical, giving rise to the formation of stabilized 1,3-pentene radicals. The latter could dimerize, leading to C_{10} hydrocarbons (known as pentene dimers), or couple with a hydroxy radical present in the medium, producing C_5 alcohols, which can be oxidized enzymatically to corresponding C_5 carbonyl compounds (Angerosa *et al.*, 2004). The level and the activity of each enzyme involved in this LOX pathway affect the concentration of each compound produced. Nevertheless, the enzymatic levels are genetically determined and the activity of LOX enzymes is influenced by many agronomic and technological factors, as will be explained subsequently.

Possible fermentations or conversion of some amino acids, either from enzymatic activities of moulds or finally from oxidative processes, are possible causes of other accumulation products, but are generally related to the VOO off-flavors. In high-quality VOOs, the LOX pathway is predominant, whereas a different relative importance of the other pathways is observed in the disagreeable aromas of defective oils according to the sensory defect.

2.1.3 Taste attributes

During organoleptic assessment in compliance with the EU Regulation EC No. 640/2008 of 4th July 2008 Amending Regulation (EEC) No. 2568/91 on the Characteristics of Olive Oil and Olive-Residue Oil and on the Relevant Methods of Analysis, L 178/11–16, "bitter" and "pungent" are listed among the positive attributes perceived in the mouth. The former is primary taste perceived throughout the mouth cavity, and the latter is a tingling sensation perceived especially in the throat.

Numerous studies have clarified the relationship between those attributes in VOO and its polyphenol content (Gutiérrez *et al.*, 1989; Gutiérrez Rosales, Rios, and Gomez-Rey, 1992; Montedoro, Baldioli, and Servili, 1992; Tsimidou, 1998; Mateos *et al.*, 2004). In general, the intensity of the bitterness and pungency in olive oil is related to the presence of the phenolic fraction consisting of secoiridoid derivatives (Servili *et al.* 2004).

From a chemical standpoint, secoiridoids are characterized by the presence in their molecules of elenolic acid (EA) or some of its derivatives; they have been found only in plants belonging to the family of Oleaceae, which includes *Olea europaea* L., hence the only natural food sources are olives and VOO. The most abundant secoiridoids in VOO are the dialdehydic form of decarboxymethyl-EA linked to hydroxytyrosol or tyrosol named

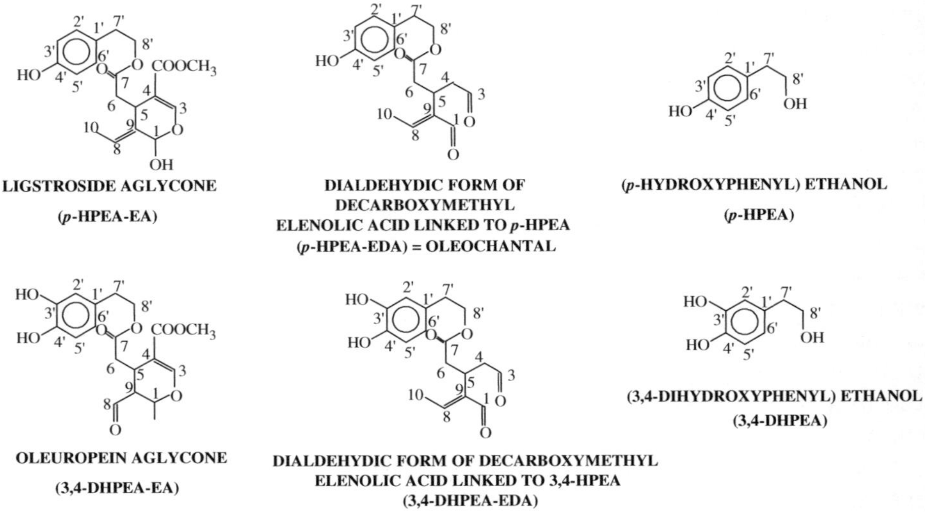

Figure 2.1 Structures of secoiridoid derivatives and phenyl alcohols of VOO. *Source:* Reprinted from Servili, M., Selvaggini, R., Esposto, S., Taticchi, A., Montedoro, G. F., and Morozzi, G. (2004) with kind permission from Springer Science and Business Media.

3,4-DHPEA-EDA and *p*-HPEA-EDA, an isomer of oleuropein aglycone (3,4-DHPEA-EA), and the ligstroside aglycone (*p*-HPEA-EA) (Figure 2.1). These compounds were found for the first time by Montedoro *et al.* (1993), who also assigned their chemical structures, later confirmed by other authors (Angerosa *et al.*, 1996a; Owen *et al.*, 2000). These substances are aglycone derivatives of secoiridoid glucosides contained in the olive fruit, originating during the oil mechanical extraction process, by the hydrolysis of oleuropein, demethyloleuropein, and ligstroside; these reactions are catalyzed by endogenous β-glucosidases (Servili *et al.*, 2004). Olives, in fact, contain a high quantities of phenolic compounds, the concentration of which can range from 1 to 3% of the fresh pulp weight. Oleuropein, demethyloleuropein, ligstroside, and nüzhenide are the most abundant secoiridoid glucosides in the olive fruit (Gariboldi, Jommi, and Verotta, 1986; Garrido Fernández, Fernández Diez, and Adamos, 1997; Servili *et al.*, 1999a), the main phenolic compounds of which also contain verbascoside, a derivative of hydroxycinnamic acid with 3,4-DHPEA in its molecular structure (Servili *et al.*, 1999b).

The health benefits of VOO are, moreover, most likely related to those compounds, since they are now considered to be the main bioactive molecules of extra virgin olive oil; these compounds show a high antioxidant activity and are also credited with the main role in the relationship between the consumption of olive oil and the prevention of cardiovascular disease and cancer in humans (Covas, 2009; Servili *et al.*, 2009a; EFSA, 2011). Furthermore, as primary antioxidants, these hydrophilic phenols can delay the VOO oxidation process, guaranteeing protection against rancidity and prolonging product shelf-life (Servili *et al.*, 2009b).

Most of the studies regarding the sensory properties of VOO phenolic compounds have focused on bitterness. In this context, several authors have suggested that secoiridoid derivatives of oleuropein and demethyloleuropein, such as 3,4-DHPEA-EDA and 3,4-DHPEA-EA, are the main contributors to VOO bitterness (Servili *et al.*, 2009b). In a study carried out in 2004, Mateos *et al.* excluded a list of simple VOO phenols from the contributors to the bitterness: hydroxytyrosol, tyrosol, vanillic acid, vanillin, *p*-coumaric acid, ferulic acid, cinnamic acid, elenolic acid, hydroxytyrosyl acetate, tyrosyl acetate, luteolin, and apigenin (Mateos *et al.*, 2004). 3,4-DHPEA-EA has been described as the most bitter, so that the authors proposed its concentration in VOO as an index of bitterness. Gutiérrez Rosales *et al.* (2003) also concluded that 3,4-DHPEA-EDA, 3,4-DHPEA-EA, and *p*-HPEA-EDA are the compounds mainly responsible for the bitter taste of VOO. However, during VOO tasting, the free endings of the trigeminal nerve located in the palate and also in the taste buds can be stimulated to enable pungency and/or other attributes, such as astringency and metallic sensations, to be perceived. These two last attributes, however, have never been investigated, whereas some evidence of a correlation with the chemical composition has been found for pungency. In particular, a slight burning sensation has been attributed to 3,4-DHPEA-EDA, whereas deacetoxyligstroside aglycone (*p*-HPEA-EDA or oleocanthal) has been described as the main compound responsible for the pungent sensation perceived mostly at the back of the tongue (Andrewes *et al.*, 2003). A study has been carried out more recently to explain why oleocanthal causes a unique kind of pungency, perceived as an unusual irritation in the pharynx when ingested. Unlike most known chemical irritants, in fact, oleocanthal does not significantly irritate the oral cavity. The sting is restricted to the upper airways and is often accompanied by throat clearing and coughing. It has been found that this rare irritation pattern is a consequence of both the specificity of oleocanthal for a single sensory receptor and the anatomical restriction of this sensory receptor to the pharynx, within the oral cavity highlighting the unusual pungency of extra virgin olive oil (Peyrot Des Gachons *et al.*, 2011).

The sensory properties elicited by secoiridoid derivatives in VOO represent a rare case of the healthy value of a food product which is directly perceptible by means of sensory stimuli. The scientific evidence for the protective effect of those compounds against oxidative damage and many implicated diseases has indeed been at the basis of increasing consumer awareness of the positive correlation between VOO bitterness and pungency and its good quality. However, those attributes in some VOOs may be perceived by consumers as excessively intense, so that many methods have been proposed to optimize the phenolic concentration, which will attenuate the bitterness and the pungency of VOOs (Bendini *et al.*, 2012; Servili *et al.*, 2003a).

2.2 Agronomic and technological aspects of production that affect sensory properties and their occurrence in olive oil

The presence and relative concentration of the substances responsible for the sensory attributes of VOOs are influenced by a large number of factors related both to the characteristics of the raw material (olives) and the conditions of extraction and conservation. The composition of pigments and of the phenolic and volatile compound is, in fact, subject in VOOs to extreme variability due to genetic, agro-climatic, and technological causes.

2.2.1 Olive composition: varieties, ripeness, and agro-climatic factors

The olive fruit, *Olea europaea*, a well-known and widespread species of the Oleacea family, is a green drupe that begins to form after flowering in May or June (Northern Hemisphere) and ripens towards the end of autumn. During this process, it darkens to purple–black and the oil content increases at the same time. When the fruits are mature, the violet or purple color is due to the formation of anthocyanins, which appear at the end of the maturation process. The pigment content of a wide variety of olive oils has been the subject of many studies, which Moyano, Heredia, and Meléndez-Martinez (2010) have summarized. They highlighted the fact that, whereas pheophytin *a* and *b*, chlorophyll *a* and *b*, lutein, and β-carotene are widely present in all the VOOs, other pigments may occur depending on the olive variety, such as pheophytin *a*, pheophorbide *a*, pyropheophytin *a*, violaxanthin, luteoxanthin, antheraxanthin, mutatoxanthin, neoxanthin, β-cryptoxanthin, and α-carotene.

Much research has been carried out with the goal of assessing color changes throughout the ripening processes, highlighting the existence of a high variability in those changes, depending on the olive cultivar, even though the level of maturation of the fruits is generally closely correlated with the amount of pigment: the harvest of all varieties of green or cherry olives guarantees a more elevated content of these substances than yields obtained from complete or late maturity (Cichelli and Pertesana, 2004).

Several studies on the volatile composition of VOOs, have demonstrated that the cultivar is the dominant factor in the classification of different varieties. In fact, in monovarietal VOOs, samples of three Italian cultivars from various Italian areas and harvested at different ripening stages showed that the genetic origin of the olives is the prevalent factor which affects their volatiles composition. However, when fruits from two different varieties are simultaneously processed, their enzymatic stores interact and some synergisms occur. The volatiles profile changes from the qualitative and, more often, quantitative point of view in relation to the percentage of fruits of each cultivar, and does not reflect the volatile composition of the cultivars considered nor of the blend of oils of the same varieties at the same percentage. A greater accumulation of volatile compounds is observed at a given percentage of fruits, which differs according to the cultivars used. When the olives are ripe, the amount of volatile compounds, especially of (*E*)-2-hexenal, increases until a maximum concentration is reached, when the skin of the fruit turns from yellow–green to purple. Subsequently, the concentration of the volatile compound decreases, owing to lower activity by the enzymes involved in their production, and the intensity of some "green" sensory notes weakens (Servili *et al.*, 2008). According to Aparicio and Morales (1998), the major reported indicators for ripeness in olive oil are (*E*)-3-hexen-1-ol, (*Z*)-3-hexen-1-ol, (*E*)-2-hexen-1-ol, hexanal, and hexyl acetate.

The phenolic content of olives has a strong genetic base and can be strongly affected by pedoclimatic production conditions, agronomic techniques, and fruit ripening. Indeed, the phenolic composition of fruit is qualitatively and quantitatively affected by the cultivar. Thus, whereas oleuropein is always present in the drupes of all olive cultivars, demethyloleuropein and verbascoside, in contrast, are cultivar dependent, since they occur only in some olive varieties. The olive cultivar also affects the absolute concentration of the specific hydrophilic phenols of VOO, whereas the phenolic profile remains almost the same (Table 2.2).

Fruit ripening also modifies the phenolic composition of the olive drupe: oleuropein decreases during maturation, whereas demethyloleuropein increases. The concentration of both compounds, however, strongly diminishes in over-ripened olives. The negative effect of fruit ripening on the phenolic concentration of VOO is particularly clear. The lowest concentrations of hydrophilic phenols occur in oils obtained from over-ripened olives (Servili *et al.*, 2004). Nevertheless, the climatic and agronomic conditions of olive growing might influence the volatile and phenolic profile of VOOs from the same cultivar.

Table 2.2 Phenolic content (mg/kg) of VOOs obtained from five Italian varieties[a].

Compound	Coratina	Moraiolo	Frantoio	Carolea	Leccino
3,4-DHPEA[b]	1.96 ± 0.30	2.08 ± 0.18	1.38 ± 0.14	2.70 ± 0.20	7.94 ± 1.08
p-HPEA	0.89 ± 0.10	0.87 ± 0.07	0.82 ± 0.09	0.72 ± 0.11	12.3 ± 1.6
3,4-DHPEA-EDA	382.4 ± 13.8	340.0 ± 26.2	154.0 ± 26.0933	268.0 ± 11.4	67.6 ± 1.6
p-HPEA-EDA	193.2 ± 6.5	99.839 ± 6.1	89.8 ± 7.9	189.6 ± 9.0	12.5 ± 0.6
3,4-DHPEA-EA	17.5 ± 9.3	157.14 ± 8.5	84.1 ± 10.3	134.5 ± 5.6	47.2 ± 0.2

[a]The results are expressed as mean ± standard deviation of 10 samples. The olives were harvested at the industrial ripening stage and malaxed at 30°C for 60 min and extracted by pressure on a laboratory scale.
[b]The concentration of hydrophilic phenols was evaluated by high-performance liquid chromatography (HPLC), as reported by Montedoro, Baldioli, and Servili (1992).
Source: Reprinted from Servili, M., Selvaggini, R., Esposto, S., Taticchi, A., Montedoro, G. F., and Morozzi, G. (2004) with kind permission from Elsevier.

In this area of expertise, the relationships between water availability during fruit ripening and the volatile and phenolic composition have been studied. Servili *et al.* (2007a) observed that in the Leccino cultivar, grown in Tuscany under three different water conditions [full irrigation (FI), deficit irrigation (DI), and stress irrigation (SI)], tree water status had a negative correlation with the phenolic concentration, since FI decreased the concentrations of total phenols and *o*-diphenols in VOO. They also found that, although the response of volatile compound concentrations depended mainly on the growing season, there were some consistent trends, which could be attributed to water status. In particular, the concentrations of (*E*)-2-hexenal, (*E*)-2-hexen-1-ol, (*Z*)-2-hexen-1-ol and 1-hexen-3-ol with the DI treatment differed from those with other irrigation regimes. However, the chemical composition of VOOs obtained from the DI treatment highlighted an interesting combination of high concentrations of phenols and similar concentrations of volatile substances responsible for VOO aroma to those of VOO from FI trees. The PLS model that was built confirmed a negative correlation between the phenolic concentration and water availability for the tree, expressed as integrated water potential. At the same time, it showed that several volatile compounds, such as hexanal, (*E*)-2-hexenal, and other lipoxygenase derivative products, were positively correlated with the irrigation rate. The correlation between the phenolic and volatiles composition of some typical VOO taste and olfactory sensations was also confirmed: the "pungent" and "bitter" sensory notes were associated with the total phenols and the *o*-diphenols, whereas "cut grass" was linked to the C_6 unsaturated aldehydes and alcohols. The behavior of phenolic compounds under different irrigation regimes was confirmed for different olive cultivars and geographical areas, whereas the results concerning the aromatic profile often proved difficult to compare (Bucelli *et al.*, 2011; Dabbou *et al.*, 2011).

The geographic origin of drupes also plays a fundamental role in defining the occurrence of the compounds responsible for the sensory attributes of VOOs. Early studies, conducted by Montedoro *et al.* (1995), showed the ability of volatiles to discriminate between oils coming from a different area of origin (Servili *et al.*, 2000). The differences in the C_6 and C_5 volatile contents of oils from various regions were described in particular. For example, the dominance of (*E*)-2-hexenal in the volatile profile of European extra virgin olive oils (Cavalli *et al.*, 2004) is not always seen in oils from other regions (Kiritsakis, 1998). The variation in levels of C_6 aldehydes and alcohols for oil samples from different regions implies that environmental growth conditions may influence the activity of ADH.

The soundness and freshness of the olive are the first criterion for a good sensory quality of VOO, which otherwise risks deterioration. The method of olive storage before the oil mechanical extraction process may cause some important changes: a decrease in oil phenols and a reduction in the volatile compounds responsible for the VOO flavor (Angerosa *et al.*, 1996b;

Kiritsakis, 1998; Kiritsakis *et al.*, 1998; Servili *et al.*, 2004, 2009a,b). The activation during olive storage of endogenous polyphenoloxidase (PPO) and peroxidase (POD), which catalyze phenol degradation, can explain the loss of phenolic compounds. The reduction of volatile compounds responsible for the oil flavor can be due to the inhibition of the LOX pathway. The moulds, yeasts, and bacterial contaminations, together with their corresponding metabolisms, are at the basis of the off-flavor biogenesis in VOO, when they form the characteristic sensory defect of "mould" (Angerosa *et al.*, 2004). Sugar fermentation leads to the formation of acetic acid and ethyl acetate, which are considered to be responsible for the "vinegar" off-flavor.

2.2.2 Technology of mechanical extraction

Virgin olive oil is extracted from the olive fruit using only mechanical processes, which include olive crushing, malaxation of the resulting pastes, and separation of the oil from the other phases essentially by pressure or centrifugation. The different technological operations have repercussions on the extraction and/or biogenesis and modification of pigments, phenols, and volatile compounds. As a result, each sensory attribute is seriously affected during the mechanical extraction process. The influence of each individual operation is discussed below.

2.2.2.1 Olive crushing The main objective of olive crushing performed in industrial oil mills is to break up plant tissues in order to liberate the oil drops contained in the pulp cells. The equipment used may be the traditional, discontinuous stone-mill or the more commonly used continuous hammer crusher, as well as some more recently introduced continuous systems. Each of these systems extracts the pigments from the olive skin in a different way: the most effective is the hammer crusher, which gives a higher concentration of β-carotene, chlorophyll *a* and pheophytin *a* to produce greener oils (Servili *et al.*, 2002).

These differing concentrations confirm the different crushing effects on the constituent parts of the olive fruit. Therefore, as almost all volatile compounds of a good quality olive oil are produced at the moment of tissue disruption of the olive pulp, the effectiveness of crushing plays an important role in the LOX activity and their production. In fact, the impact of crushing in the VOO volatile compounds (together with phenols) can be related to the differentiated distribution of the endogenous oxidoreductases in the constituent parts of the olive fruit (pulp, stone, and seed). The use of a hammer mill crusher (Angerosa and Di Giacinto, 1995), which leads to a more violent grinding of pulp tissues, causes an increase in the temperature of the olive paste and a reduction in HPL activity (Salas and Sánchez, 1999a,b). The new generation of mild seed crushers, which include the blade crusher, teeth crusher, and double stoker crusher, improve the concentration of volatile compounds, especially hexanal, (*E*)-2-hexenal, and C_6 esters, with a consequent positive increase in the intensity of "cut grass" and "floral" sensory notes (Angerosa *et al.*, 2004).

Since the occurrence of hydrophilic phenols in VOO is strictly related to the activities of various endogenous enzymes of olive fruit, crushing and malaxation are the most important critical points for their concentration in the oil mechanical extraction process (Servili *et al.*, 2004). Secoiridoid aglycones such as 3,4-DHPEA-EDA, *p*-HPEA-EDA, *p*-HPEAEA, and 3,4-DHPEA-EA result from the hydrolysis of oleuropein, demethyloleuropein, and ligstroside during crushing. The reaction is catalyzed by the endogenous β-glucosidases, according to a partially known mechanism (Obied *et al.*, 2008). VOO, moreover, does not contain oleuropein and demethyloleuropein, owing to the low solubility of these glycoside substances in the oil phase. Although the hammer crushers have some advantages, they also show some disadvantages, due to the high degradation of the seed tissues, which can encourage phenolic oxidation (Angerosa and Di Giacinto, 1995; Servili *et al.*, 1999c). In fact, endogenous oxidoreductases responsible for phenolic oxidation during processing (such as POD and PPO) are liberated from the olive tissue during the crushing operation. A high amount of POD occurs in the olive seed. The phenolic compounds, in contrast, are largely concentrated in the pulp, whereas the stone and seed contain only small quantities of these substances (Servili *et al.*, 2004, 2007b). As a result, the crushing methods, such as the olive stoning process or the use of mild seed crushers, which enable or reduce the degradation of the seed tissues by limiting the release of POD in the pastes, prevent the oxidation of hydrophilic phenols during malaxation, thus improving their concentration in the VOO (Servili *et al.*, 1999b, 2004, 2007b). Additionally, olive stoning during the mechanical extraction process of VOO modifies the composition of volatile compounds produced by the LOX pathway and increases the concentration of those volatile substances correlated with the "green" sensory notes These results are particularly important, because they appear to demonstrate that the enzymes involved in the LOX pathway have a different activity in the pulp and in the seed of the olive (Servili *et al.*, 2007b).

2.2.2.2 Olive paste malaxation The malaxing stage consists of slow, continuous mixing of the olive paste at a moderate temperature between 25 and 35 °C. This helps the small droplets of oil formed during crushing to aggregate into larger drops, which can be easily separated by means of mechanical systems, and to break up the oil/water emulsions formed during the crushing operation. As in crushing, malaxation results in significant modifications to the minor components of VOO, particularly the volatile and phenolic composition. These two fractions, which are very important in determining the VOO flavor, are most affected by temperature and oxygen management. The effect of the temperature and its management during malaxation has been the subject of over 20 years of in-depth investigations, which have demonstrated a substantial, negative relationship between the processing temperature and the quality of the VOO (Garcia *et al.*, 2001; Di Giovacchino, Sestili, and Di Vincenzo, 2002; Servili *et al.*, 2003b, 2004; Kalua *et al.*, 2006). However, some aspects of the relationship between the operating conditions of malaxation and oil quality must be better defined.

The loss of phenolic compounds during malaxation may be explained by the optimal temperature of activity for PPO and POD, which catalyze the oxidative degradation of phenolic compounds (approximately 50 and 45 °C for PPO and POD, respectively). Thus, the oxidation of phenols within the range 20–35 °C would be progressively greater with increase in malaxation temperature. These results are generally obtained by performing malaxation with the pastes in continuous contact with air, as in the case of open-top malaxing machines, which have been mostly phased out in recent decades (Servili *et al.*, 1998, 2003a).

However, the new generation of malaxing machines operate in a sealed state, to allow for the control of gaseous exchanges ("covered malaxer"), and under these conditions the results obtained in terms of relationships between phenol concentrations in VOO and the processing temperature are completely different. In fact, during processing, the olive pastes release CO_2 and the dissolved O_2 is quickly consumed by the oxidoreductase activities. As a result, the reduction of the O_2 content obtained in the covered malaxer inhibits PPO and POD activities, and increases the concentration of hydrophilic phenols (mainly the derivatives of secoiridoids responsible for the bitter, pungent notes) in the olive pastes and in the corresponding VOO (Servili *et al.*, 2008). Nevertheless, the O_2 depletion can be achieved by using nitrogen or argon. A systematic relationship between high temperatures and VOO phenolic loss is observed in traditional malaxing machines, where a large amount of O_2 is dissolved in the paste during the process due to contact with the air. Small amounts of O_2, in contrast, inhibit the oxidative reactions of phenols during malaxation and, in this case, their concentration in the VOO increases according to the temperature, due to a major release of phenols into the oil (Servili *et al.*, 2008, 2009b). In addition to its beneficial effect on the healthy quality, the decrease in phenolic oxidation also improves the organoleptic properties of the oil.

Thus, the O_2 concentration during malaxation, in combination with the traditional process parameters (time and temperature), may be managed to optimize the VOO phenolic and volatile concentrations (Servili *et al.*, 2004, 2008, 2009a,b). Proposed models to optimize the time of exposure of the olive pastes to air contact (TEOPAC) during malaxation operations aim to reach the most desirable sensory and healthy properties of VOOs according to the olive cultivar and its phenolic and aroma potentiality.

Malaxation time and temperature also affect not only the volatiles profile, but also the sensory characteristics of the resulting VOOs. The main effect of malaxation time is an increase in C_6 and C_5 carbonyl compounds, especially (*E*)-2-hexenal and hexanal, which represent the main contributors to the cut-grass olfactory note, whereas high malaxation temperatures cause a loss of esters and (*Z*)-3-hexen-1-ol and an accumulation of hexan-1-ol and (*E*)-2-hexen-1-ol, both considered by some authors as not completely agreeable odors (Angerosa *et al.*, 2004; Servili *et al.*, 2009a,b).

The enzymes involved in the LOX pathway show optimal temperatures between 15 and 25 °C, whereas their activity diminishes above 30 °C. Hence the malaxing process carried out at temperatures above 35 °C can produce a reduction in the generation of volatile compounds during malaxation and less "green" and "fruity" VOOs. In addition, high temperatures in the malaxation process activate the amino acid conversion pathway and produce considerable amounts of 2- and 3-methylbutanal, without accumulating any corresponding alcohols, associated with the "fusty" defect (Angerosa *et al.*, 1996b, 2001). In contrast, no effects have been noted between the O_2 concentration and aroma production during malaxation.

From the aforementioned considerations, it is clear that a malaxation temperature choice of lower than 28–30 °C generally represents a good compromise in the extraction process in order to obtain a high-quality VOO, in terms of phenols and volatiles (Angerosa *et al.*, 2001, 2004; Servili *et al.*, 2009a), in the new covered malaxing machines. It should be noted that a decrease in the solubility of phenolic compounds and chlorophylls has been found at temperatures below 22 °C.

2.2.2.3 Separation system Centrifugation and pressing are the most widespread systems for the separation of oil from olive pastes. The effects on the sensory attributes differ according to the method chosen and depend on several factors.

A general decrease in volatile compounds responsible for the positive aroma notes occurs. The losses of volatile compounds depend on the importance of the interactions between the oil and solids on the one hand and between oil and the minimal vegetation waters produced by the traditional pressing plant on the other. However, it must be stressed that to obtain high-quality VOOs, the pressing plant needs to work continuously on fruits of the same good quality to prevent possible fermentations and/or degradation phenomena of the pulp residues and vegetation waters on the filtering diaphragms (Angerosa *et al.*, 2004), which could give rise to the defect termed "pressing mats" (IOOC, 1987).

The decanters for separation by centrifugation can be classified as follows:

1. traditional three-phases decanters, featuring water addition ranging from 0.5 to 1 m^3/ton;
2. two-phases decanters, which can operate without the addition of water and do not produce vegetation water as a by-product of the extraction oil process;
3. new three-phases decanters, working with low water consumption (ranging from 0.2 to 0.3 m^3/ton).

The factor involved in the deterioration of the VOO quality is linked to the water used for dilution and the processing temperature. The addition of warm water to dilute the olive pastes in three-phases centrifugal decanters can explain the decrease in C_6 alcohols, hexan-1-ol and (*E*)-2-hexen-1-ol with respect to oils separated by pressing (Angerosa *et al.*, 2004).

In addition to volatile compounds, there is also a dramatic loss of phenolic substances from this important antioxidant fraction, washed away by the action of the water. This problem has been limited by using decanters, which can separate the oil phase from the malaxed pastes without requiring any additional warm water (Servili *et al.*, 2004).

2.2.2.4 Olive oil storage The VOO profile changes during its storage, when all the aspects related to the quality of the oil deteriorate. The endogenous enzymatic activities contained in the cloudy phase modify the phenolic composition. These enzymes may reduce the "pungent" and "bitter" sensory notes, the intensity of which is strictly linked to the content of secoiridoid aglycones, and, at the same time, can produce olfactory and taste defects. Simultaneously, there is a drastic reduction in compounds from the LOX pathway and the neo-formation of volatile compounds, responsible for some common defects referred to as "rancid," "cucumber," and "muddy sediment" (Morales and Aparicio, 1997; Servili *et al.*, 2009a). This runs parallel to the increase in saturated aldehydes, such as nonanal and mainly hexanal, coming from in the oxidation process (Angerosa *et al.*, 2004). Furthermore, under suitable temperature conditions, the presence of sediment from the decanting of unfiltered olive oil can determine the production of unpleasant compounds, responsible for the typical "muddy sediment" defect due to fermentation probably producing butyric-type compounds (Servili *et al.*, 2009a).

Oil filtration partially removes the water and enzymes from VOOs, and keeps the phenolic content and the aroma profile more stable during storage (Montedoro *et al.*, 2005).

2.3 Conclusion

The sensory attributes of VOO may be summarized as visual, olfactory, gustatory, tactile, and chemesthetic sensations. The qualitative and quantitative composition in some minor components, contributing through various roles, determines the global sensory quality and the consumer appreciation. The color nuances going from yellow–green to greenish gold are due to different proportions of the green (chlorophyll pigments) and yellow (carotenoids) components. The olfactory properties are correlated with the concentrations of several volatile compounds, mainly arising from the activity of the lipoxygenase pathways, developed during the first processing operations. In particular, the cut-grass aromas are due to (*E*)-2-hexenal, hexanal, and (*Z*)-3-hexenal, the floral notes to some esters, and the hay-like sensory notes to some C_5 and C_6 alcohols. The positive attributes perceived in the mouth are described as "bitter" and "pungent" and are related to the presence of the phenolic fraction consisting principally of secoiridoid derivatives. These compounds are effective against the oxidative damage and implicated diseases, and they are responsible for the positive correlation between oil bitterness and pungency and its good healthy quality.

All the components involved in the determination of the VOO sensory properties are affected qualitatively and quantitatively by several factors. The oils produced from different olive cultivars are highly variably in terms of composition in pigments, aromas, and phenols; nevertheless, the climate and the agronomic management of the olive cultivations are able to exert strong influences. Other sources of variability are represented by the numerous technological parameters that occur during the mechanical extraction process and the way in which they have to modulate the extraction and the modification of the sensory active components and to promote or control the enzymatic activities on which their concentration depends. Finally, the storage conditions may greatly affect the sensory attributes of VOOs and the speed of the several modifications to which they are subjected.

References

Andrewes, P., Busch, J.L.H.C., de Joode, T., Groenewegen, A., and Alexandre, H. (2003) Sensory properties of virgin olive oil polyphenols: identification of deacetoxy-ligstroside aglycon as a key contributor to pungency. *J. Agric. Food Chem.*, **51**, 1415–1420.

Angerosa, F. (2002) Influence of volatile compounds on virgin olive oil quality evaluated by analytical approaches and sensor panels. *Eur. J. Lipid Sci. Technol.*, **104**, 639–660.

Angerosa, F., and Di Giacinto, L. (1995) Quality characteristics of virgin olive oil in relation to crushing method. Note II. *Riv. Ital. Sostanze Grasse*, **72**, 1–4

Angerosa, F., d'Alessandro, N., Corana, F., and Mellerio, G. (1996a) Characterization of phenolic and secoiridoid aglycons present in virgin olive oil by gas chromatography–chemical ionization mass spectrometry. *J. Chromatogr. A*, **736**, 195–203.

Angerosa, F., Lanza, B., and Marsilio, V. (1996b) Biogenesis of fusty defect in virgin olive oils. *Grasas Aceites*, **47**, 142–150.

Angerosa, F., Mostallino, R., Basti, C., and Vito, R. (2000) Virgin olive oil odour notes: their relationships with volatile compounds from the lipoxygenase pathway and secoiridoid compounds. *Food Chem.*, **68**, 283–287.

Angerosa, F., Mostallino, R., Basti, C., and Vito, R. (2001) Influence of malaxation temperature and time on the quality of virgin olive oils. *Food Chem.*, **72**, 19–28.

Angerosa, F., Servili, M., Selvaggini, R., Taticchi, A., Esposto, S., and Montedoro, G.F. (2004) Volatile compounds in virgin olive oil: occurrence and their relationship with the quality. *J. Chromatogr. A*, **1054**, 17–31.

Aparicio, R., and Morales, M.T. (1998) Characterization of olive ripeness by green aroma compounds of virgin olive oil. *J. Agric. Food Chem.*, **46**, 1116–1122.

Bendini, A., Valli, E., Barbieri, S., and Toschi, T.G. (2012) Sensory analysis of virgin olive oil, in *Olive Oil – Constituents, Quality, Health Properties and Bioconversions* (ed. B. Dimitrios), InTech, Rijeka, Croatia, pp 109–130.

Boskou, D. (ed.) (1996) *Olive Oil: Chemistry and Technology*, AOCS Press, Champaign, IL.

Bucelli, P., Costantini, E.A.C., Barbetti, R., and Franchini, E. (2011) Soil water availability in rainfed cultivation affects more than cultivar some nutraceutical components and the sensory profile of virgin olive oil. *J. Agric. Food Chem.*, **59**, 8304–8313.

Cavalli, J.-F., Fernandez, X., Lizzani-Cuvelier, L., and Loiseau, A.-M. (2004) Characterization of volatile compounds of French and Spanish virgin olive oils by HS-SPME: identification of quality-freshness markers. *Food Chem.*, **88**, 151–157.

Cichelli, A., and Pertesana, G.P. (2004) High-performance liquid chromatographic analysis of chlorophylls, pheophytins and carotenoids in virgin olive oils: chemometric approach to variety classification. *J. Chromatogr. A*, **1046**, 141–146.

Covas, M. I. (2009) Bioactive effects of olive oil phenolic compounds in humans: reduction of heart disease factors and oxidative damage. *Inflammopharmacology*, **16**, 216–218.

Criado, M.-N., Romero, M.-P, Casanovas, M., and Motilva. M.-J. (2008) Pigment profile and colour of monovarietal virgin olive oils from Arbequina cultivar obtained during two consecutive crop seasons. *Food Chem.*, **110**, 873–880.

Dabbou, S., Brahmi, F., Selvaggini, R., Chehab, H., Dabbou, S., Taticchi, A., Servili, M., and Hammami, M. (2011) Contribution of irrigation and cultivars to volatile profile and sensory attributes of selected virgin olive oils produced in Tunisia. *Int. J. Food Sci. Technol.*, **46**, 1964–1976.

Di Giovacchino, L., Sestili, S., and Di Vincenzo, D. (2002) Influence of olive processing on virgin olive oil quality. *Eur. J. Lipid Sci. Technol.*, **104**, 587–601.

EFSA Panel on Dietetic Products, Nutrition and Allergies (NDA) (2011) Scientific opinion on the substantiation of health claims related to polyphenols in olive and protection of LDL particles from oxidative damage (ID 1333, 1638, 1639, 1696, 2865) pursuant to Article 13(1) of Regulation (EC) No. 1924/2006. *EFSA J.*, **9**, 1–25.

Gandul-Rojas, B., and Minguez-Mosquera, M.I (1996) Chlorophyll and carotenoid composition in virgin olive oils from various Spanish olive varieties. *J. Sci. Food Agric.*, **72**, 31–39.

Garcia, J.M., Yousfi, K., Mateos, R., Olmo, M., and Cert, A. (2001) Reduction of oil bitterness by heating of olive (*Olea europaea*) fruits. *J. Agric. Food Chem.*, **49**, 4231–4235.

Gariboldi, P., Jommi, G., and Verotta, L. (1986) Secoiridoids from *Olea europea*, *Phytochemistry*, **25**, 865–869.

Garrido Fernández, A., Fernández Díez, M.J., and Adamos, M.R. (1997) Physical and chemical characteristics of the olive fruit, in *Table Olives* (eds A. Garrido Fernández, M.J. Fernández Díez, and M.R. Adams), Chapman and Hall, London, pp. 67–109.

Gordillo, B., Ciaccheri, L., Mignani, A.G., Gonzalez-Miret, M.L., and Heredia. F.J. (2011) Influence of turbidity grade on color and appearance. *J. Am. Oil Chem. Soc.*, **88**, 1317–1327.

Gutiérrez, F., Albi, M.A., Palma, R., Rìos, J.J., and Olìas, J.M. (1989) Bitter taste of virgin olive oil: correlation of sensory evaluation and instrumental HPLC analysis. *J. Food Sci.*, **54**, 68–70.

Gutiérrez-Rosales, F., Perdiguero, S., Gutiérrez, R., and Olías, J.M. (1992) Evaluation of. the bitter taste in virgin olive oil. *J. Am. Oil Chem. Soc.*, **69**, 394–395.

Gutiérrez Rosales, F., Rios, J.J., and Gomez-Rey, M.L. (2003) Main polyphenols in the bitter taste of virgin olive oil. Structural confirmation by on-line high-performance liquid chromatography–electrospray ionization mass spectrometry. *J. Agric. Food Chem.*, **51**, 6021–6025.

IOOC (1987) *Organoleptic Assessment of Virgin Olive Oil*, COI/T.20/Doc. No. 3, International Olive Oil Council, Madrid.

Kalua, C.M., Bedgood, D.R., Jr, Bishop, A.G., and Prenzler, P.D. (2006) Changes in volatile and phenolic compounds with malaxation time and temperature during virgin olive oil production. *J. Agric. Food Chem.*, **54**, 7641–7651.

Kalua, C.M., Allen, M.S., Bedgood, D.R., Jr, Bishop, A.G., Prenzler, P.D., and Robards, K. (2007) Olive oil volatile compounds, flavour development and quality: a critical review. *Food Chem.*, **100**, 273–286.

Kiritsakis, A.K. (1998) Flavor components of olive oil – a review. *J. Am. Oil Chem. Soc.*, **75**, 673–681.

Kiritsakis, A., Nanos, G.D., Polymenopoulos, Z., Thomai, T., and Sfakiotakis, E.M. (1998) Effect of fruit storage conditions on olive oil quality. *J. Am. Oil Chem. Soc.*, **75**, 721–724.

Mateos, R., Cert, A., Pérez-Camino, M.C., and Garcìa, J.M. (2004) Evaluation of virgin olive oil bitterness by quantification of secoiridoid derivatives. *J. Am. Oil Chem. Soc.*, **81**, 71–75.

Mínguez-Mosquera, M.I., and Garrido-Fernández, J. (1989) Chlorophyll and carotenoid presence in olive fruit (*Olea europaea*). *J. Agric. Food Chem.*, **37**, 1–7.

Montedoro, G.F., Baldioli, M., and Servili, M. (1992) I composti fenolici dell'olio di oliva e la loro importanza sensoriale, nutrizionale e merceologica. *G. Ital. Nutriz. Clin. Prev.*, **1**, 19–32.

Montedoro, G.F., Servili, M., Baldioli, M., Selvaggini, R., Miniati, E., and Macchioni, A. (1993) Simple and hydrolyzable compounds in virgin olive oil. Note 3: spectroscopic characterization of the secoiridoids derivatives. *J. Agric. Food Chem.*, **41**, 2228–2234.

Montedoro, G.F., Servili, M., Baldioli, M., Selvaggini, R., Perretti, G., Magnarini, C., Cossignani, L., and Damiani, P. (1995) Caratterizzazione di alcuni oli vergini di oliva italiani in relazione al luogo di origine. *Riv. Ital. Sostanze Grasse*, **72**, 403–414.

Montedoro, G.F., Selvaggini, R., Begliomini, A.L., Baldioli, M., Esposto, S., and Servili, M. (2005) Questa filtrazione s'ha da fare. *Olivo Olio*, **5**, 32–40.

Morales, M.T., Alonso, M.V., Rios, J.J., and Aparicio, R. (1995) Virgin olive oil aroma: relationship between volatile compounds and sensory attributes by chemometrics. *J. Agric. Food Chem.*, **43**, 2925–2931.

Morales, M.T., and Aparicio, R. (1997) Changes in the volatile composition olive oil during oxidation: flavors and off-flavors. *J. Agric. Food Chem.*, **45**, 2666–2673.

Moyano, M.J., Heredia, F.J., and Meléndez-Martinez. A.J. (2010) The color of olive oils: the pigments and their likely health benefits and visual and instrumental methods of analysis. *Compr. Rev. Food Sci. Food Saf.*, **9**, 278–291.

Obied, H.K., Prenzler, P.D., Ryan, D., Servili, M., Taticchi, A., Esposto, S., and Robards, K. (2008) Biosynthesis and biotransformations of phenol-conjugated oleosidic secoiridoids from *Olea europaea* L. *Nat. Prod. Rep.*, **25**, 1167–1179.

Owen, R.W., Giacosa, A., Hull, W.E., Haubner, R., Würtele, G., Spiegelhalder, B., and Bartsch, H. (2000) Olive-oil consumption and health: the possible role of antioxidants. *Lancet Oncol.*, **1**, 107–112.

Peyrot Des Gachons, C., Uchida, K., Bryant, B., Shima, A., Sperry, J.B., Dankulich-Nagrudny, L., Tominaga, M., Smith, A.B., III, Beauchamp, G.K., and Breslin, P.A.S. (2011) Unusual pungency from extra-virgin olive oil is attributable to restricted spatial expression of the receptor of oleocanthal. *J. Neurosci.*, **31**, 999–1009.

Ranalli, A., Malfatti, A., Lucera, L., Contento, S., and Sotiriuo, E. (2005) Effects of the processing techniques on the natural colourings and the other functional constituents in virgin olive oil. *Food Res. Int.*, **38**, 873–878.

Salas, J.J., and Sánchez, J. (1999a) Hydropeoxyde lyase from olive (*Olea europaea*) fruit. *Plant Sci.*, **143**, 19–26.

Salas, J.J., and Sánchez, J. (1999b) The decrease of virgin olive oil flavor produced by high malaxation temperature is due to inactivation of hydroperoxide lyase. *J. Agric. Food Chem.*, **47**, 809–812.

Servili, M., Baldioli, M., Selvaggini, R., Mariotti, F., Federici, E., and Montedoro, G.F. (1998) Effect of malaxation under N_2 flush on phenolic and volatile compounds of virgin olive oil, in *13th International Symposium on Plant Lipids*, Seville, July 1998, pp. 307–310.

Servili, M., Baldioli, M., Selvaggini, R., Macchioni, A., and Montedoro, G.F. (1999a) Phenolic compounds of olive fruit: one- and two-dimensional nuclear magnetic resonance characterization of nüzhenide and its distribution in the constitutive parts of fruit. *J. Agric. Food Chem.*, **47**, 12–18.

Servili, M., Baldioli, M., Selvaggini, R., Miniati, E., Macchioni, A., and Montedoro, G.F. (1999b) High-performance liquid chromatography evaluation of phenols in olive fruit, virgin olive oil, vegetation waters and pomace and 1D and 2D nuclear magnetic resonance characterization. *J. Am. Oil Chem. Soc.*, **76**, 873–882.

Servili, M., De Stefano, G., Piacquadio, P., Di Giovacchino, L., and Sciancalepore, V. (1999c) Effect of extraction systems on the phenolic composition of virgin olive oils. *Eur. J. Lipid Sci. Technol.*, **101**, 328–332.

Servili, M., Selvaggini, R., Taticchi, A., and Montedoro, G.F. (2000) Volatile compounds of virgin olive oil evaluated by solid phase microextraction: an application in the discrimination of virgin olive oil according to the cultivar and area, in *Flavour and Fragrance Chemistry* (eds V. Lanzotti and O. Taglialatela-Scafati), Kluwer, Dordrecht, pp. 211–220.

Servili, M., Selvaggini, R., Taticchi, A., and Montedoro, G.F. (2001) Headspace composition of virgin olive oil evaluated by solid phase microextraction: relationship with the oil sensory characteristics, in *Food Flavors and Chemistry: Advances of the New Millennium* (eds A.M. Spanier, F. Shahidi, T.H. Parliment, C. Mussiman,

C.-T. Ho, and E. Tratras Contis), Special Publication No. 274, Royal Society of Chemistry, Cambridge, pp 236–247.

Servili, M., Piacquadio, P., De Stefano, G., Taticchi, A., and Sciancalepore. V. (2002) Influence of a new crushing technique on the composition of the volatile compounds and related sensory quality of virgin olive oil. *Eur. J. Lipid Sci. Technol.*, **104**, 483–489.

Servili, M., Selvaggini, R., Taticchi, A., Esposto, S. and and Montedoro, G.F. (2003a) Volatile compounds and phenolic composition of virgin olive oil: optimization of temperature and time of exposure of olive pastes to air contact during the mechanical extraction process. *J. Agric. Food Chem.*, **51**, 7980–7988.

Servili, M., Selvaggini, R., Taticchi, A., Esposto, S., and Montedoro, G.F. (2003b) Volatile compounds and phenolic composition of virgin olive oil: optimization of temperature and time of exposure of olive pastes to air contact during the mechanical extraction process. *J. Agric. Food Chem.*, **27**, 7980–7988.

Servili, M., Selvaggini, R., Esposto, S., Taticchi, A., Montedoro, G. F., and Morozzi, G. (2004) Health and sensory properties of virgin olive oil hydrophilic phenols: agronomic and technological aspects of production that affect their occurrence in the oil. *J. Chromatgr. A*, **1054**, 113–127.

Servili, M., Esposto, S., Lodolini, E., Selvaggini, R., Taticchi, A., Urbani, S., Montedoro, G.F., Serravalle, M., and Gucci, R. (2007a) Irrigation effects on quality, phenolic composition and selected volatiles of virgin olive oils cv. Leccino. *J. Agric. Food Chem.*, **55**, 6609–6618.

Servili, M., Taticchi, A., Esposto, S., Urbani, S., Selvaggini, R., and Montedoro, G.F. (2007b) Effect of olive stoning on the volatile and phenolic composition of virgin olive oil. *J. Agric. Food Chem.*, **55**, 7028–7035.

Servili, M., Taticchi, A., Esposto, S., Urbani, S., Selvaggini, R., and Montedoro, G.F. (2008) Influence of the decrease in oxygen during malaxation of olive paste on the composition of volatiles and phenolic compounds in virgin olive oil. *J. Agric. Food Chem.*, **56**, 10048–10055.

Servili, M., Esposto, S., Taticchi, A., Urbani, S. Di Maio, I., Sordini, B., Selvaggini, R., Montedoro, G.F., and Angerosa, F. (2009a) Volatile compounds of virgin olive oil: their importance in the sensory quality, in *Advances in Olive Resources* (eds L. Berti and J. Maury), Transworld Research Network, Trivandrum, pp. 45–77.

Servili, M., Esposto, S., Fabiani, R., Urbani, S., Taticchi, A., Mariucci, F., Selvaggini, R., and Montedoro, G.F. (2009b) Phenolic compounds in olive oil: antioxidant, health and organoleptic activities according to their chemical structure. *Inflammopharmacology*, **17**, 76–84.

Sotiroudis, T.G., Sotiroudis, G.T., Varkas, N., and Xenakis. A. (2005) The role of endogenous amphiphiles on the stability of virgin olive oil-in-water emulsions. *J. Am. Oil Chem. Soc.*, **82**, 415–420.

Tsimidou, M. (1998) Polyphenols and quality of virgin olive oil in retrospect. *Ital. J. Food Sci.*, **10**, 99–116.

Vichi, S., Pizzale, L., Conte, L.S., Buxaderas, S., and Lopez-Tamames, E. (2003) Solid-phase microextraction in the analysis of virgin olive oil volatile fraction: modifications induced by oxidation and suitable markers of oxidative status. *J. Agric. Food Chem.*, **51**, 6564–6571.

3 Sensory Perception and Other Factors Affecting Consumer Choice of Olive Oil

Hely Tuorila[1] **and Annamaria Recchia**[2]
[1] *Department of Food and Environmental Sciences, University of Helsinki, Helsinki, Finland*
[2] *Dipartimento di Gestione dei Sistemi Agrari, Alimentari e Forestali (GESAAF), Università degli Studi di Firenze, Florence, Italy*

3.1 Introduction

Human senses mediate information from the environment. One of their key functions is to help identify edible material and prevent the ingestion of materials with possibly adverse effects on health and well-being. Eating is necessary for survival, but caution is needed before ingestion. To encourage eating, evolution has provided supportive mechanisms. The motivation to eat is organized by the homeostatic (hunger/satiety) and hedonic (rewarding) systems (Sape, Chou, and Elmquist, 2002). Thus, perceived hunger forces us to find something to eat, and expected or actual satisfaction from pleasant perceptions encourages eating and drinking.

Besides the inherent (physiological) inclinations, the human array of experience-based and cognitive factors that potentially have an impact on food choice is extensive. In the first years of life, in particular, individuals learn to eat and enjoy foods that are available in their environment. The preferences shift, change, and update continuously. Sensory quality must not fail in attracting consumers (Tuorila and Cardello, 2002), but the vast range of other factors finally dictating the choice also need attention. With increasing

Olive Oil Sensory Science, First Edition. Edited by Erminio Monteleone and Susan Langstaff.

experience and information, socio-cognitively based inclinations to choose certain foods and drinks over others take place in the choice process (Costell, Tarrega, and Bayarri, 2010). They may be elusive when it comes to measurement, but they are extremely crucial to choice and therefore not to be neglected.

This chapter deals with the role of sensory perception and socio-cognitive predictors of olive oil choice and consumption. General principles and instruments that possibly predict the choice are introduced and discussed for their suitability for this special category of products. Concerning the functions of senses in food acceptance, several reviews are available (e.g., Cardello, 1996; Tuorila, 2007). Therefore, the following text focuses on the special case of extra virgin olive oil (EVOO), which we refer to as "olive oil" or sometimes just "oil."

3.2 The sensory system

All sensory modalities are involved with the perception of food (Table 3.1). The perceived quality of olive oil is heavily determined by odor and flavor, while hearing probably has no role in defining the quality or in differentiating among various olive oils. Different modalities are important at different phases of perception.

Visual perception or olfactory signals are often the first to provide information on the qualities of food. Other sensory modalities come into play when the food is touched, tasted, or eaten, and later inputs complement or revise the early visual and olfactory information. Although senses are categorized

Table 3.1 Sensory perception of food attributes, and the special case of olive oil.

Category of attributes	Participating sense(s)	Attributes relevant for olive oil
Appearance	Vision	Green, yellow (color) Glossy (other visual appearance) Fluid, viscous (visual texture)
Aroma	Smell (orthonasal)	Green olive, ripe olive, grassy, almond, artichoke, tomato leaf, citrus, apple
Flavor	Taste Smell (retronasal) Chemesthesis	Bitter (taste) Retronasally perceived smells (see above, orthonasal smells) Burning, pungent, peppery (mouth feel) Astringent (mouth feel)
Temperature	Somesthesis – Cold and warm perceptions	Perceived heat, warmth, cold
Texture	Somesthesis – Cutaneous touch – Muscle tension, joint position	Oily (mouth feel) Viscous (oral texture)

into five or six modalities, their action is multimodal. Thus, several modalities participate in the identification of a food, supporting each other's perceptual outcome. Multimodality is also evident in the human categorization of food attributes: texture and flavor, two major categories of food quality, are perceived via several sense modalities in a complementary or integrative manner – flavor by chemical senses taste, smell, and chemesthesis, and texture visually and also by hearing and sense of touch (somesthesis).

Oral sensations experienced when eating involve several sensory modalities: taste, retronasal olfaction, and somatosensory pathways responsible for temperature, mouth feel/texture, and chemesthetic (burning, cooling, and astringent) sensations. Several attributes are frequently used to describe the sensory properties of olive oil (Table 3.1). Odor and flavor (green olive, ripe olive, grassy, almond, artichoke, tomato leaf, apple, and citrus), taste (bitterness), tactile (astringency, thickness, and viscosity), and chemesthetic (pungency, peppery) attributes define olive oil (Mojet and de Jong, 1994; Caporale *et al.*, 2006; Cerretani *et al.*, 2007; Delgado and Guinard, 2011).

3.2.1 Vision

The viewer of olive oil notes a green or yellowish ("olive green"), viscous and glossy liquid that may have turbidity if kept in a chilly place.

The human eye responds to a certain part of electromagnetic radiation of the light, resulting as the perception of color and other appearance attributes. The latter are a broad group of properties, such as transparency, turbidity, and shine, and properties related to visually perceived texture, such as smooth, lumpy, rough, flaky, crystalline, and viscous.

The appearance of foods and beverages has always been considered an important part of the quality, but brain/learning research (e.g., Gottfried and Dolan, 2003) helps to understand the crucial power of visual cues in the identification of a product and in anticipation of the remaining attributes.

3.2.2 Odor

A person sniffing olive oil finds one or several attributes, depending on the origin and quality of the product and their experience of odors and their naming. In addition to qualities directly referring to olive (green and ripe olive) they may include, for example, grassy, almond/nutty, apple, artichoke, and citrus notes. The same qualities may be observed when the oil is tasted and ingested.

Breathing in through the nostrils brings to the olfactory epithelium evaporated odorous chemical compounds triggering the perception of orthonasal odor. During mastication and swallowing, further odorous compounds pass through nasopharynx channel to the olfactory epithelium, thus causing the perception of retronasal odor. Traditionally, most research attention has been directed to the orthonasal perception of odor (e.g., Firestein, 2001), yet the retronasal function is of the utmost importance for food flavor (e.g., Small and Prescott, 2005). Volatile compounds often act as a group to produce a certain perception – the aroma of a certain fruit, for example, can be composed of tens or even hundreds of compounds. This diversity translates into rich and individually varying patterns of odor perception. Individual capabilities of differentiating among odors vary by exposures and training (e.g., Larsson, 2002). Odor description can be remarkably aided by the provision of descriptive words and by training (Cain, 1982; de Wijk and Cain, 1994).

The human olfactory system is not prepared to perceive odors as individual compounds, but rather as patterns based on a group of odorants. Detecting a single odorous compound from a mixture is not substantially improved by training: odor compounds are blended in a manner that cannot be broken down by the human odor detection and recognition device (Livermore and Laing, 1998). The olfactory cortex interprets signals deriving from mixtures of compounds as unitary (Zou and Buck, 2006).

The retronasal odor stimulates the olfactory epithelium in the nose, similarly to orthonasal smell, but preceding the stimulation odorous compounds have evaporated from the food in the mouth to the epithelium via the nasopharynx. An individual locates the perception of retronasal odor in their mouth, where the source of smell is during tasting (Small and Prescott, 2005). Thereby, retronasal smell, along with taste and chemesthetic sensations, becomes an important part of flavor.

The proper and precise description of odors requires systematic learning based on stimuli representing different odor qualities. The fact that even trained sensory panelists perceive a composite odor rather than several separate odorous compounds, as demonstrated by Livermore and Laing (1998), stresses the importance of humans as instruments in odor and flavor analyses. Research on retronasal odor perception is new and there is a clear need to understand better the similarities and differences in ortho- and retronasal perception.

3.2.3 Taste

Among the five "basic tastes" (sweet, salty, sour, umami, and bitter), only the bitter note is readily perceived in olive oil.

Taste perception is initiated on the tongue, on which taste buds are located in three types of taste papillae. Taste buds contain receptor cells with microvilli that have active reception sites. Stimulation of receptors by a bitter substance leads to transduction to nerve impulses via a G protein-coupled cascade of reactions. Nerve impulses mediate the information to the brain (Lindemann, 2001). Bitter substances are probably the largest group of taste-eliciting substances. Sensitivities to various bitter substances are not, or are only weakly, correlated (Cubero-Castillo and Noble, 2004).

Bitterness is elicited by a large range of chemical compounds, and responses to these differ. For example, detection thresholds for seven bitter compounds among 41 panelists varied widely and were correlated from 0.07 to 0.75 (Pearson's *r*). Sensitivity to the extensively studied compound PROP (*n*-propylthiouracil) – the perception of which is known to have a genetic basis – did not predict sensitivity to other bitter compounds (Cubero-Castillo and Noble, 2004), demonstrating the diversity and complexity of bitterness at the perceptual level. The polyphenols causing the bitterness of olive oil are, again, different from the purely bitter substances such as PROP, quinine, and caffeine, since they also elicit astringency and sourness (Monteleone *et al.*, 2004).

Several factors make the evaluation of perceived bitterness in olive oils complex. First, bitterness is a persistent sensation and involves adaptation during repeated exposures. Furthermore, it is enhanced by the presence of typical odor of cut grass note (Caporale, Policastro, and Monteleone, 2004). The odor–taste interaction was observed in an experiment in which model oils were modulated by adding quinine hydrochloride and *cis*-3-hexen-1-ol at different levels. Both the duration and the intensity of the sensation of bitterness in oil, measured by the time–intensity (TI) technique, were affected by the smell of cut grass.

3.2.4 Chemesthesis

At the point of ingesting a sample of EVOO, a peppery note and pungency in the throat will be noticed.

Chemesthesis, in earlier literature often referred to as trigeminal stimulation, is based on the chemical stimulation of free nerve endings of mainly trigeminal, but also other, nerves in the mouth, nose, and eyes (Green, 2002). Oral chemesthetic sensations are described as burning, stinging, tingling, feeling of numbness, hot, warm, cool or cold, and pain. Typical foods causing these sensations include a range of spices (e.g., mustard, chili, cinnamon, pepper, and cloves contain compounds capable of irritation), alcoholic beverages (containing the irritant ethanol), and cooling agents such as eucalyptus and menthol (Prescott and Stevenson, 1995). The sensation often emerges slowly and dissipates even more slowly. The irritation is either based on direct stimulation of the free nerve endings or occurs via enzymatic reaction of an irritant (such as carbon dioxide) in the epithelium (Alimohammadi and Silver, 2002).

Pungency refers to the sharp or biting quality of long-lasting sensation evoked by some foods, spices, or volatile compounds in contact with trigeminal and other free nerve endings of the nasal or oral epithelium (Viana, 2011). Repeated exposures to pungent stimuli with short (<2 min) interstimulus intervals induce sensitization, whereas long (>3 min) interstimulus intervals produce desensitization (Brand and Jacquot, 2002). Subjects normally exposed to pungent stimuli show a chronic desensitization. Panelists usually evaluate several samples in the same session, and the sensitization/desensitization effect induced by repeated exposures to pungent stimuli makes the evaluation of this attribute particularly complex. Chronic desensitization due to dietary habits could explain differences in liking for oils with different levels of perceived pungency.

3.2.5 Tactile perceptions

Olive oil tasting and ingestion lead to the perception of viscous oiliness and astringency in the mouth.

Tactile perceptions elicited in the mouth by olive oil include astringency (discussed below), oral viscosity, and oiliness. According to classic categories devised by Szczesniak (1963), viscosity belongs to mechanical properties and is perceived via the kinesthetic system that relates to joint positions and using force in manipulation. In Szczesniak's categories, oiliness belongs to "Other textural properties" and is perceived by mechanoreceptors responding to pressure, stretching, and vibration. In contrast to odors, there are plenty of words to describe tactile perceptions (e.g., Drake, 1989).

Astringency, primarily a complex tactile sensation involving dryness of the oral surface and tightening and puckering sensations of the mucosa and muscles around the mouth, is induced by mechanoreceptors (Lee and Lawless, 1991; Gawel, Oberholster, and Francis, 2000). It arises from the interaction of dietary tannins with lubricating salivary proteins, thus inducing the loss of saliva lubricating properties (de Wijk and Prinz, 2006; Nayak and Carpenter, 2008; Dinnella *et al.*, 2010). Astringency is a long-lasting sensation that exhibits a carryover effect upon repeated ingestion of astringent stimuli (Guinard, Pangborn, and Lewis, 1986, Dinnella *et al.*, 2010).

Bitterness, pungency, and astringency can persist for a fairly long time after swallowing, showing a clear after-effect that can vary strongly among olive oils in intensity and duration and might affect consumer acceptance (Caporale *et al.*, 2006; Delgado and Guinard, 2011). The dynamic sensory evaluation of bitterness and pungency has been studied in monovarietal autochthonous olive oil stored for up to 18 months utilizing the TI technique (Sinesio, Moneta, and Esti, 2005). Bitterness dominated in oils just after production, but in aged samples the bitter taste rapidly decreased (maximum intensity after 16–20 s), and pungency became the prevailing sensation (maximum intensity after 26–29 s) of the TI measurement. This result has been confirmed in a more recent study using the TDS (temporal dominance of sensations) technique applied to two Italian EVOOs produced in two different regions, Tuscany and the area of Lake Garda–Veneto. TDS curves computed on frequency values during 90 s clearly showed the different dynamics, with taste (bitterness) and flavor (grassy flavor in Tuscany oil and unripe flavor in Veneto oil) perceived as dominant when the oils were in the mouth (during the first 10 s of the evaluation) and the chemesthetic and tactile sensations (pungency and astringency) dominating from swallowing until the end of evaluation (Dinnella *et al.*, 2012).

3.2.6 Chemical basis of perceptions

The color of olive oils ranges from light yellow to a more or less deep green, depending on the content of liposoluble pigments (Roca and Minguez-Mosquera, 1991; Ayuso, Haro, and Escolar, 2004).

The physical characteristics of oils are determined by their chemical compositions, in particular by the chain length and degree of unsaturation of fatty acid residues, and also the distribution of fatty acid residues along the glycerol backbone (Criad, Hernandez-Martin, and Otero, 2007). The triacylglycerol fraction and its capacity to modify texture and carry many taste and aroma compounds affect the increased palatability of fat-containing foods (Mattes, 2009). Fat plays an important role in the flavor perception of food, by affecting both taste and aroma (Pripp, Busch, and Vrecker, 2004). The addition of monoglycerides to a food matrix induces a significant effect on the perception of fattiness and other attributes (Benjamins *et al.*, 2009).

The phenolic composition is largely responsible for the chemosensory profile of a particular oil. The relationships between bitterness, pungency, and total phenol concentration have been investigated in several studies (Gutiérrez-Rosales *et al.*, 1992; Montedoro, Baldioli, and Servili, 1992; Tsimidou, 1998). Only relatively recently have the phenolic fractions directly responsible for these sensations been identified. In particular, secoiridoid derivatives of oleuropein and demethyloleuropein (3,4-DHPEA-EDA and 3,4-DHPEA-EA) as the main contributors to the bitterness have been reported (Kiritsakis, 1998; García *et al.*, 2001). Good linear correlations between bitterness and oleuropein and ligstroside derivatives such as *p*-HPEA-EDA and 3,4-DHPEA-EDA have been found (Tovar, Motilva, and Romero, 2001; Gutiérrez-Rosales, Ríos, and Gómez-Rey, 2003). The secoiridoid derivative fractions containing *p*-HPEA-EDA produced a strong burning, pungent sensation and those containing 3,4-DHPEA-EDA produced a slight burning sensation perceived on the tongue (Andrewes *et al.*, 2003). These results were confirmed in a more recent study (Esti *et al.*, 2009) in which oleuropein and ligstroside derivatives predicted bitterness and pungency in static and dynamic (TI) measurements of bitterness, whereas *p*-HPEA-EDA was effective only in predicting pungency. Beauchamp *et al.* (2005) reported that the perceived pungency intensity in olive oils was positively correlated with the concentration of oleocanthal, a compound with anti-inflammatory and antioxidant properties. More recently, a significant decrease in phenolic content during the storage of oils was associated with a slight decrease in their perceived bitterness and pungency (Romani *et al.*, 2007; Gambacorta *et al.*, 2010).

Volatile compounds represent another important source of flavor in olive oil (Doleschall *et al.*, 2001; Psomiadou *et al.*, 2003; Bubola *et al.*, 2012). They belong to several chemical families such as acids, alcohols, esters, and carbonyls, and they are mainly produced by oxidation of fatty acids. It is generally agreed that endogenous plant enzymes, through the lipoxygenase pathway, are responsible for the positive aroma perceptions (Angerosa, 2002; Kiritsakis, 1998), whereas chemical oxidation and exogenous enzymes, usually from microbial activity, are associated with sensory defects. The major aromatic components in high-quality olive oils are C_6 volatile compounds, mainly responsible for the fruity notes of oils (Aparicio, Morales, and Alonso, 1996; Angerosa, 2002). In particular, fruity was highly correlated with several components: linolenic acid through the lipoxygenase pathway, 1-penten-3-one, (*Z*)-2-hexenal, and individual alcohols, such as hexanol, (*Z*)-3-hexen-1-ol, (*E*)-2-penten-1-ol, (*Z*)-2-penten-1-ol, 1-penten-3-ol, and 3-methylbutan-1-al (Morales *et al.*, 1995; Luna, Morales, and Aparicio, 2006; Bubola *et al.*, 2012). The peculiar note "green" was associated with (*Z*)-3-hexenal (Kuentzel and Bahri, 1991), (*Z*)-3-hexen-1-ol (Hatanaka *et al.*, 1992), (*E*)-2-hexenal, hexanal, ethyl 2-methylbutyrate, hexyl acetate, and (*Z*)-3-hexenyl acetate (Angerosa *et al.*, 2004; Morales, Luna and Aparicio, 2005; Luna, Morales, and Aparicio, 2006; Kotti *et al.*, 2011; Bubola *et al.*, 2012). According to Reiners and Grosh (1998), octanal is associated with citrus aroma of olive oil.

The oxidation of the triglycerides and their derivatives in olive oils causes changes in the chemical, sensory, and nutritional properties of the oil. Several volatile compounds from oxidation lead to an undesirable flavor (Vichi *et al.*, 2003; Angerosa *et al.*, 2004; Kanavouras *et al.*, 2004; Luna, Morales, and Aparicio, 2006; Kalua *et al.*, 2007). Butyl acetate and 1-butanol contributed to heated defects, and ethyl propionate, methyl propionate, 3-methyl-1-butanol, and 1-octen-3-ol were responsible for fusty/muddy sediment defect (Morales, Luna and Aparicio, 2005). Ethylguaiacol and guaiacol were strongly associated with the perception of musty defect (Vichi *et al.*, 2009). Rancid was related to (*E*)-2-heptenal, 2,4-heptadienal, 2,3 nonadienal, (*Z*)-2-octenal, 2-heptanol, and nonanal. Acetic acid was responsible for the vinegary defect (Morales, Luna and Aparicio, 2005).

3.2.7 Integration of perceptions

The sensory system consists of (1) receptors in periphery, as described above, (2) nerves mediating impulses from receptors, and (3) brain areas which receive the impulses and interpret them. The information reaches the brain as impulses, and only the brain is capable of making sense of the message. Specialized brain regions receive and interpret the information from different sense modalities, and the orbitofrontal cortex has a central role in integrating the information (Rolls, 2005). The multimodality of the information means that anatomically separated units (receptor sites and mediating nerves) provide a holistic perception. Hence, although the flavor of food consists of impulses that enter the perceptual system via different senses, the learning process unites the perception into a seemingly single experience from which the parts are difficult to sort out. Whether multimodal processes unify perceptions of flavor, texture, or entire foods, they facilitate the perceptual process by producing "missing links." For example, individual thresholds for the odor of benzaldehyde (bitter almond) were lower when a person tasted sweet solution while smelling the compound, compared with a situation in which no sweet solution was in the mouth (Dalton *et al.*, 2000). The contribution of sweetness to the perception of odor was assumed to be derived from the learned association between the two sensations.

Flavor perception is based on chemical stimulation during the time food is in the mouth, whether this stimulation derives from taste, retronasal odor, or chemesthesis (Table 3.1). The brain integrates the information through learning and builds relevant networks (Small and Prescott, 2005). For example, the brain learns to combine vanilla odor with sweet taste, and therefore the brain response to this combination occurs more strongly, and even in a different brain region, compared with unlearned and thus an incongruent combination of vanilla and saltiness. With the same principle, the bitter, burning, and astringent notes become an entity of perceptions that an experienced user perceives as the sign of high-quality EVOO.

Texture is another multimodally perceived food attribute, the perception being based on visual, tactile, and sometimes auditory stimulation. The interplay among the sense modalities is well agreed on, but the neural integration of textural perceptions has attracted less research interest than flavor. Until now, research on brain responses to texture has focused on the perception of fat and oral viscosity (Rolls, 2005), both mainly tactile perceptions.

3.3 Affective responses to salient sensory attributes of olive oil

The appearance of olive oil gives the first clue to the identity and a powerful expectation of the quality of the product. In a study in which an outstanding EVOO had a bright green color, this unusual color was considered to be a flaw by consumers and this assumption led to negative hedonic ratings (Recchia, Monteleone, and Tuorila, 2012).

The aroma notes of olive oil refer directly to olives and to other fruit and vegetable odors. Among these, citrus, for example, is universally well liked (Pangborn, Guinard, and Davis, 1988). High plasticity is characteristic of the hedonic responses to food odors, and there is meager evidence that odors would be inherently unpleasant. Instead, cognitive cues such as information on the source of odor are important for the hedonic value: the odor of isovaleric acid may be pleasant if interpreted as the smell of cheese, but unpleasant if interpreted as a body smell (de Araujo *et al.*, 2005).

Unlike appearance, texture, and aroma, the flavor attributes of EVOOs are a challenge when it comes to affective responses. Innate human preferences for sensory stimuli are few (Rozin and Vollmecke, 1986, Birch, 1999), and none of the three salient flavor qualities (bitter taste, pungent and astringent mouthfeel sensations) is considered inherently pleasant to humans.

Rejection of bitterness has been observed in the facial expression of newborn babies whose tongue has been exposed to quinine hydrochloride (Ganchrow, Steiner, and Daher, 1983). Bitterness is often deemed unpleasant in foods and beverages. This is seen as part of a biological warning system, as bitterness predicts the presence of poisonous substances. For example, various harmful effects have been reported for tannins. These compounds have a strong bitterness effect and it has been proposed that bitterness represents a warning cue, discouraging animals from ingesting foods too high in tannin content that have various harmful effects (Mueller-Harvey, 2006). However, at the same time, bitterness is an essential quality attribute in many products, such as a range of beverages.

In spite of the initial rejection, later in life the bitter quality is accepted as part of many food products, for example, coffee, beer, many wines, grapefruit, and dairy products. A recent study on early exposures of young infants to bitter sour milk formula suggested that infants may learn to appreciate a remarkably bitter taste if it is part of the essential source of nutrition (Mennella *et al.*, 2011). Research carried out so far is only at the beginning of answering the question concerning the critical age or context leading to long-term generalized food preference, and the mechanisms of learning to like initially rejected sour and bitter tastes await their precise scientific description.

Similar remarks can be made concerning the perception of pungency and astringency. These perceptions are initially disliked, but later in life they become part of the desired qualities in spicy food and, for example, matured wines. Aspects of pungency perception and preference have been reviewed by Rozin (1990) and by Prescott and Stevenson (1995), and preferences for different astringent products have been reviewed by Dinnella *et al.* (2011). A genetic component appears to be present in individually varying preferences to both perceptions (Törnwall *et al.*, 2011, 2012). Initially unpleasant sensations or warning signals, these perceptions eventually transform to culturally accepted sources of sensory enjoyment.

Based on the discussion above, a successful learning process is required of an individual to develop a taste and appreciation for olive oil. Possible mechanisms leading to liking of an originally unacceptable quality, originating from Zellner (1991), are listed in Table 3.2, in which the learning categories have been explained and applied to olive oil. (1) A necessary condition for learning to like the flavor of olive oil is repeated exposures. (2) Social/environmental structures allow coming in contact with oil and

Table 3.2 Learning mechanisms in the acquisition of liking for food qualities, and their possible relevance and necessary conditions for learning to like olive oil.

Mechanism[a]	Explanation	Case of olive oil
(1) Mere exposure	Repeated exposures, tasting, and ingestion	Oil is available and served
(2) Influence of social environment	Making the product available and providing a social model for consumption	As above + other family/group members use oil
(3) Social conditioning	Reinforcement of learning in social situations associated with the consumption of the product	Support and positive remarks that encourage trying and consuming oil
(4) Flavor–flavor conditioning	Associating the unfamiliar flavor of a product with a well-liked familiar flavor	Serving oil with, e.g., tomato and basil
(5) Flavor–calorie conditioning	The body learns to appreciate the flavor because of the post-ingestive effect (satiety)	High calorie content of oil supports this
(6) Flavor–pharmacological conditioning	The body learns to appreciate the flavor because the ingestion contributes to health and well-being	Oil pungency may signal the medicine-like effect
(7) Benign masochism	Enjoying the strong and seemingly dangerous, but harmless, experience	Bitter, pungent, and astringent sensations are capable of contributing
(8) Violation of expectations	Product not as bad as expected, leading to leniency in responses	May work if very negative expectations of oil exist

[a]Categories from Zellner (1991).

permit model learning. (3) Momentary situational and social reinforcement lead to conditioning and thereby to a shift in liking. (4) Paired with favorable flavors, (5) energy and satiety effects, and (6) pharmacological influences further support shifts in liking. As to the pharmacological influence, Beauchamp *et al.* (2005) found that the throat-burning sensation from olive oil derives from a medicinal compound that may protect the organism from the disease, the finding supporting the true medicinal effect. (7) Benign masochism refers to harmless excitement or satisfaction from a painful experience, and is loosely related to Rozin's findings on correlations between amusement park rides and gambling with a liking for chili pepper: excitement induced by danger (Rozin, 1990). (8) Violation of expectations may work in some instances and thereby improve the final perception. However, negative expectations may also lead in the opposite direction, resulting in generalized negativity about the product.

Overall, the pathways leading to hedonic shifts in response to olive oil are as numerous as for any other complex food product. The pathways are likely to function in parallel and overlapping, and different mechanisms may act on different individuals. Not all people will shift their hedonic views concerning chili pepper or coffee – the products so far most extensively studied and discussed in the context of hedonic shifts – and the same is probably true for olive oil. Zellner (1991) suggested that the hedonic shifts of chili and coffee are due to their cultural popularity and consequently frequent exposures that allow space for all the remaining mechanisms to operate. Again, the same is probably true for olive oil in the increasing number of countries in which Mediterranean food has gained in popularity.

3.4 Nonsensory aspects of consumer behavior

3.4.1 Food choice

Traditional sensory science, when expanding to consumer responses, is focused on pleasantness and liking of food samples (Meiselman and MacFie, 1996). This is fair, because liking for a food and food preferences are known to predict the choice and consumption of a product. A rough estimation is that liking predicts 25–50% or the variation observed in the consumption measure (for the discussion and for the instruments used for the measurement of liking, see Tuorila, 2007). However, this means that 50–75% of the variation in the consumption is explained by factors other than affection.

For a producer and marketer, affective responses to a product are an instrument to measure market potential, while the ultimate interest of these players is to understand or predict the choices and the consumed quantities relative to competitors. This, in turn, requires an understanding of different approaches to food choice and consumption. Numerous instruments are aimed at quantifying the choice or consumption (Table 3.3), and they reflect different aspects of the concept. Also, their determinants may be different. For example, a single choice of a certain brand of olive oil in a supermarket may be clearly rational in that particular choice situation, but not at all representative of a person's regular olive oil consumption. The regular consumption of olive oil may be better described by the frequency of past use, or amounts used within a time unit. Intended, reported, and actual behavior, as measuring devices, have their advantages and limitations. Reported past or intended behaviors may reflect an urge to overall consistency in reporting and may contain social desirability bias, whereas actual choice behavior is subject to situational biases, such as the lack or presence of alternatives, cognitive input, or other momentary motives. Nevertheless, factors other than sensory attributes *per se* and their appreciation have a great impact on the choice, and understanding the role of nonsensory factors broadens the picture when attempts are made to understand the success of the product.

Table 3.3 Procedures for measuring food use or consumption.

Task	Operationalization
Actual choice	Which one(s) do you choose? Which one do you want to eat now?
Intended choice	Which one would you prefer to taste/eat?
Frequency of current or future (intended) use	Verbally anchored response options, e.g., from never to daily
Purchase interest	Bipolar, verbally anchored 7- or 9-point scale (from not at all to extremely interested)
Likelihood of buying/using	Unipolar 7- or 9-point scale, verbally anchored at the ends (from unlikely to likely)
Consumed quantity	Estimated by observation of the amount or from waste (or even self-reports)

3.4.2 Perceived quality of olive oil

These factors include extrinsic cues, such as the price, brand, labels, production methods, and country or region of origin. Numerous attempts have been made to identify the drivers of consumer acceptance and the characteristics associated with the perception of olive oil quality. Most of the studies have been conducted on volatile compounds and their possible relationship to olive oils (Morales *et al.*, 1995; Aparicio, Morales and Alonso, 1996, 1997; Caporale *et al.*, 2006). Some authors developed a quality index based on chemical parameters that are related to microbiological, chemical, nutritional, and technological aspects of olive oil, without considering sensory characteristics (Finotti, Bersani, and Bersani, 2007).

Nonsensory factors affecting choice behavior have been illustrated in consumer models (Mullen and Johnson, 1990; Connors *et al.*, 2001) and selected aspects of how these nonsensory factors (e.g., price, production technology, convenience, personal health, branding, and societal issue) affect people's decision in food choice have been reviewed (Jaeger, 2006). A study of wine back labels, for example, found that certain information contained in the label had a positive effect on consumer choice (Mueller and Rungie, 2009). Other authors have focused on the health benefits and flavor of EVOOs as the main motivators for consumption, without considering packaging, price, and size (Martínez, Aragonés, and Poole, 2002; Sandalidou and Baourakis, 2002; Krystallis and Ness, 2003). Perceived quality was defined in terms of sensory properties by both distributors and olive oil consumers (Matsatsinis, Grigoroudis, and Samaras, 2007) or in terms of customer satisfaction (Sandalidou and Baourakis, 2002; Krystallis and Ness, 2005). More recently, the effect of information about the origin of the product on the perceived quality of EVOO has been studied (Caporale *et al.*, 2006; Dekhili and d'Hauteville, 2009; Dekhili, Sirieix, and Cohen, 2011). Some authors have emphasized the region of origin, focusing on the influence of the Protected Designation of Origin (PDO) as the main consumption driver. Region has been considered a key factor that drives purchasing and preferences only by consumers who are experienced, local, or familiar with a particular region of origin (Fotopoulos and Krystallis, 2001; Caporale *et al.*, 2006; Stefani, Romano, and Cavicchi, 2006). Flavor, taste, and culinary use represent factors motivating olive oil purchase and consumption for both experienced and emergent market consumers (Krystallis and Ness, 2003; Caporale *et al.*, 2006; Delgado and Guinard, 2011; Santosa and Guinard, 2011).

3.4.3 Psychographic segmentation of consumers

During the past two decades, instruments, usually consisting of multiple statements or direct questions, each one to be answered on a rating scale, have been developed and validated to help clustering (or segmenting) of consumers based on their socio-cognitive profiles. Such instruments define individual dispositions to a product, such as EVOO. Certain potential instruments that were found relevant for oil choices and acceptance are discussed below. They include involvement, innovativeness, familiarity with, and knowledge of EVOO. Consumer involvement has been defined as "a state of energy (arousal) that a person experiences in regard to a consumption-related activity" (Wilkie, 1986). Involvement is related to how relevant consumers perceive a product to be within their personal needs, values, interests, ego, and motivation for a given situation (Bell and Marshall, 2003; Marshall and Bell, 2004). High involvement may lead to making an active effort to find product information, making detailed comparisons of product attributes, and perception of differences between brands (Zaichkowsky, 1985). For the product category wine, for example, product involvement has been conceptualized as the interest, enthusiasm, and excitement that consumers exhibit towards wine (Goldsmith, 1998). Increasing involvement has been found to correlate with the pleasantness and buying probability of several foods (Ares *et al.*, 2010; Hollebeek *et al.*, 2010; Meillon *et al.*, 2010). The more involved the consumers or the more knowledge consumers had about the product, the larger was the number of consumption motivations that would be elicited by the subjects (Gengler and Reynolds, 1995).

With growing experience and knowledge, a deep understanding is developed of the nature of a product. Perceptual abilities develop in the course of exposures to stimuli in those who have reasonable sensory acuity, who are motivated to spend time and effort, and who are capable of focusing their attention on specific sensory inputs (Lawless, 1984). Holbrook, Lehman, and O'Shaughnessy (1986) pointed out that the consumption experience (using) may affect future purchase (choosing) decisions. Previous experience positively affects food preferences (Pliner, 1982; Baron and Penfield, 1993) and also categorization of the products, as more experienced people evoke finer categorization than novices (Tanaka and Taylor, 1991; Johnson and Mervis,

1997; Rota and Zellner, 2007). Subjects' familiarity with EVOOs affected the outcome of a sorting task based on visual assessments of the oil bottles (Santosa, Abdi, and Guinard, 2010). Sorting was more successful in staging finer product discriminations for familiar products imported into California than for unfamiliar local EVOOs. The experiential background of respondents was found to be the main factor affecting wine categorization (Torri *et al.*, 2013). Only highly experienced and dedicated respondents (measured in terms of involvement, innovativeness, and knowledge) produce informative projective maps reflecting perceived similarities/dissimilarities among samples with subtle sensory differences. When performing the categorization task, consumers refer to their liking while experts refer to a common memorized wine prototype (Torri *et al.*, 2013).

Commitment to EVOO, defined as involvement, familiarity with, and subjective knowledge of EVOO, and knowledge about the appropriateness of sensory descriptors of the EVOOs, was recently used to define subgroups of Finnish subjects ($n = 74$) with an educational background in culinary preparation or in food science (Recchia, Monteleone, and Tuorila, 2012). Individual involvement with oils was quantified using 15 items based on the scales of Marshall and Bell (2004) and Zaichkowsky (1985) (Table 3.4). Subjects were instructed as follows according to Marshall and Bell (2004): "Below are pairs

Table 3.4 Personal Involvement Inventory Scale, ratings from 1 to 7 depending on which phrase better reflects the opinion of the respondent.

Important	1	2	3	4	5	6	7	Unimportant	®[a]
Of no concern to me	1	2	3	4	5	6	7	Of concern to me	
Irrelevant	1	2	3	4	5	6	7	Relevant	
Means a lot to me	1	2	3	4	5	6	7	Means nothing to me	®
Doesn't matter to me	1	2	3	4	5	6	7	Matters to me	
Think about often	1	2	3	4	5	6	7	Think about seldom	®
Appealing	1	2	3	4	5	6	7	Unappealing	®
Boring	1	2	3	4	5	6	7	Interesting	
Unexciting	1	2	3	4	5	6	7	Exciting	
Fun	1	2	3	4	5	6	7	Not fun	®
A good use of my money	1	2	3	4	5	6	7	A bad use of my money	®
Hard to make a good choice	1	2	3	4	5	6	7	Easy to make a good choice	
Options within category are very similar	1	2	3	4	5	6	7	Options within category are very different	
Reveals something about me	1	2	3	4	5	6	7	Reveals nothing about me	®
Tells me something about another person	1	2	3	4	5	6	7	Tells me nothing about another person	®

[a]Ratings of the marked item ® have been reversed for the mean scores.

of phrases that you are asked to use to describe your perceptions concerning extra virgin olive oils. Each phrase defines the scale point 1 or 7. Value 1 indicates that the phrase on the left is adequate in expressing how you feel about the product. The value 7 indicates that the phrase on the right is adequate. The values from 2 to 6 indicate intermediate evaluations. For each pair of phrases we ask that you circle a number from 1 to 7 according to your feelings the extra virgin olive oil."

Subjects rated their subjective or self-perceived knowledge of extra virgin olive oil by rating nine statements from 1 = "strongly disagree" to 7 = "strongly agree" with a neutral point at 4 = "neither agree nor disagree" (Table 3.5) (Flynn and Goldsmith, 1999). Subjects were instructed as follows according to Flynn and Goldsmith (1999): "Please indicate the extent to which you agree or disagree with the following statements about extra virgin olive oils."

Finally, respondents rated the appropriateness of 24 sensory attributes in describing the properties of EVOOs. Thirteen of these (green olive, ripe

Table 3.5 Self-perceived knowledge scale.

		Strongly disagree			Neither agree nor disagree			Strongly agree
I know pretty much about extra virgin olive oils	®[a]	1	2	3	4	5	6	7
I know how to judge the quality of extra virgin olive oils	®	1	2	3	4	5	6	7
I think I know enough about extra virgin olive oils to enjoy them	®	1	2	3	4	5	6	7
I do not feel very knowledgeable about extra virgin olive oils		1	2	3	4	5	6	7
Among my circle of friends, I'm one of the "experts" on extra virgin olive oils	®	1	2	3	4	5	6	7
Compared to most other people, I know less about extra virgin olive oils		1	2	3	4	5	6	7
I have heard of most of the new extra virgin olive oils that are available	®	1	2	3	4	5	6	7
When it comes to extra virgin olive oils, I really don't know a lot		1	2	3	4	5	6	7
I can tell if an extra virgin olive oil is worth the price or not	®	1	2	3	4	5	6	7

[a]Ratings of the marked items ® have been reversed for the mean scores.

olive, cut grass, tomato leaf, apple, citrus, artichoke, almond, bitter, pungent, sweet, stringent, and viscous) were selected from those generally used for EVOOs (Mojet and de Jong, 1994; Caporale *et al.*, 2006) and the remaining 11 were added as inappropriate attributes (not reported here). The inappropriate attributes were included since a development of an appropriateness scale consisting of ratings of both appropriate and inappropriate attributes was initially planned. The appropriateness was rated on a seven-point scale ranging from 1 = "not at all appropriate" to 7 = "extremely appropriate" (Schutz, 1994; Caporale *et al.*, 2006), with the option "I do not know." Subjects were instructed: "Please read carefully the following list of sensory attributes and rate the appropriateness of each one on the scale 1–7 in describing the odor, flavor, taste, tactile sensations or consistency of extra virgin olive oils. If you do not know whether an attribute is appropriate or not, please do not hesitate selecting the option 8 ('I don't know')."

Based on these constructs, three clusters were identified: Cl1 (n = 12), Cl2 (n = 35), and Cl3 (n = 27), characterized by a low, a medium, and a high commitment level, respectively. Empirical construct scores of data were well dispersed (Figure 3.1) (Recchia, Monteleone, and Tuorila, 2012). Subjects from Cl3 were characterized by the highest involvement with EVOO and they considered themselves to be the most knowledgeable. Furthermore, they reported the highest frequency of buying and consumption of EVOO. Cl3 was the most familiar with and the most willing to buy EVOOs. Even though only the most important attributes typically used in the description of the sensory profile of EVOO (Mojet and de Jong, 1994; Caporale *et al.*, 2006; IOC, 2007; Delgado and Guinard, 2011) were used in the questionnaire, the knowledge of the sensory properties of the product category was low for all subjects who participated in the study, regardless of their commitment to the product. This scale, in fact, may reflect "real" or "concrete" knowledge (what the consumers actually know), as opposed to another distinct knowledge category, the subjective knowledge of the product (what the consumers think they know) (Brucks, 1985).

Furthermore, the experiential background of subjects did not affect either categorization or liking for four Italian EVOOs with very different color, bitterness, and pungency, and with different states of oxidation. All clusters, in fact, disliked the most bitter and pungent oils with a low oxidation state and they categorized all EVOOs in the same way. Bitterness and pungency were the main preference drivers rather than subjects' commitment, confirming what has been reported in the literature (Bertuccioli, 1994; McEwan, 1994; Garcia *et al.*, 2001; Caporale, Policastro, and Monteleone, 2004; Delgado and Guinard, 2011). Since humans learn to accept bitterness and pungency as adults and in response to cultural processes (see Section 3.3 and Table 3.2; Drewnowski, 1997; Kim *et al.*, 2004), consumers in emergent markets may not have enough exposures to the product to be able to appreciate bitterness and pungency in olive oil (Delgado and Guinard, 2011). The "commitment,"

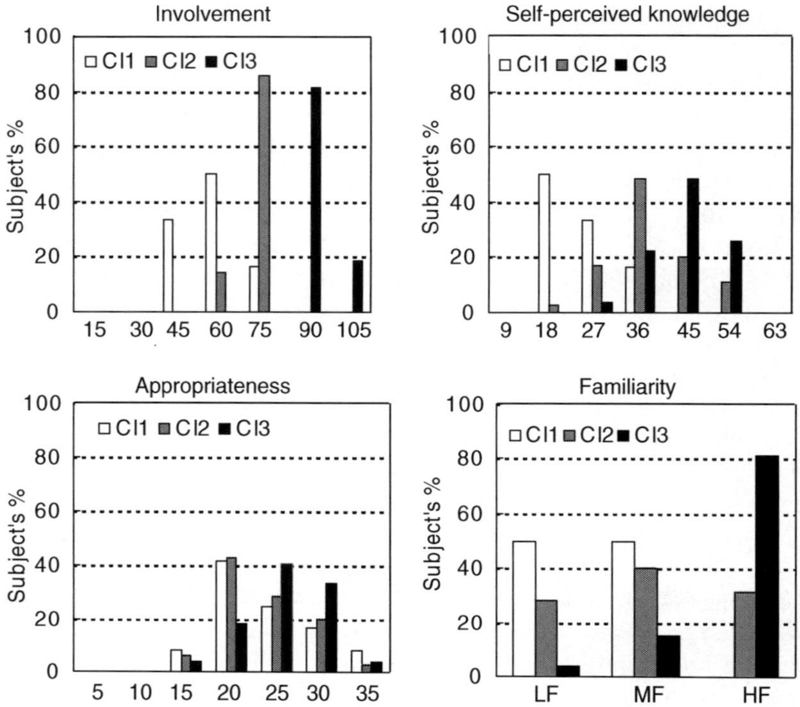

Figure 3.1 Frequency distributions of involvement, self-perceived knowledge, perceived appropriateness, and familiarity with EVOO in consumer clusters Cl1 ($n = 12$), Cl2 ($n = 35$), and Cl3 ($n = 27$). LF = low familiarity (scores 1, 2, 3); MF = medium familiarity (score 4); HF = high familiarity (score 5).

defined as the level of involvement and subjective knowledge, and knowledge of the sensory descriptors appropriate for EVOOs, does not necessarily encompass the sensory experience.

EVOO could function as both a low- and a high-involvement product depending on how it is being used (Santosa and Guinard, 2011). Certain product attributes associated with EVOOs as a low-involvement product were cheaper/on sale, reasonable price, large quantity/bulk size, and less flavor. These attributes allowed consumers to follow a budget or save money so that money would be available for other things. Product attributes associated with EVOOs as a high-involvement product were more flavor, more expensive, and smaller size. EVOOs in this category were associated mostly with providing pleasure.

EVOOs are usually classified into quality categories based on the intensity of flavors in compliance with national regulations or defined by experts in award events. These methods often do not relate to consumer hedonic responses. In fact, liking expressed by consumers from emergent markets and consequently with low experience do not match experts' quality (Delgado and Guinard, 2011; Recchia, Monteleone, and Tuorila, 2012). Because of expertise and exposure to different olive oil profiles, experts are aware that bitterness and pungency are positive characteristics, in contrast to the majority of consumers, who did not favor these sensory properties in "new" food products. It is fair to assume that as the number of official competitions and awards for EVOOs continues to grow worldwide, consumers will increasingly come to follow the opinions of the experts behind those events, much in the same way as they do for wine (Delgado and Guinard, 2011). However, sensory expectations of EVOO's contribution to the sensory profile of the food matrix to which it is added should also be taken into account. Information about conditions of use that allow an oil to express its best sensory potential would greatly help in optimizing both production and market positioning strategies for this ingredient (Dinnella *et al.*, 2012).

3.5 Conclusion

In sensory and consumer terms, EVOO is complex and rich in nuance. In spite of its demanding sensory nature, its popularity is increasing. Discrepancies between consumer preferences and expert quality ratings of EVOOs point to the need to educate consumers about the different styles of oils and the wide range of sensory properties associated with them. For success in the market, it is important to communicate properly the valuable sensory properties to the potential consumer segments. Likewise, it is important to identify the existing and forthcoming segments of consumers who are inclined to like and use EVOO, and to provide information to support their interest in the product. Information about conditions of use that allow an oil to express its best sensory potential would greatly help in optimizing both production and market positioning strategies for this ingredient. Indeed, the complex interactions between food components represent one of the most important drivers of perceived differences among oils. Therefore, the actual conditions in which consumers are exposed to oils, that is, mixed with other foods rather than oils on their own, and how they affect hedonic responses of potential consumer segments should be taken into account.

Acknowledgment

This work was supported by the Tuscany Regional Government within the framework of the project "Oleosalusistem – Validazione di Protocolli per la Produzione di Oli ad Elevato Valore Nutrizionale ed a Ridotto Impatto Ambientale."

References

Alimohammadi, H., and Silver, W.L. (2002) Chemesthesis: hot and cold mechanisms. *Chemosense*, **4**, 1–9.

Andrewes, P., Busch, J.L.H.C., de Joode, T., Groenewegen, A., and Alexandre, H. (2003) Sensory properties of virgin olive oil polyphenols: identification of deacetoxy-ligstroside aglycon as a key contributor to pungency. *J. Agric. Food Chem.*, **51**, 1415–1420.

Angerosa, F. (2002) Influence of volatile compounds on virgin olive oil quality evaluated by analytical approaches and sensory panels. *Eur. J. Lipid Sci. Technol.*, **104**, 639–660.

Angerosa, F., Servili, M., Selvaggini, R., Taticchi, A., Esposto, S., and Montedoro, G. (2004) Volatile compounds in virgin olive oil: occurrence and their relationship with the quality. *J. Chromatogr. A*, **1054**, 17–31.

Aparicio, R., Morales, M.T., and Alonso, M.V. (1996) Relationship between volatile compounds and sensory attributes of olive oils by the sensory wheel. *J. Am. Oil Chem. Soc.*, **73**, 1253–1264.

Aparicio, R., Morales, M.T., and Alonso, V. (1997) Authentication of European virgin olive oils by their chemical compounds, sensory attributes, and consumers' attitudes. *J. Sci. Food Agric.*, **45**, 1076–1083.

Ares, G., Besio, M., Gimenez, A., and Deliza, R. (2010) Relationship between involvement and functional milk desserts intention to purchase. Influence on attitude towards packaging characteristics. *Appetite*, **55**, 298–304.

Ayuso, J., Haro, M.R., and Escolar, D. (2004) Simulation of the visible spectra for edible virgin oils: potential uses. *Appl. Spectrosc.*, **58**, 474–480.

Baron, R.F., and Penfield, M.P. (1993) Panelist texture preferences affect sensory evaluation of green bean cultivars (*Phaseolus vulgaris* L.). *J. Food Sci.*, **58**, 138–139.

Beauchamp, G.K., Keast, R.S.J., Morel, D., Lin, J., Pika, J., Han, Q., Lee, C.H., Smith, A.B., and Breslin, P.A.S. (2005) Phytochemistry: ibuprofen-like activity in extra-virgin olive oil. *Nature*, **437**, 45–46.

Bell, R., and Marshall, D.W. (2003) The construct of food involvement in behavioural research: scale development and validation. *Appetite*, **40**, 235–244.

Benjamins, J., Vingerhoedsa, M.H., Zoeta, F.D., De Hooga, E.H.A., and Van Aken, G.A. (2009) Partial coalescence as a tool to control sensory perception of emulsions. *Food Hydrocolloids*, **23**, 102–115.

Bertuccioli, M. (1994) A study of sensory and nutritional quality of virgin olive oil in relation to variety, ripeness and extraction technology. Overview of three year study and conclusion. *Grasas Aceites*, **45**, 55–59.

Birch, L.L. (1999) Development of food preferences. *Annu. Rev. Nutr.*, **19**, 41–62.

Brand, G., and Jacquot, L. (2002) Sensitization and desensitization to allyl isothiocyanate (mustard oil) in the nasal cavity. *Chem. Sens.*, **27**, 593–598.

Brucks, M. (1985) The effects of product class knowledge on information search behavior. *J. Consumer Res.*, **12**, 1–16.

Bubola, K.B., Koprivnjak, O., Sladonja, B., and Luki, I. (2012) Volatile compounds and sensory profiles of monovarietal virgin olive oil from Buza, Crna and Rosinjola cultivars in Istria (Croatia). *Food Technol. Biotechnol.*, **50**, 192–198.

Cain, W.S. (1982) Odor identification by males and females: predictions *vs.* performance. *Chem. Senses*, **7**, 129–142.

Caporale, G., Policastro, S., and Monteleone, E. (2004) Bitterness enhancement induced by cut grass odorant (*cis*-3-hexen-1-ol) in a model olive oil. *Food Qual. Pref.*, **15**, 219–227.

Caporale, G., Policastro, S., Carlucci, A., and Monteleone, E. (2006) Consumer expectations for sensory properties in virgin olive oils. *Food Qual. Pref.*, **17**, 116–125.

Cardello, A.V. (1996) The role of human senses in food acceptance, in *Food Choice, Acceptance, and Consumption* (eds H.L. Meiselman and H.J.H. MacFie), Blackie, London, pp. 1–82.

Cerretani, L., Biasini, G., Bonoli-Carbognin, M., and Bendini, A. (2007) Harmony of virgin olive oil and food pairing: a methodological proposal. *J. Sens. Stud.*, **22**, 403–416.

Connors, M., Bisogni, C.A., Sobal, J., and Devine, C.M. (2001) Managing values in personal food systems. *Appetite*, **36**, 189–200.

Costell, E., Tarrega, A., and Bayarri, S. (2010) Food acceptance: the role of consumer perception and attitudes. *Chemosens. Percept.*, **3**, 42–50.

Criad, M., Hernandez-Martin, E., and Otero, C. (2007) Optimized interesterification of virgin olive oil with a fully hydrogenated fat in a batch reactor: effect of mass transfer limitations. *Eur. J. Lipid Sci. Technol.*, **109**, 474–485.

Cubero-Castillo, E., and Noble, A.C. (2004) Relationship of 6-*n*-propylthiouracil status to bitterness sensitivity, in *Genetic Variation in Taste Sensitivity* (eds J. Prescott and B.J. Tepper), Marcel Dekker, New York pp. 105–116.

Dalton, P., Doolittle, N., Nagata, H., and Breslin, P.A.S. (2000) The merging of the senses: integration of subthreshold taste and smell. *Nat. Neurosci.*, **3**, 431–432.

de Araujo, I.E., Rolls, E.T., Velazco, M.I., Margot, C., and Cayeux, I. (2005) Cognitive modulation of olfactory processing. *Neuron*, **46**, 671–679.

de Wijk, R.A., and Cain, W.S. (1994) Odor quality: discrimination versus free and cued identification. *Percept. Psychophys.*, **56**, 12–18.

de Wijk, R.A., and Prinz, J.F. (2006) Mechanisms underlying the role of friction in oral texture. *J. Texture Stud.*, **37**, 413–427.

Dekhili, S., and d'Hauteville, F. (2009) Effect of the region of origin on the perceived quality of olive oil: an experimental approach using a control group. *Food Qual. Pref.*, **20**, 525–532.

Dekhili, S., Sirieix, L., and Cohen, E. (2011) How consumers choose olive oil: the importance of origin cues. *Food Qual. Pref.*, **22**, 757–762.

Delgado, C., and Guinard, J.X. (2011) How do consumer hedonic ratings for extra virgin olive oil relate to quality ratings by experts and descriptive analysis ratings? *Food Qual. Pref.*, **22**, 213–225.

Dinnella, C., Recchia, A., Vincenzi, S., Tuorila, H., and Monteleone, E. (2010) Temporary modification of salivary protein profile and individual responses to repeated phenolic astringent stimuli. *Chem. Sens.*, **35**, 75–85.

Dinnella, C., Recchia, A., Tuorila, H., and Monteleone, E. (2011) Individual astringency responsiveness affects the acceptance of phenol-rich foods. *Appetite*, **56**, 633–642.

Dinnella, C., Masi, C., Zoboli, G., and Monteleone, E. (2012) Sensory functionality of extra-virgin olive oil in vegetable foods assessed by temporal dominance of sensations and descriptive analysis. *Food Qual. Pref.*, **26**, 141–150.

Doleschall, F., Kemény, Z., Recseg, K., and Kővári, K. (2001) Monitoring of lipid degradation products by solid-phase microextraction. *J. Microcol. Sep.*, **13**, 215–220.

Drake, B. (1989) Sensory textural/rheological properties – a polyglot list. *J. Texture Stud.*, **20**, 1–27.

Drewnowski, A. (1997) Taste preferences and food intake. *Annu. Rev. Nutr.*, **17**, 237–253.

Esti, M., Contini, M., Moneta, E., and Sinesio, F. (2009) Phenolics compounds and temporal perception of bitterness and pungency in extra-virgin olive oils: changes occurring throughout storage. *Food Chem.*, **113**, 1095–1100.

Finotti, E., Bersani, A.M., and Bersani, E. (2007) Total quality indexes for extra-virgin olive oils. *J. Food Qual.*, **30**, 911–931.

Firestein, S. (2001) How the olfactory system makes sense of scents. *Nature*, **413**, 211–218.

Flynn, L.R., and Goldsmith, R.E. (1999) A short, reliable measure of subjective knowledge. *J. Bus. Res.*, **46**, 57–66.

Fotopoulos, C., and Krystallis, A. (2001) Are quality labels a real marketing advantage? A conjoint application on Greek PDO protected olive oil. *J. Int. Food Agribus. Mark.*, **12**, 1–22.

Gambacorta, G., Faccia, M., Previtali, M.A., Pati, S., La Notte, E., and Baiano, A. (2010) Effects of olive maturation and stoning on quality indices and antioxidant content of extra virgin oils (cv. *Coratina*) during storage. *Food Chem.*, **75**, 229–235.

Ganchrow, J.R., Steiner, J.E., and Daher, M. (1983) Neonatal facial expressions in response to different qualities and intensities of gustatory stimuli. *Infant Behav. Dev.*, **6**, 473–484.

Garcıa, J.M., Yousfi, K., Mateos, R., Olmo, M., and Cert, A. (2001) Reduction of oil bitterness by heating of olive (*Olea europea*) fruits. *J. Agric. Food Chem.*, **49**, 4231–4235.

Gawel, R., Oberholster, A., and Francis, I. (2000) A mouth-feel wheel: terminology for communicating the mouth-feel characteristics of red wine. *Aust. J. Grape Wine Res.*, **6**, 203–207.

Gengler, C.E., and Reynolds, T.J. (1995) Consumer understanding and advertising strategy: analysis and strategic translation of laddering data. *J Advert. Res.*, **35**, 19–33.

Goldsmith, R.E. (1998) Some characteristics of wine innovators. *Int. J. Wine Mark.*, **10**, 5–12.

Gottfried, J.A., and Dolan, R.J. (2003) The nose smells what the eye sees: crossmodal visual facilitation of human olfactory perspeption. *Neuron*, **39**, 375–386.

Green, B.G. (2002) Psychophysical measurement of oral chemesthesis, in *Methods in Chemosensory Research* (eds S.A. Simon and M.A.L. Nicolelis), CRC Press, Boca Raton, FL, pp. 3–19.

Guinard, J.X., Pangborn, R.M., and Lewis, M.J. (1986) The time-course of astringency in wine upon repeated ingestion. *Am. J. Enol. Vitic.*, **37**, 184–189.

Gutiérrez-Rosales, F., Perdiguero, S., Gutiérrez, R., and Olias, J.M. (1992) Evaluation of the bitter taste in virgin olive oil. *J. Am. Oil Chem. Soc.*, **69**, 394–395.

Gutiérrez-Rosales, F., Ríos, J.J., and Gómez-Rey, M.L. (2003) Main polyphenols in the bitter taste of virgin olive oil. Structural confirmation by on-line high-performance liquid chromatography electrospray ionization mass spectrometry. *J. Agric. Food Chem.*, **51**, 6021–6025.

Hatanaka, A., Kajiwara, T., Horino, H., and Inokuchi, K. (1992) Odor–structure relationships in *n*-hexenols and *n*-hexenals. *Z. Naturforsch.*, **47**, 183–189.

Holbrook, M.B., Lehman, D.R., and O'Shaughnessy, J. (1986) Using versus choosing: the relationship of the consumption experience to the reasons for purchasing. *Eur. J. Mark.*, **20** (8), 49–62.

Hollebeek, L.D., Jaeger, S.R., Brodie, R.J., and Balemi, A. (2010) The influence of involvement on purchase intention for new world wine. *Food Qual. Pref.*, **18**, 1033–1049.

IOC (2007) *Sensory Analysis of Olive Oil – Method – Organoleptic Assessment of Virgin Olive Oil*, COI/T.20/Doc. No. 15/Rev. 2, International Olive Council, Madrid.

Jaeger, S.R. (2006) Non-sensory factors in sensory science research. *Food Qual. Pref.*, **17**, 132–144.

Johnson, K.E., and Mervis, C.B. (1997) Effects of varying levels of expertise on the basic level of categorization. *J. Exp. Psychol.*, **126**, 248–277.

Kalua, C.M., Allen, M.S., Bedgood, D.R., Bishop, A.G., Prenzler, P.D., and Robards, K. (2007) Olive oil volatile compounds, flavor development and quality: a critical review. *Food Chem.*, **100**, 273–286.

Kanavouras, A., Hernandez-Munoz, P., Coutelieris, F., and Selke, S. (2004) Oxidation-derived flavor compounds as quality indicators for packaged olive oil. *J. Am. Oil Chem. Soc.*, **81**, 251–257.

Kim, U.K., Breslin, P.A.S., Reed, D., and Drayna, D. (2004) Genetics of human taste perception. *J. Dent. Res.*, **83**, 448–453.

Kiritsakis, A.K. (1998) Flavor components of olive oil – a review. *J. Am. Oil Chem. Soc.*, **75**, 673–681.

Kotti, F., Cerretani, L., Gargouri, M., Chiavaro, E., and Bendini, A. (2011) Evaluation of the volatile fraction of commercial virgin olive oils from Tunisia and Italy: relation with olfactory attributes. *J. Food Biochem.*, **35**, 681–698.

Krystallis, A., and Ness, M. (2003) Motivational and cognitive structures of Greek consumers in the purchase of quality food products. *J. Int. Consumer Mark.*, **16**, 7–36.

Krystallis, A., and Ness, M. (2005) Consumer preferences for quality foods from a South European perspective: a conjoint analysis implementation on Greek olive oil. *Int. Food Agribus. Manage. Rev.*, **8**, 62–91.

Kuentzel, H., and Bahri, D. (1991) Synthetic ingredients of food flavourings, in *Food Flavourings* (ed. P.R. Ashurst), Blackie, Glasgow, pp. 115–157.

Larsson, M. (2002) Odor memory: a memory systems approach, in *Olfaction, Taste, and Cognition* (eds C. Rouby, B. Schaal, D. Dubois, R. Gervais, and A. Holley), Cambridge University Press, Cambridge, pp. 231–245.

Lawless, H.T. (1984) Flavor description of white wine by "expert" and nonexpert wine consumers. *J. Food Sci.*, **49**, 120–123.

Lee, C.B., and Lawless, H.T. (1991) Time-course of astringent materials. *Chem. Sens.*, **16**, 225–238.

Lindemann, B. (2001) Receptors and transduction in taste. *Nature*, **413**, 219–225.

Livermore, A., and Laing, D.G. (1998) The influence of odor type on the discrimination and identification of odorants in multicomponent odor mixtures. *Physiol. Behav.*, **65**, 311–320.

Luna, G., Morales, M.T., and Aparicio, R. (2006) Characterization of varietal virgin olive oils by their volatile compositions. *Food Chem.*, **98**, 243–252.

Marshall, D., and Bell, R. (2004) Relating the food involvement scale to demographic variables, food choice and other constructs. *Food Qual. Pref.*, **15**, 871–879.

Martínez, M.G., Aragonés, Z., and Poole, N. (2002) A repositioning strategy for olive oil in the UK market. *Agribusiness*, **18**, 163–180.

Matsatsinis, N.F., Grigoroudis, E., and Samaras, A.P. (2007) Comparing distributors' judgements to buyers' preferences: a consumer value analysis in the Greek olive oil market. *Int. J. Retail Distrib. Manage.*, **35**, 342–362.

Mattes, R.D. (2009) Is there a fatty acid taste? *Annu. Rev. Nutr.*, **29**, 305–327.

McEwan, J. (1994) Consumer attitudes and olive oil acceptance, the potential consumer. *Grasas Aceites*, **45**, 9–15.

Meillon, S., Urbano, C., Guillot, G., and Schlich, P. (2010) Acceptability of partially dealcoholized wines – measuring the impact of sensory and information cues on overall liking in real-life settings. *Food Qual. Pref.*, **21**, 763–773.

Meiselman, H.L., and MacFie, H.J.H. (eds) (1996) *Food Choice, Acceptance and Consumption*, Blackie, London.

Mennella, J.A., Lukasewycz, L.D., Castor, S.M., and Beauchamp, G.K. (2011) The timing and duration of a sensitive period in human flavor learning: a randomized trial. *Am. J. Clin. Nutr.*, **93**, 1019–1024.

Mojet, J., and de Jong, S. (1994) The sensory wheel of virgin olive oil. *Grasas Aceites*, **45**, 42–47.

Montedoro, G., Baldioli, M., and Servili, M. (1992) Sensory and nutritional relevance of phenolic compounds in olive oil. *G. Ital. Nutr. Clin. Prev.*, **1**, 19–32.

Monteleone, E., Condelli, N., Dinnella, C., and Bertuccioli, M. (2004) Prediction of perceived astringency induced by phenolic compounds. *Food Qual. Pref.*, **15**, 761–769.

Morales, M.T., Alonso, M.V., Rios, J.J., and Aparicio, R. (1995) Virgin olive oil aroma: relationship between volatile compounds and sensory attributes by chemometrics. *J. Agric. Food Chem.*, **43**, 2925–2931.

Morales, M.T., Luna, G., and Aparicio, R. (2005) Comparative study of virgin olive oil defects. *Food Chem.*, **91**, 293–301.

Mueller, S., and Rungie, C. (2009) Is there more information in best–worst choice data? Using the attitude heterogeneity structure to identify consumer segments. *Int. J. Wine Bus. Res*, **21**, 24–40.

Mueller-Harvey, I. (2006). Unravelling the conundrum of tannins in animal nutrition and health. *J. Sci. Food Agric.*, **86**, 2010–2037.

Mullen, B., and Johnson, C. (eds) (1990) *The Psychology of Consumer Behaviour*, Lawrence Erlbaum Associates, Hillsdale, NJ.

Nayak, A., and Carpenter, G.H. (2008) A physiological model for tea induced astringency. *Physiol. Behav.*, **95**, 290–294.

Pangborn, R.M., Guinard, J.X., and Davis, R.G. (1988) Regional aroma preferences. *Food Qual. Pref.*, **1**, 11–19.

Pliner, P. (1982) The effects of mere exposure on liking for edible substances. *Appetite*, **3**, 283–290.

Prescott, J., and Stevenson, R.J. (1995) Pungency in food perception and preference. *Food Rev. Int.*, **11** 665–698.

Pripp, A.H., Busch, J., and Vreeker, R. (2004) Effect of viscosity, sodium caseinate and oil on bitterness perception of olive oil phenolics. *Food Qual. Pref.*, **15**, 375–382.

Psomiadou, E., Konstantinos, X., Blekas, K.G., Tsimidou, M.Z., and Boskou, D. (2003) Proposed parameters for monitoring quality of virgin olive oil (*koroneiki* cv). *Eur. J. Lipid Sci. Technol.*, **105**, 403–409.

Recchia, A., Monteleone, E., and Tuorila, H. (2012) Responses to extra virgin olive oils in consumers with varying commitment to oils. *Food Qual. Pref.*, **24**, 153–161.

Reiners, J., and Grosh, W. (1998) Odorants of virgin olive oils with different flavor profiles. *J. Agric. Food Chem.*, **46**, 2754–2763.

Roca, M., and Minguez-Mosquera, A. (1991) Change in the natural ratio between chlorophylls and carotenoids in olive fruit during processing for virgin olive oil. *J. Am. Oil Chem. Soc.*, **78**, 133–138.

Rolls, E.T. (2005) Taste, olfactory, and food texture processing in the brain, and the control of food intake. *Physiol. Behav.*, **85**, 45–56.

Romani, A., Lapucci, C., Cantini, C., Ieri, F., Mulinacci, N., and Visioli, F. (2007) Evolution of minor polar compounds and antioxidant capacity during storage of bottled extra virgin olive oil. *J. Agric. Food Chem.*, **55**, 1315–1320.

Rota, L.M., and Zellner, D.A. (2007) The categorization effect in hedonic contrast: experts differ from novices. *Psychon. Bull. Rev.*, **14**, 179–183.

Rozin, P. (1990) Getting to like the burn of chili pepper: biological, psychological, and cultural perspectives, in *Chemical Senses. Vol 2. Irritation* (eds B.G. Green, J.R. Mason, and M.R. Kare), Marcel Dekker, New York, pp. 231–299.

Rozin, P., and Vollmecke, T.A. (1986) Food likes and dislikes. *Annu. Rev. Nutr.*, **6**, 433–456.

Sandalidou, E., and Baourakis, G. (2002) Customers' perspectives on the quality of organic olive oil in Greece. A satisfaction evaluation approach. *Br. Food J.*, **104**, 391–406.

Santosa, M., and Guinard, J.X. (2011) Means–end chains analysis of extra virgin olive oil purchase and consumption behaviour. *Food Qual. Pref.*, **22**, 304–316.

Santosa, M., Abdi, H., and Guinard, J.X. (2010) Modified sorting task to investigate consumer perceptions of extra virgin olive oils. *Food Qual. Pref.*, **21**, 881–892.

Saper, C.B., Chou, T., and Elmquist, J.K. (2002) The need to feed: homeostatic and hedonic control of eating. *Neuron*, **36**, 199–211.

Schutz, H.G. (1994) Appropriateness as a measure of the cognitive-contextual aspects of food acceptance, in *Measurement of Food Preferences* (eds H.J.M. MacFie and D.M.H. Thomson), Blackie, London, pp. 25–50.

Sinesio, F., Moneta, E., and Esti, M. (2005) The dynamic sensory evaluation of bitterness and pungency in virgin olive oil. *Food Qual. Pref.*, **16**, 557–564.

Small, D.M., and Prescott, J. (2005) Odor/taste integration and the perception of flavor. *Exp. Brain Res.*, **166**, 345–357.

Stefani, G., Romano, D., and Cavicchi, A. (2006) Consumer expectations, liking and willingness to pay for specialty foods: do sensory characteristics tell the whole story? *Food Qual. Pref.*, **17**, 53–62.

Szczesniak, A.S. (1963) Classification of textural characteristics. *J. Food Sci.*, **28**, 385–389.

Tanaka, J.W., and Taylor, M. (1991) Object categories and expertise: is the basic level in the eye of the beholder? *Cogn. Psychol.*, **23**, 457–482.

Törnwall, O., Dinnella, C., Keskitalo-Vuokko, K., Silventoinen, K., Perola, M., Monteleone, E., Kaprio, J., and Tuorila, H. (2011) Astringency perception and heritability among young Finnish twins. *Chemosens. Percept.*, **4**, 134–144.

Törnwall, O., Silventoinen, K., Kaprio, J., and Tuorila, H. (2012) Why do some like it hot? Genetic and environmental contributions to the pleasantness of oral pungency. *Physiol. Behav.*, **107**, 381–389.

Torri, L., Recchia, A., Dinnella, C., Næs, T., Tuorila, H., and Monteleone, E. (2013) Projective Mapping for interpreting wine aroma differences as perceived by naïve and experienced assessors. *Food Qual. Pref.*, **29**, 6–15.

Tovar, M.J., Motilva, M.J., and Romero, M.P. (2001) Changes in the phenolic composition of virgin olive oil from young trees (*Olea europaea* L. cv. *Arbequina*) grown under linear irrigation strategies. *J. Agric. Food Chem.*, **49**, 5502–5508.

Tsimidou, M. (1998) Polyphenols and quality of virgin olive oil in retrospect. *Ital. J. Food Sci.*, **10**, 99–116.

Tuorila, H. (2007) Sensory perception as a basis of food acceptance and consumption, in *Consumer-Led Food Product Development* (ed. H.J.H. MacFie), Woodhead Publishing, Cambridge, pp. 34–65.

Tuorila, H., and Cardello, A.V. (2002) Consumer responses to an off-flavour in juice in the presence of specific health claims. *Food Qual. Pref.*, **13**, 561–569.

Viana, F. (2011) Chemosensory properties of the trigeminal system. *ACS Chem. Neurosci.*, **2**, 38–50.

Vichi, S., Pizzale, L., Conte, L.S., Buxaderas, S., and Lopez-Tamames, E. (2003) Solid-phase microextraction in the analysis of virgin olive oil volatile fraction: modifications induced by oxidation and suitable markers of oxidative status. *J. Agric. Food Chem.*, **51**, 6564–6571.

Vichi, S., Romero, A., Gallardo-Chacón, J., Tous, J., López-Tamames, E., and Buxaderas, S. (2009) Volatile phenols in virgin olive oils: influence of olive variety on their formation during fruits storage. *Food Chem.*, **116**, 651–656.

Wilkie, W.L. (1986) *Consumer Behaviour*, John Wiley & Sons, Inc., New York, pp. 350–353.

Zaichkowsky, J. (1985) Measuring the involvement construct. *J. Consumer Res.*, **12**, 341–352.

Zellner, D.A. (1991) How foods get be liked: some general mechanisms and some special cases, in *The Hedonics of Taste* (ed. R.C. Bolles), Laurence Erlbaum Associates, Hillsdale, NJ, pp. 199–217.

Zou, Z., and Buck, L.B. (2006) Combinatorial effects of olfactory mixes in olfactory cortex. *Science*, **311**, 1477–1481.

4
Sensory Quality Control

Susan Langstaff
Applied Sensory LLC, Fairfield, CA, USA

4.1 Introduction

Flavor and changes to flavor due to the chemistry in olive oil play a major role in dictating standards and the market value of olive oil. These flavors can be amplified, diminished, or disturbed by changing the quality of the fruit, the way it was handled, the manipulation of the paste, the extraction process, and finally the cleaning and storage of the oil (Vossen, 2005). Premium quality, fresh, virgin olive oil is characterized by a fruity aroma with a bitter and pungent finish. For such oil, it is common for consumers to pay high prices.

The development of sensory quality standards for olive oil began in the 1970s to address five main issues (Mailer, 2012):

- Trade – to comply with nutritional guidelines and tolerances.
- Safety – to verify that the oil is safe and fit for use.
- Production – to ensure that deviation from expected quality is kept within an acceptable tolerance.
- Fraud – to ensure that legal requirements are met and that the oil is genuine.
- Freshness – to provide an indication of the age of the oil.

4.2 Historical perspective

In the ancient Mediterranean area, olive oil was an important product and methods of measuring quality were a matter of concern to communities

Olive Oil Sensory Science, First Edition. Edited by Erminio Monteleone and Susan Langstaff.

that traded it. High-quality olive oil was used as nourishment and in the preparation of cosmetics and perfumes, where it was mixed with other ingredients. Inferior-quality oil was equally valuable since it was used in lamps for lighting and to make soap (Zampounis, 2006). Later, the use of steam and electricity boosted the production of higher quality oils from olive mills. These oils cost more to produce and therefore demanded a higher price than other vegetable oils (Angerosa, 2006).

In 1959, the United Nations founded the International Olive Oil Council (renamed the International Olive Council, or IOC, in 2006), an intergovernmental agency composed of 18 olive-growing and oil-producing countries from around the Mediterranean. The purpose of the IOC was to supply financial and technical assistance to growers and millers, fund research in oil quality and chemistry, and promote olive oil consumption throughout the world.

The IOC took upon itself the task of creating a scientific "taste test" for olive oil. A group of chemists and sensory scientists from Italy and Spain experimented with olive oil for 20 years and developed a tasting method in the early 1980s. They identified various flaws in taste and aroma which can arise in oil, and instituted a rigorous and reproducible tasting protocol to determine the quality grade of an olive oil (Mueller, 2012).

The resulting IOC tasting methodology, together with a set of chemical requirements, became the basis of a new European Union (EU) olive oil law, which was passed in 1991 and, with some modifications, is still in force today. To be called extra virgin, an oil must have zero "organoleptic" flaws and some perceptible fruitiness. With this law, olive oil became the first food in the world whose quality was legally determined at least in part by its sensory properties (Mueller, 2012).

In the mid- to late-1990s, olive oil became one of the most frequently adulterated agricultural products in the EU (Mueller, 2012). The substitution or adulteration of food products is not only an economic fraud but may also have severe health implications to consumers. This was the case with the Spanish toxic oil syndrome (TOS), which resulted from the ingestion of aniline-denatured rapeseed oil that involved more than 20 000 people (Angerosa, 2006).

Today, more than 85% of the world's olives are grown in IOC member nations (IOC, 2012a). The IOC's duties include adopting standards and quality-management guides for industry, developing chemical and sensory testing methods to assess olive oil quality and authenticity, and providing official recognition to laboratories that demonstrate proficiency in employing the IOC's recommended testing methods (IOC, 2012b).

In recent years, various individual countries have been developing their own olive oil specifications. The United States and Australia, for example, have based their specifications on IOC methodology but have made some modifications (Mailer, 2012).

4.3 Standard methods

Sensory testing plays an important role in maintaining olive oil quality. To establish a control program, methods of measuring the characteristics must be identified and limits of variability around the target level must be set. Standard methods for the evaluation of olive oil have been published by different organizations (Mailer, 2012):

- American Oil Chemists' Society (AOCS)
- International Olive Council (IOC) (formerly International Olive Oil Council)
- International Standards Organization (ISO)
- AOAC International (formerly the Association of Official Agricultural Chemists and then the Association of Official Analytical Chemists) (AOAC)
- European Committee for Standardization (CEN)

Additionally, the American Society for Testing and Materials (ASTM) has a standard practice in place for the sensory evaluation of edible oils and fats (ASTM, 2011).

4.4 Legislative standards

The evaluation of olive oil quality and the monitoring of product authenticity are accomplished by analytical techniques designed to provide qualitative and quantitative data on olive oil composition. Several different standardizing bodies – IOC, European Community (EC), and Codex Alimentarius (Codex, 2009) – have evaluated and adopted many of the methods listed above (Mailer, 2012). These organizations have also set limit values for many of the physical, chemical, and sensory parameters in olive oil (Table 4.1). Various national governments have also set limits based on those set by the

Table 4.1 Organizations that develop methods and/or standard limits for olive oil evaluation.

Organization	Methods	Standard limits?
American Oil Chemists' Society (AOCS)	AOCS	No
AOAC International	AOAC	No
International Standards Organization (ISO)	ISO	No
International Olive Council (IOC)	COI[a], AOAC, ISO, IUPAC[b]	Yes
European Community (EC)	CEN	Yes
Codex Alimentarius	COI[a], AOAC, ISO, IUPAC[b]	Yes
National governments	Based on IOC, EC, Codex Alimentarius, others	Yes

[a]COI = IOC.
[b]IUPAC = International Union of Pure and Applied Chemistry.

above standardizing bodies. Olive oil is classified by how it was produced, by its chemistry, and by its flavor. The classification of olive oil grades is discussed later in this chapter.

4.5 Parameters used to evaluate olive oil quality

The IOC, EC, and Codex Alimentarius share five parameters to evaluate olive oil quality:

- free acidity
- peroxide value
- absorbance in the UV region
- organoleptic assessment
- volatile halogenated solvents.

Table 4.2 displays standard limit values and explanations of free acidity, peroxide value, and UV absorbance. Median values for sensory defects and

Table 4.2 Virgin olive oil quality criteria and standard limits (IOC, EC, and USDA).

Parameters		Extra virgin	Virgin	Lampante
Chemical parameters	Free acidity[a] (FA%)	≤0.8	≤2.0	>3.3 (IOC) >2.0 (EC, USDA)
	Peroxide value[b] (meq.O_2/kg)	≤20	≤20	>20 (IOC, EC) No limit (USDA)
UV absorbance[c]	$K_{232\ nm}$	≤2.5	≤2.6	–
	$K_{270\ nm}$	≤0.22	≤0.25	–
	ΔK	≤0.01	≤0.01	–
Sensory attributes (median values across panelists)	Defects	0	≤3.5 (IOC) ≤2.5 (EC, USDA)	>6 (IOC) >2.5 (EC, USDA)
	Fruitiness	>0	>0	–

[a]This is a very broad indicator of the quality of the fruit and handling procedures prior to milling. It is a measurement of hydrolytic breakdown of the fatty acid chains from triglycerides into diglycerides and monoglycerides, liberating free fatty acids. It is usually expressed as percent free fatty acids (free acidity, FA%) on the basis of the oleic acid, because that is the predominant fatty acid in olive oil. The higher the FA% (usually when >0.5%), the greater is the probability that the oil will have a taste defect. Free acidity remains stable as the oil ages.

[b]This measurement is a very broad indicator of the amount of primary oxidation that has occurred, forming peroxide compounds within the oil. The peroxide value is determined by conducting a titration that liberates iodine from potassium iodide and is expressed as a value in milliequivalents of free oxygen per kilogram of oil (meq. O_2/kg). A high value indicates that olives or paste were likely handled improperly, the oil could be defective, and the oil might not keep well.

[c]This measurement is a more specific indicator of oxidation, especially in oils that have been heated in the refining process. It measures the quantity of certain oxidized compounds that resonate at wavelengths of 232 and 270 nm in the UV spectrum in a spectrophotometer. ΔK detects oil treatments with color-removing substances and the presence of refined or pomace oil by measuring the difference between the absorbances at 270 and 266–274 nm. A high peroxide value in young oils suggests a short shelf life but it is more difficult to interpret in older oils since the value decreases as the available oxygen is used up and the oil starts to become rancid. Peroxide value declines with oil age.

Table 4.3 Additional quality criteria for virgin olive oils and standard limits (IOC).

Parameters		Extra virgin, Virgin	Lampante
Halogenated solvents[a] (mg/kg)	Each solvent	$\leq$0.1	–
	Sum of solvents	$\leq$0.2	–
Metal traces[b] (mg/kg)	Iron	$\leq$3.0	$\leq$3.0
	Copper	$\leq$0.1	$\leq$0.1
Insoluble impurities[c] (%)		$\leq$0.1	$\leq$0.2
Unsaponifiable material[d] (g/kg)		Max. 15	Max. 15
α-Tocopherol as an additive[e]		None	None

[a]This measures the quantity of chloroform, trichloroethylene, and tetrachloroethylene that may be present as residuals in solvent-extracted oils. Headspace volatile gases are measured in a gas chromatograph and expressed in mg/kg.
[b]This is a measure of the amount of iron and copper in a sample taken by burning the oil in a special high-temperature graphite oven and analyzing the ash by atomic absorption spectrometry.
[c]This is a measurement of the presence of dirt, minerals, resins, oxidized fatty acids, alkaline soaps of palmitic and stearic acids, and proteins that are suspended in the oil. It is determined by dissolving some oil in petroleum ether and filtering out the impurities and is expressed as a percentage of the total.
[d]This test measures the content of components that do not turn to soap under the process of saponification (addition of lye). It is a measure of the number of ester bonds in the fatty acid. A high number of ester bonds suggests that the fat molecule is intact, which assumes that the oil has been correctly processed from sound fruit.
[e]None is permitted for virgin olive oils. For refined olive oil, olive oil, refined pomace oil, and pomace oil, a maximum of 200 mg/kg of α-tocopherol is permitted to aid stability.

fruity aroma in the various olive oil categories are also shown and are discussed in more detail in Section 4.6. Standard limits for halogenated solvents and their definition are shown in Table 4.3.

Additionally, the IOC and Codex Alimentarius include the following physical and chemical tests:

- α-tocopherol
- Cu, Fe, Pb, As determination
- insoluble impurities
- unsaponifiable matter content.

Standard limit values and explanations of the above parameters are shown in Table 4.3.

Finally, only the EC specifies oil content in pomace residue.

In addition to the above parameters, there are several other analyses that are helpful to complete the evaluation of olive oil quality. These are measurements related to the:

- level of antioxidants
- state of oxidation
- hydrolysis
- shelf life
- presence of contaminants and volatile compounds.

4.6 Organoleptic assessment – aroma and flavor

Organoleptic assessment – also called sensory evaluation or sensory analysis – is a process by which trained panelists evaluate the sight, aroma, taste, texture, and aftertaste of products. The methodology for evaluating the sensory attributes of virgin olive oils, known as the Panel Test method, was developed in the 1980s by the IOC and later adopted by the European Commission (Regulation No. 640/2008). The main goal of the Panel Test, also known as the IOC test, is to detect sensory defects and their intensities for the classification of various grades of virgin olive oils.

IOC methods define the:

- selection and training of tasters (IOC, 2011a)
- minimum number of tasters (IOOC, 1996)
- basic olive oil vocabulary (IOC, 2007a)
- tasting technique (IOOC, 1996; IOC, 2011b)
- tasting glass (IOC, 2007b)
- test booth (IOC, 2007c)
- profile sheet (IOOC, 1996).

4.7 IOC taste panel development

Potential panel members are selected and screened for their ability to detect small differences in the intensity of aromas from standard diluted samples of common olive oil defects. To determine the prospective panelists' sensitivity to bitterness, a caffeine–water solution is evaluated. These two tests are intended to measure the innate abilities of a taster rather than their previous experience with olive oil.

For evaluating the sense of smell, an arrangement test is commonly used. The candidate is asked to place a sample glass containing a specific concentration of a defect back into a series of 12 glasses with dilution arranged from strongest (left) to weakest (right). If correctly placed, the sample should have an intensity that is slightly weaker than the one on the left and slightly stronger than the one on the right. The potential tasters are given 0 points for a correct placement, 1 point if off by one position, 4 points if off by two positions, 9 points if off by three positions and 16 points, including disqualification, if off by four positions. After four replications of each defect arrangement test, for a total of 16 tests, panelists are given a final score, which may not exceed 34 points.

For tasting bitterness, candidates are rated for their absolute detection threshold and recognition threshold. To determine the difference threshold, tasters are given four samples and they must arrange the oils in order of increasing bitterness. The ideal taster would have a great ability to discriminate between different stimuli and their intensities.

The IOC-recognized taste panel consists of a minimum of eight tasters who have been screened, trained, and tested for their skill level. Sensory evaluations are conducted in a controlled environment (IOOC, 1996) using special blue glasses so that the color of the oil is concealed (IOC, 2007b). Three-digit random number codes are used to identify the oils and to prevent bias. The oils are warmed to 80°F (26.5° C) and judges evaluate them using a prescribed methodology (IOOC, 1996).

4.8 IOC terminology for virgin olive oils

A special vocabulary has been developed to characterize the flavors of virgin olive oils. These oils, as described more in detail below, are produced only from fruit from the olive tree using solely mechanical or other physical means and can be consumed without any further treatment. In the IOC methodology, panelists do not rate aroma and flavor attributes separately; rather, they evaluate each perception resulting from the entire olfactory–gustatory–tactile sensation. Research has indicated that the sensory characterization of olive oil was not improved by the independent evaluation of all sensations that form the flavor (Angerosa, 2000).

4.8.1 Negative attributes (defined by the IOC)

These defects are due mainly to the deterioration of olives that result from prolonged retention and from the fermentation or oxidation process to which the oils are subjected during storage. Factors influencing olive oil sensory defects are listed in Table 4.4. Olive oil sensory defects associated with processing and storage are shown in Figure 4.1. A useful tool for learning to recognize off-aromas and flavors in olive oil is the Defects Wheel for Olive Oil (Figure 4.2) (Langstaff *et al.*, 2011).

Table 4.4 Factors influencing olive oil sensory defects.

- Condition of fruit (pest infestation and diseases)
- Speed to crush
- Period of storage in final package prior to use
- Cleanliness of tanks, piping and bottling equipment
- Method and period of oil storage prior to packing
- Oil storage temperature
- Exposure to oxygen/light
- Harvesting method
- Fruit transport method
- Lack of good manufacturing practices (GMPs)

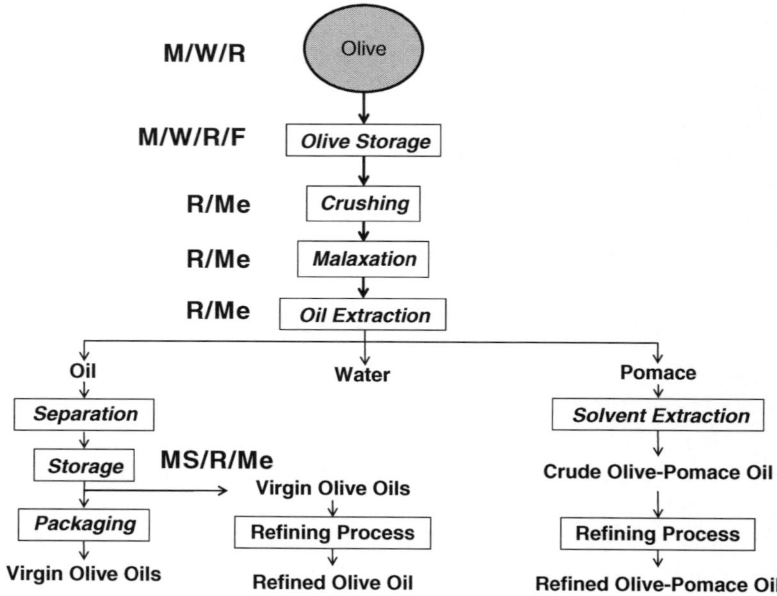

Defects (see text for definitions):
- M = musty – humid – earthy
- W = winey – vinegary – acid – sour
- R = rancid
- F = fusty
- MS = muddy sediment
- Me = metallic

Figure 4.1 Olive oil sensory defects associated with processing/storage.

4.8.1.1 Fusty/muddy sediment This is the characteristic flavor of oil obtained from olives piled or stored in conditions such as to have undergone an advanced stage of anaerobic fermentation, or of oil that has been left in contact with the sediment that settles in underground tanks and vats which has also undergone a process of anaerobic fermentation.

4.8.1.2 Musty–humid–earthy This is the characteristic flavor of oils obtained from fruit in which large numbers of fungi and yeasts have developed as a result of its being stored in humid conditions for several days or of oil obtained from olives that have been collected with earth or mud on them and that have not been washed.

4.8.1.3 Winey–vinegary–acid–sour This is the characteristic flavor of certain oils reminiscent of wine or vinegar. It is mainly due to a process of aerobic fermentation in the olives or in olive paste left on processing mats which have not been properly cleaned and leads to the formation of acetic acid, ethyl acetate, and ethanol.

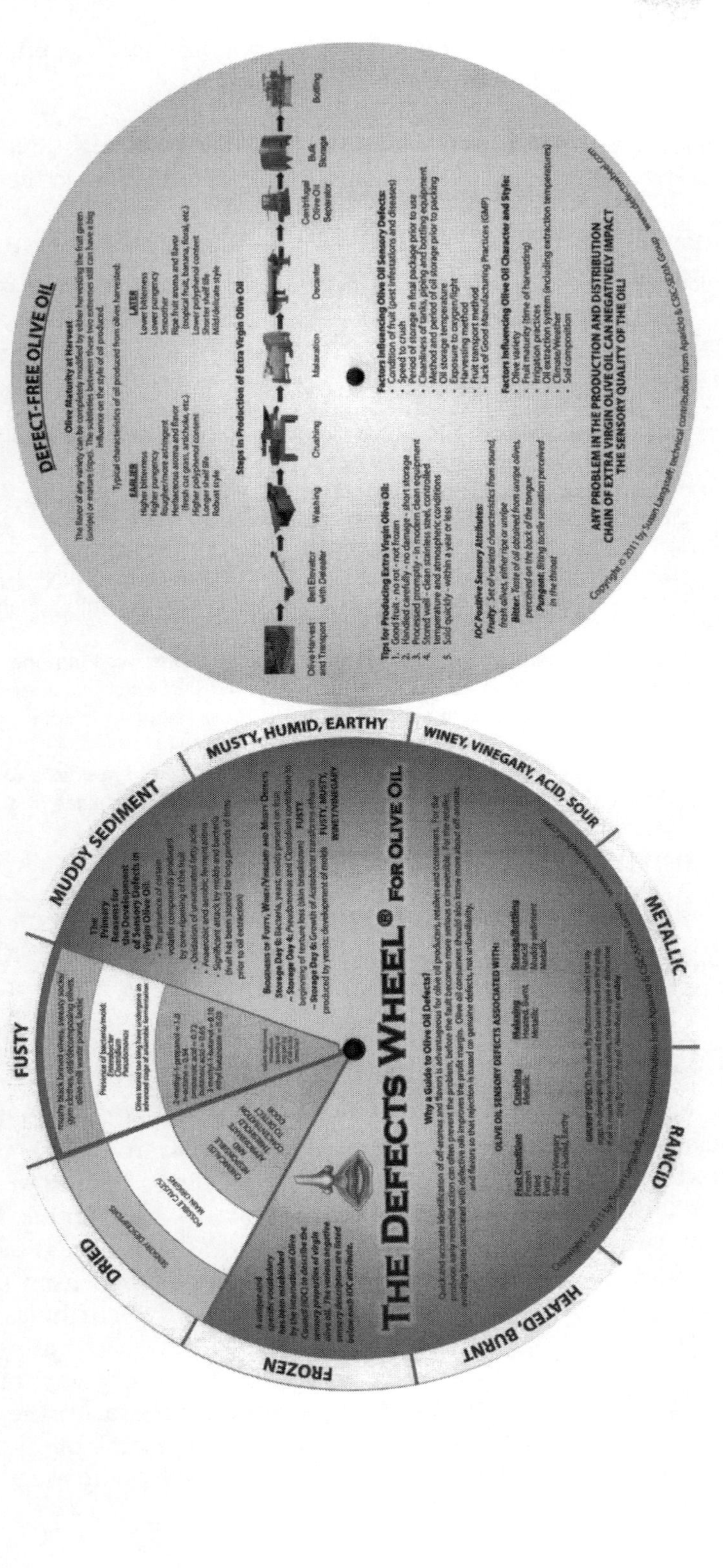

Figure 4.2 The Defects Wheel for Olive Oil. For a color version of this figure, see the color plates section.

4.8.1.4 Rancid This is the characteristic flavor of oils that have undergone an intense process of oxidation.

4.8.1.5 Frostbitten olives (wet wood) This is the characteristic flavor of oils extracted from olives that have been injured by frost while on the tree.

4.8.1.6 Other negative attributes (IOC) Ten additional negative attributes identified and defined by the IOC are listed in Table 4.5.

Table 4.5 Other IOC negative attributes.

Attribute	Definition
Heated or burnt	Characteristic flavor of oils caused by excessive and/or prolonged heating during processing, particularly when the paste is thermally mixed, if this is done under unsuitable thermal conditions
Hay–wood	Characteristic flavor of certain oils produced from olives that have dried out
Rough	Thick, pasty mouthfeel sensation produced by certain old oils
Greasy	Flavor of oil reminiscent of that of diesel oil, grease, or mineral oil
Vegetable water	Flavor acquired by the oil as a result of prolonged contact with vegetable water which has undergone fermentation processes
Brine	Flavor of oil extracted from olives which have been preserved in brine
Metallic	Flavor reminiscent of metals. It is characteristic of oil which has been in prolonged contact with metallic surfaces during crushing, mixing, pressing, or storage
Esparto	Characteristic flavor of oil obtained from olives pressed in new esparto mats. The flavor may differ depending on whether the mats are made of green esparto or dried esparto
Grubby	Flavor of oil obtained from olives which have been heavily attacked by the grubs of the olive fruit fly (*Bactrocera oleae*)
Cucumber	Flavor produced when an oil is hermetically packed for too long, particularly in tin containers, and which is attributed to the formation of 2,6-nonadienal

Source: IOC (2011b).

4.8.2 Positive attributes (defined by the IOC)

The aroma descriptions of defect-free virgin olive oils fall generally into two main categories: olive fruity and green notes (Angerosa, 2000). Olive fruity refers to the aroma/flavor of healthy, fresh olives harvested at the correct degree of ripeness. Green notes suggest leaves, green fruits, unripe olives, freshly cut grass, bitterness, pungency, and astringency. Bitterness, pungency, and astringency are characteristic of oils obtained from unripe, healthy fresh olives, and are therefore considered positive attributes. They will mellow as the oils age.

4.8.2.1 Fruity This is the set of olfactory sensations characteristic of the oil, which depends on the variety and comes from sound, fresh olives, either ripe or unripe. It is perceived directly or through the back of the nose.

4.8.2.2 Bitter This is the characteristic primary taste of oil obtained from green olives or olives turning color. It is perceived in the circumvallate papillae on the "V" region of the tongue.

4.8.2.3 Pungent This is the biting tactile sensation characteristic of oils produced at the start of the crop year, primarily from olives that are still unripe. It is perceived throughout the whole of the mouth cavity, particularly in the throat.

4.9 IOC profile sheet

The IOC profile sheet (Figure 4.3) emphasizes the main negative attributes defined above. Fruity, bitter, and pungent are the only positive attributes considered. The intensity of each attribute is scored on an unstructured 10 cm line scale and converted to numerical values for statistical analysis. The median of each attribute is calculated. Although the mean has traditionally been a popular measure of a mid-point in a set of sensory scores, it has the disadvantage of being affected by any single value being too high or too low compared with the rest of the scores. This is why a median is sometimes taken as a better measure of a mid-point.

A majority of five out of eight tasters must agree in their profile sheet characterization of the tasted oil. If the coefficient of variation (CV) of the main defect is greater than 20% in defective oil or the CV is greater than 10% in an extra virgin oil for the fruitiness character, the test must be repeated. This means that the tasters must be quite close in indentifying the primary defect in each oil, if it has one, and the intensities of the defect must be within two points on the 10 cm scale. For fruitiness, the intensity has to be within one point on the intensity scale (Vossen, 2010).

Each grade of olive oil (see Section 4.11) is identified by a sensory standard that defines the intensity level of both positive and negative characteristics. Table 4.2 displays the median values for sensory defects and fruity aroma in the various olive oil categories, along with several physical and chemical limits related to quality.

Ideally, a taste panel should rate oils with very little discrepancy between individuals within the panel and rate the oils very similarly to other taste panels around the world. Statistically they should have a within-panel repeatability of 5–20%.

4.10 "Ring tests"

Across multiple locations, there is a need to standardize sensory quality control (QC) procedures. "Ring tests," also called "round-robin tests," compare different sensory panels that perform the same evaluations with their own individual panel and are often called *proficiency tests* (Hund, Massart, and

PROFILE SHEET FOR VIRGIN OLIVE OIL

INTENSITY OF PERCEPTION OF DEFECTS:

Fusty/ muddy sediment

Musty–humid–earthy

Winey–vinegary–acid–sour

Frostbitten olives (wet wood)

Rancid

Others (specify)

INTENSITY OF PERCEPTION OF POSITIVE ATTRIBUTES:

Fruity

greenly ripely

Bitter

Pungent

Name of taster:

Sample code:

Date:

Comments:

Figure 4.3 IOC sensory profile sheet for virgin oil.

Smeyers-Verbeke, 2000). They are an essential part of the requirements for laboratory accreditation. The participating sensory panels have to prove their technical competence regularly, so proficiency tests are usually performed several times a year. They are often managed by a central body, often a governmental organization.

Twice each year, all IOC-accredited sensory panels around the world receive the same five oils for evaluation. Results from past ring tests indicated poor reproducibility of overall grading scores (Angerosa, 2000). The researchers attributed this mainly to panelists using different parts of the intensity scale in their evaluation of oil samples with slightly perceptible defects because of ineffective training. They also speculated that the poor reproducibility could be due to differences in the evaluations of positive quality attributes, since judges on the various panels have dissimilar origins, food habits, and cultures (Angerosa, 2000).

The AOCS has instituted a program of olive oil sensory panel proficiency tests. The testing and scoring will abide with the IOC guideline COI/T.20/Doc. no. 15/Rev. 4 (November 2011) (IOC, 2011b) and qualifying panels will be promoted as an AOCS recognized olive oil sensory panel (AOCS, 2012).

A more detailed discussion of proficiency testing of sensory panels may be found in the book edited by Kilcast (Hyldig, 2010).

4.11 IOC classification of olive oil grades

The International Standards under IOC resolution COI/T.15/NC no. 3–25 (revised June 2003) lists nine grades of olive oil in two primary categories – (1) Olive Oil and (2) Olive Pomace Oil. A hierarchy of the olive grades is shown in Figure 4.4. The following are the official definitions of each of the nine grades:

- Olive Oil category
  Oil obtained solely from the fruit of the olive tree (*Olea europaea sativa*) to the exclusion of oils obtained using solvents or re-esterification processes and of any mixture with oils of other kinds (seed or nut oils).

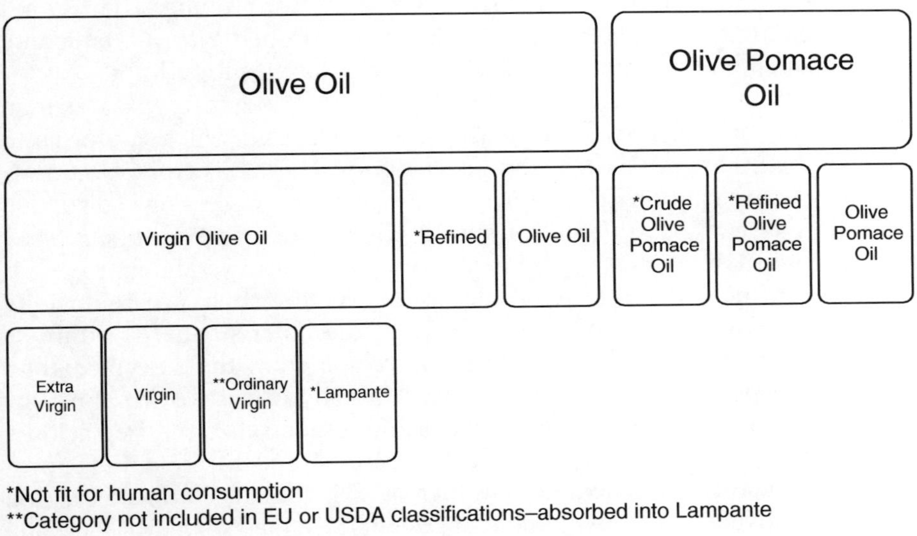

Figure 4.4 Hierarchy of olive oil grades.

- Virgin Olive Oils
  Oils obtained solely by mechanical or physical means under thermal conditions that do not lead to alterations in the oil, using only treatments such as washing, decantation, centrifugation, and filtration. Those fit for human consumption are as follows:
  - Extra Virgin Olive Oil. This oil, as evaluated numerically by the mean of a certified taste panel, contains zero defects and greater than zero positive attributes. In other words, more than half of the tasters indicated that it is not defective and has some fruitiness. Extra virgin olive oil also must have a free acidity of less than 0.8% and conform to all the standards listed in its category. This is the highest quality rating for an olive oil. Extra virgin olive oil should have clear flavor characteristics that reflect the fruit from which it was made. In relation to the complex matrix of variety, fruit maturity, growing region, and extraction technique, extra virgin olive oils can be very different from one another.
  - Virgin Olive Oil. Oil with a sensory analysis rating of the mean of tasters, having defects from 0 to less than 3.5, a free acidity of less than 2%, and conforms to all the other standards in its category. These are oils with analytical and sensory indices that reflect slightly lower quality than extra virgin olive oil.
  - Ordinary Virgin Olive Oil. Oil with a lower sensory rating (defects from the mean of tasters 3.5 to less than 6.0), a free acidity of less than 3.3%, and conformity within its category for all other standards. This is inferior oil with notable defects that is not permitted to be bottled under EU laws, so it is sent for refining. The EU has eliminated this category (it has been absorbed into the lampante category) and other regulating agencies have followed suit.
  - Virgin Olive Oil–Lampante. Oil with severe defects (greater than 6.0) or free acidity of greater than 3.3%, and which conforms to the other standards within its category. These oils come from low-quality fruit or from improper handling and processing. This grade is designated as not fit for human consumption and must be refined
- Refined Olive Oil. Oil obtained from virgin oils by refining methods that do not alter the initial glyceride structure. It has a free acidity of less than 0.3% and must conform to the other standards within its category. Refined olive oil must not come from the solvent extraction of pomace. The refining process usually consists of treating virgin oil/lampante with sodium hydroxide to neutralize the free acidity, washing, drying, odor removal, color removal, and filtration. In the process, the oil can be heated to as high as 430 °F (220 °C) under vacuum to remove all of the volatile components. Refined olive oil is usually odorless, tasteless, and colorless. It is designated as not fit for human consumption.

  - Olive Oil. Oil that is a blend of refined and unrefined virgin olive oils. It must have a free acidity of not more than 1% and conform to the other standards within its category. This grade of oil actually represents the bulk of the oil sold to consumers on the world market. Blends are made in proportions to create specific styles and prices. Oils in the United States labeled "Extra Light" would most likely be a blend dominated by refined olive oil. Other blends with more color and flavor would contain more virgin or extra virgin olive oil.
- Olive Pomace Oil category
  Oil obtained by treating olive pomace with solvents. It does not include oils obtained in the re-esterification processes or any mixture with oils of other kinds (seed or nut oils).
  - Crude Olive Pomace Oil – Not Fit for Human Consumption. This is the solvent-extracted crude oil product as it comes out of the pomace extractor after distillation to separate and recover most of the solvent. EU law also defines any oil containing 300–350 mg/kg of waxes and more than 350 mg/kg of aliphatic alcohols to be crude pomace oil. It is designated as not fit for human consumption, but is intended for refining.
  - Refined Olive Pomace Oil – Not Fit for Human Consumption. Oil obtained from crude pomace oil by refining methods that do not alter the initial glyceride structure. It has a free acidity of not more than 0.3% and its other characteristics must conform to the standard in its category. Refining includes the same methods used for "refined olive oil" except that the source of the raw product comes from pomace by means of solvent extraction. It is designated as not fit for human consumption.
  - Olive Pomace Oil. A blend of refined olive-pomace oil and virgin olive oil that is fit for human consumption. It has a free acidity of not more than 1% and must conform to the other standards within its category. In no case shall this blend be called "olive oil."

4.12 Other certification systems

4.12.1 EU denominations

There is a system of certification for specific geographic designations of origin that guarantee that oil is from a particular region, typical of that region, and meets certain quality standards. These are referred to as denomination of controlled origin (DOC) or oils with protected denomination of origin (PDO) designation. This EU quality guarantee is based on Italian national laws and is variously known in other countries. These oils are produced in licensed geographic areas according to conditions which specify the particular variety or varieties, the designated growing region, and that the olives are harvested and milled to a certain style that is typical of that region. The EU has a seal that appears on the labels of protected designation products.

4.12.2 United States Department of Agriculture (USDA)

In 1948, the USDA outlined definitions for olive oil. This was prior to the creation of the IOC and therefore the definitions were not aligned with the international descriptions or linked to well-defined testing methods. Since the United States is not a member of the IOC, the USDA does not legally recognize its classifications. However, in 2010, after more than 60 years, the USDA updated its standard to follow closely the IOC guidelines, and it now has the facilities in place to test for compliance at the USDA laboratory in Fresno, CA (USDA, 2010).

The USDA defines a voluntary standard for companies that want USDA grading certifications. In order to be certified as US Extra Virgin Olive Oil, the product must be inspected by the USDA. This means that a USDA inspector will pull samples according to a statistically designed protocol and submit them for analysis. The standard requires that the product that is sampled and certified is verifiably the same product that is sold as such. This means that one sample cannot be tested and the determined grade regarded as covering the entire production of that year. If a tank of oil is inspected and graded, it is only that tank of oil that can bear the USDA grade. Because of the perishable nature of olive oil, there is also likely to be some protocol providing for updating of an inspection (Kicenik Devarenne and Vossen, 2010).

The USDA sensory definition of an extra virgin olive oil is one in which the median score of the eight panel members is zero defects and fruitiness greater than zero. For the grades with defects, the median of the eight panel members must have noted the same defect. The grade of virgin olive oil can have a defect intensity up to a median of 2.5 and lampante is defined as having a defect intensity of over 2.5 (Table 4.6).

4.12.3 Miscellaneous certifying organizations

In California, the California Olive Oil Council (COOC) has a seal program that grants members a “COOC Certified Extra Virgin” seal for California-produced olive oil that meets the COOC standard for chemical analysis and is found free of defects by the COOC taste panel. Their 0.5% standard for free fatty acids is stricter that the IOC’s 0.8% standard (California Olive Oil Council, 2012).

The North American Olive Oil Association (NAOOA) has a quality seal program for its members. The NAOOA Certified Quality seal is awarded to imported products that meet the IOC standards as determined by an IOC-accredited laboratory. The seal may be used on virgin and refined olive oils, but not olive pomace oil. In the case of extra virgin grade oils, the

Table 4.6 USDA olive oil grades.

USDA grade	Definition
US Extra Virgin Olive Oil	Virgin olive oil which has excellent flavor and odor (median of defects equal to zero and median of fruitiness greater than zero) and a free fatty acid content, expressed as oleic acid, of not more than 0.8 g per 100 g
US Virgin Olive Oil	Virgin olive oil which has reasonably good flavor and odor (median of defects between zero and 2.5 and median of fruitiness greater than zero) and a free fatty acid content, expressed as oleic acid, of not more than 2.0 g per 100 g
US Lampante Virgin Olive Oil	Virgin olive oil which has poor flavor and odor (median of defects between 2.5 and 6.0 or when the median of defects is less than or equal to 2.5 and the median of fruit is zero), a free fatty acid content, expressed as oleic acid, of more than 2.0 g per 100 g. This oil is intended for refining or for purposes other than food use
US Olive Oil	Oil consisting of a blend of refined olive oil and virgin olive oils fit for consumption without further processing. It has a free fatty acid content, expressed as oleic acid, of not more than 1.0 g per 100 g, and has acceptable odor and flavor characteristic of "virgin olive oil." The maximum level permitted of total α-tocopherol in the final product is 200 mg/kg
"US Refined Olive Oil"	Olive oil obtained from virgin olive oils by refining methods that do not lead to alterations in the initial glyceridic structure (basic glycerine–fatty acid structure). It has a free fatty acid content, expressed as oleic acid, of not more than 0.3 g per 100 g and is flavorless and odorless. The addition of α-tocopherol is permitted to restore natural tocopherol lost in the refining process. The maximum level is 200 mg/kg of total α-tocopherol in the final product

Source: USDA (2010).

NAOOA seal requires sensory analysis by an IOC-recognized panel (North American Olive Oil Association, 2012).

The Australian Olive Association (AOA) has an "Australian Certified Extra Virgin" seal which is found only on AOA member olive oils produced in Australia. In order to receive the seal, a producer must sign a Code of Practices that addresses olive oil quality, food safety, ethical marketing, environmental standards, and other issues. The oil must meet most of the same chemical standards as the IOC, and be certified free from sensory defects and possessing some olive fruitiness by three AOA-accredited tasters (Australian Olive Association, 2012).

The Olives New Zealand (ONZ) certification program is open to both nonmembers and members. It includes bottling and labeling standards and chemical and sensory criteria. The ONZ certification requires that the oil be bottled in dark glass or, if in clear glass, that it is in a box or other light-excluding package. A pressing date is also required. The sensory requirements are the same as the IOC's: no defects and some olive fruitiness (Olives New Zealand, 2012).

4.13 Designing a sensory quality control program

For whatever reasons, an olive oil production facility may choose not to follow IOC protocols when establishing a sensory quality control program. There are many resources which describe specific methods to initiate a sensory QC or quality assurance (QA) program and the issues related to this (Yantis, 1992; Stone and Sidel, 1993; European Co-operation for Accreditation, 2003; Meilgaard, Civille, and Carr, 2006; Lawlesss and Heymann, 2010). ISO and ASTM also provide guidelines for panel set-up, training, and monitoring (ASTM, 1981; ISO, 1985, 1991, 1993, 2005, 2006a–c, 2008).

The components required for a commercial sensory program with low cost and optimum conditions are summarized in Table 4.7. Good sensory practices can be implemented in the olive oil production facility by adjustments to current practices such as quality control, blending, competitive tasting, and new market development. Tasting should be organized by a third party in an appropriate tasting room, oils should be served blind, tasters should not be informed of the purpose of the test, and the individual tasting data and comments should be collected independently before any group discussion. Decisions emerging from the tasting should be based on the data and not on the opinion of the group leader, regardless of experience. Tasters should maintain their tasting skills through regular training sessions. Managing such programs requires a dedicated position and must be fully endorsed by managers and employees (Lesschaeve, 2007).

4.13.1 Practical application of sensory QC

Once a trained sensory panel is in place, QC checks should be performed at various critical points in the production. These are shown in Table 4.8 along with the reason why the check is needed and whether the sensory testing is mandatory or optional.

4.14 New developments and future opportunities

Several areas of growth for sensory evaluation in quality control have been identified (Prenzler, Robards, and Bedgood, 2007). Many of these can be applied to the olive oil industry.

4.14.1 Increased involvement of sensory scientists in in-plant QC/sensory programs

All olive oil production companies need to have in place some form of in-house sensory testing. Demands for routine sensory QC come from many

Table 4.7 Key components in implementing a sensory program in commercial olive oil operations.

Component	General requirements	Cost-effective implementation for occasional sensory activities	Optimum implementation for regular sensory activities
Tasting room	Quiet environment: free form noise, odor, visual distractions	Conference room away from production plant	Dedicated sensory laboratory with preparation laboratory
	Individual tasting booth	Individual tables to minimize panelist interactions or temporary booth with odor-free cardboard dividers	Permanent booth with dividers
Data collection	Questionnaire featuring test instruction and scorecard to record individual assessments	Paper questionnaire	Computerized questionnaire from sensory software
Panel	*Recruitment*	*Internal panel*	*Preferably an external panel*
	Screening	Screening of volunteers on basic taste and olive oil defects and olive oil typical flavor recognition; availability and motivation	Two- to three-step screening of volunteers recruited from outside the olive oil plant; screening based on sensory acuity, odor description ability, olfactory memory, motivation, interests, and long-term availability
	Training	Short training session on methodology	Comprehensive training or a basic training
	Control of panel performance	Occasional on some duplicated sample evaluations	Ongoing with reference identification, duplicate sample evaluations
Methods	Sensory tests	Difference testing, ranking, sorting tasks, descriptive analysis on limited predefined descriptors	Difference testing, ranking, sorting tasks, descriptive analysis of full sensory profile
	Analysis of sensory data	Acquire inexpensive statistical package to perform univariate [analysis of variance (ANOVA)] and multivariate [principal component analysis (PCA)] analyses	Acquire packages to run predictive modeling or hire statistician
Sensory personnel	Laboratory technician to prepare samples; sensory specialist to plan, conduct, and analyze sensory tests	Same person often assumes both roles	Hire consultant to implement a program to train on-site staff

Source: Adapted from Lesschaeve (2007); reprinted with permission.

Table 4.8 Production points for sensory control check, aims, and requirements.

Operation/phase	Aim of Control System	Requirement
Olive production Harvest Olive transportation and storage	1. To preserve olive integrity and limit lipolytic and oxidative degradation 2. To obtain optimal chemical composition in terms of volatile compounds, phenols, and pigments 3. To detect any sensory defects which may develop 4. To evaluate quantitative and qualitative changes in color and the aroma/flavor profile	Mandatory
Crushing Mixing Extraction Separation	1. To limit lipolytic and oxidative degradation 2. To prevent the loss of volatile compounds and phenols 3. To detect any sensory defects which may develop 4. To evaluate quantitative and qualitative changes in the aroma/flavor profile	Mandatory
Filtration	To test if operation is effective	Optional
Blending	1. To evaluate each component of the blend for the presence of defects and to profile the positive attributes 2. To evaluate the final blend to determine if blending goals were met	Mandatory
Oil storage Packaging	1. To limit lipolytic and oxidative degradation 2. To prevent the loss of volatile compounds and phenols 3. To detect any sensory defects which may develop 4. To evaluate quantitative and qualitative changes in the aroma/flavor profile	Mandatory

directions, in particular from national or governmental bodies, through accreditation procedures, and from customer demands higher up the production chain.

4.14.2 Establishment of new or improved QC/sensory training programs

When training panelists, it is often difficult to find a set of reference standards capable of defining each possible sensory characteristic of virgin olive oils. This is a major handicap because unfortunately only a few such standards are available. In the past, the IOC had made an effort to provide some reference standards to all panels that had been selected and trained according to their procedures. Despite this effort, the number of reference standards available from the IOC is not sufficient for all possible attributes of virgin oils.

Every year panel leaders must look for oils with some dominant attributes to be used as references due to the poor stability of oil over time.

There are several companies that provide flavor standards to various industries. The beer industry has relied upon Cara Technology for taster validation schemes which employ stabilized reference beer flavor standards to generate beers with consistent sensory properties (Cara Technologies, 2012). The flavor is delivered into a neutral beer by means of nanoencapsulated reference flavor standards (chemicals) which are safe to smell and taste. Performance assessment is determined by the ability of tasters to identify flavor attributes presented at low levels in beer. A second type of assessment relates to scaling ability. Samples are presented to assessors which span a range of intensities for a single flavor attribute. Both replicates and control samples containing no added flavor are included in the test set. Assessors first have to rank the samples in order of intensity before rating the intensity of the added flavor in the beers. In the case of both types of assessment, the data can be analyzed in a variety of ways to provide information relating to multinational companies, country operations, individual taste panels, and individual assessors.

For the olive oil industry, flavor reference standards such as those described above could be used to:

- screen and select new tasters;
- measure the performance of individual tasters and trainees;
- measure the performance of sensory panels;
- measure the effectiveness of training programs.

4.14.3 Collaboration between R&D and QC functions to produce high and consistent quality

The goal of the study "Quality Enhancement of Australian Extra Virgin Olive Oils" was to investigate more thoroughly the types of compounds – volatiles and phenolics – that are more closely linked with flavor (Prenzler, Robards, and Bedgood, 2007). Results from this study led to a more fundamental understanding of the chemistry of flavor development and at the same time provided processors with information on how to optimize production conditions to maximize oil quality. The researchers saw a need to understand how production conditions could be fine-tuned to improve and maintain quality. Among the production conditions that can be manipulated, they determined that time and temperature were important in:

- increasing extracting more desirable compounds;
- retarding extraction of less desirable compounds;
- influencing chemical changes that can result in off-flavors and less stable oils.

4.14.4 Establishment of new or improved sensory specifications in techniques and the integration of consumer/management input

Although it allows an oil to be placed in a given category rather than in another, the current official IOC profile sheet of flavor description and quality grading does not provide a complete description of olive oil flavors. Panelists mainly evaluate defects and only fruity, bitterness, and pungency among positive attributes. Small but meaningful changes of oil sensory profiles attributed to genetic factors, ripening stages of fruits, or technological processing of oil extraction cannot be monitored using the current IOC scorecard. In its latest revision of the sensory assessment protocol, the IOC has included a section on "Optional terminology for labeling purposes." The positive attributes of fruity, bitter, and pungent may be classified as:

- intense, when the median of the attribute is more than 6;
- medium, when the median of the attribute is between 3 and 6;
- light, when the median of the attribute is less than 3.

A new official characteristic "well balanced" describes oils that show no lack of balance between olfactory, gustatory, and tactile sensations. In other words, none of the medians for bitterness and pungency are allowed to be two points higher than that for fruitiness. However, olive oils with low-intensity flavors can be legally described as "well balanced," which may lead to consumer confusion.

German and Swiss researchers looked for a better way to discriminate olive oils in the extra virgin category (Bongartz and Oberg, 2011). They wanted to characterize differences between extra virgin olive oils of only standard quality and "excellent" extra virgin olive oils (see Chapter 1). They defined "harmony" as the degree of balance of all positive characteristics. Harmony includes both the presence and intensity of positive attributes (fruitiness, bitterness, pungency, and the presence of aromatic components – perceived both ortho- and retronasally) and the complexity of these positive characteristics. This definition of "harmony" is not the same as the IOC characteristic of "well balanced."

These researchers claimed that "harmony" is an objective descriptor and is well suited as a parameter for describing different quality levels in olive oil competitions in the segment of excellent premium oils.

Other researchers in the wine field used the term "harmonious" as one of several integrated terms and found it to be the most subjective attribute (Sivertsen and Risvik, 1994). While all judges showed relatively good repeatability, which indicated a consistent personal opinion of the term, different judges seemed to have a different opinion of what the term meant.

The researchers also noted a distinct grouping of judges by gender for "harmonious."

The profile sheet used by the University of California at Davis (UC Davis) olive oil taste panel includes concepts of balance, complexity, and freshness. This scorecard, shown in Figure 4.5, was developed to grant IOC sensory

UC Davis Olive Oil Taste Panel **Olive Oil Scorecard** **Name** ______ **Date** ______ **Oil Code** ______

Total aroma intensity: ______

Total flavor intensity ______

Intensity of defects:

Rancid ______
Fusty/muddy sediment ______
Winey–vinegary–acid–sour ______
Musty–humid–earthy ______
Frostbitten olives (wet wood) ______
Other ______
Specify other defect(s): ______

Additional defects/negative attributes (mark all that apply):

❑ Heated/Burnt ❑ Rough/Coarse MF ❑ Vegetable water ❑ Grubby ❑ Metallic
❑ Hay–wood ❑ Greasy/mineral oil flavor ❑ Brine ❑ Cucumber

Intensity of positive attributes:

Fruit ______
Ripe fruit ______
Green fruit ______
Bitter ______
Pungent ______

Ripe fruit aroma and flavor descriptors (mark all that apply):

❑ Ripe olive ❑ Ripe banana ❑ Ripe apple ❑ Floral ❑ Other:
❑ Buttery ❑ Tropical ❑ Apricot/peach ❑ Nutty

Green fruit aroma and flavor descriptors (mark all that apply):

❑ Green grass/freshly cut grass ❑ Green apple ❑ Green banana ❑ Green olive ❑ Green almond ❑ Tomato leaf ❑ Artichoke
❑ Green tea ❑ Herbaceous ❑ Minty ❑ Eucalyptus ❑ Pine ❑ Other:

Other aroma and flavor descriptors (mark all that apply):

❑ Citrus ❑ Black pepper ❑ Spice: ______ ❑ Cinnamon ❑ Straw ❑ Woody/olive pit ❑ Other:

Taste:

Sweetness: ______

Mouthfeel:

Astringency: ______

Integrative measures:

Balance: ______
Complexity: ______
Freshness: ______

Style/level of intensity (select one): ❑ Not applicable–defective ❑ Delicate ❑ Medium ❑ Robust

Overall comments

Figure 4.5 UC Davis olive oil sensory scorecard.

certification to submitted oils, to provide olive oil producers with a full descriptive profile of their oils, and to give university personnel detailed sensory feedback on oils for research projects. The scorecard contains a large number of terms to describe the various flavor nuances of oils.

4.14.5 Internet applications in the QC field

With the decrease in costs associated with personal computers, a number of data entry systems have become very readily available. The advantages of computerization of the sensory booth include:

- the speed of receiving test results;
- a ready interface between the data entry system and statistical and graphing programs;
- a reduction in the errors involved in manual data entry.

The UC Davis scorecard (above) has been configured for Internet use.

4.14.6 Continued use of instrumental measures: their establishment and validation

Although instrumental measurements of quality parameters complement sensory information, they cannot replace sensory evaluation for many important product characteristics. Sometimes the instrumental parameters do not indicate unacceptable changes to the oil as judged by sensory perception, and this was seen in a study conducted by UC Davis (Frankel *et al.*, 2011). They found that the top-selling imported brands of "extra virgin" olive oil sold in the United States and purchased at retail locations throughout California often failed the IOC's sensory standards for extra virgin olive oil. Sensory analysis indicated that these failed samples had objectionable descriptors such as rancid and fusty. They recommended that it is essential to support sensory evaluations by chemical tests for volatile compounds that are known to be produced by lipid oxidation. Their chemical tests indicated that the samples usually passed the IOC's chemical tests even when those samples failed two IOC-accredited sensory panels. Their report showed that chemical confirmation of the negative sensory results was strongest with the German/Australian DAGs and PPP tests (see below), followed by IOC tests for UV absorption. It was concluded that the IOC and USDA standards would be more effective in assessing and enforcing olive oil quality by including the German/Australian DAGs and PPP standards. Their testing also indicated that the olive oil samples failed extra virgin olive oil standards for reasons that included one or more of the following: (a) oxidation by exposure to elevated temperatures, light, and/or aging; (b) adulteration with cheaper refined olive oil; and (c) poor quality oil made from damaged and over-ripe olives, processing flaws, and/or improper oil storage.

Two standards developed by the German Fat and Oil Society (DGF) and adopted by the German government and the Australian Olive Association (AOA) to detect old and illegally deodorized oils are the following:

- *1,2- and 1,3-diacylglycerol (DAGs)*
  DAGs are lipid compounds that are found at different levels in fresh and refined olive oil. During the breakdown of DAGs, diglycerols are formed. Fresh extra virgin oil contains a high proportion of 1,2-DAG whereas olive oil from poor quality fruits and refined olive oil have elevated levels of 1,3-DAG. The ratio of 1,2-DAG to 1,2- + 1,3-DAGs should be greater than 40% and a higher percentage indicates fresher oil. Lower percentages indicate that the olive oil was hydrolyzed, oxidized, of poor quality, and/or adulterated with refined oil (DFG method C-VI 16(06) – ISO 29822:2009).
- *Pyropheophytins (PPP)*
  These are chlorophyll pigment breakdown compounds from the thermal degradation of olive oil. Heat and long storage time break down chlorophyll into pheophytins and then to pyropheophytins. An elevated PPP level is an indicator of oil that is oxidized and/or adulterated with refined oil.

The DAGs and PPP standards must be met by members of the AOA to receive AOA certification for extra virgin olive oil (Australian Olive Association, 2012).

4.14.7 Global quality control

The ultimate goal for the olive oil industry is a more effective approach to delivering consistent quality products worldwide.

4.15 Conclusion

Olive oil producers need to meet consumer demands by generating and improving the supply chain to deliver olive oils of appropriate quality to consumers in different world market segments. It is increasingly important to produce olive oils consistently to definable aroma and flavor specifications. With this comes the need to integrate better oil and olive research and to focus on the management of sensory features of olive oils that attract the consumer.

The consumer of today is quality focused, image conscious, and price sensitive, and this has led to a change in the rules of the marketplace to a degree where quality can be defined as "sustainable customer and consumer satisfaction." The process of transforming the olive oil industry from a production-oriented to a market-driven industry results in an increasing dependence on analytical sensory evaluation techniques.

References

Angerosa, F. (2000) Sensory quality of olive oils, in *Handbook of Olive Oil: Analysis and Properties* (eds J.L. Harwood and R. Aparicio), Aspen Publishers, Gaithersburg, MD, pp. 355–372.

Angerosa, F. (2006) Analysis and authentication, in *Olive Oil: Chemistry and Technology*, 2nd edn (ed. D. Boskou), AOCS Press, Champaign, IL, pp. 113–172.

AOCS (2012) *Olive Oil Sensory Panel Proficiency Testing Series*, American Oil Chemists' Society, Urbana, IL, http://www.aocs.org/LabServices/content.cfm?ItemNumber=18252 (last accessed 26 May 2013).

ASTM (1981) *Guidelines for the Selection and Training of Sensory Panel Members*, Special Technical Publication 758, American Society for Testing and Materials, West Conshohocken, PA.

ASTM (2011) *Standard Practice for Sensory Evaluation of Edible Oils and Fats,* ASTM Standard E1627-11, ASTM International, West Conshohocken, PA, DOI: 10.1520/E1627-11, www.astm.org (last accessed 26 May 2013).

Australian Olive Association (2012) *Australian Certified Extra Virgin*, Australian Olive Association, Coomandook, SA, http://www.australianolives.com.au (last accessed 26 May 2013).

Bongartz, A., and Oberg, D.G. (2011) Sensory evaluation of extra virgin olive oil (EVOO) extended to include the quality factor "harmony." *Journal of Agricultural Science and Technology A*, **1**, 422–435.

California Olive Oil Council (2012) *Certification Process*, California Olive Oil Council, Berkeley, CA, http://www.cooc.com/about_certification.html (last accessed 26 May 2013).

Cara Technologies (2012) *Taste Panel Management*, Cara Technologies, Macclesfield, http://www.cara-online.com/tastepanelmanagement (last accessed 26 May 2013).

Codex (2009) *Codex Standard for Olive Oils and Pomace Oils*, Codex Stan 33-1981, Codex Alimnentarius, Rome, www.codexalimentarius.org/input/download/standards/88/CXS_033e.pdf (last accessed 26 May 2013).

European Co-operation for Accreditation (2003) *Accreditation for Sensory Testing Laboratories*, EA-4/09 G, European Co-operation for Accreditation, Paris, http://www.european-accreditation.org/publication/ea-4-09-g (last accessed 26 May 2013).

Frankel, E.N., Mailer, R.J., Wang, S.C., Shoemaker, C.F., Guinard, J.-X., Flynn, J.D., and Sturzenberger, N.D. (2011) *Evaluation of Extra-Virgin Olive Oil Sold in California*, UC Davis Olive Center, Robert Mondavi Institute for Wine and Food Science, University of California Davis, Davis, CA.

Hund, E., Massart, D.L., and Smeyers-Verbeke, J. (2000) Review. Inter-laboratory studies in analytical chemistry. *Analytica Chimica Acta*, **423**, 145–165.

Hyldig, G. (2010) Proficiency testing of sensory panels, in *Sensory Analysis for Food and Beverage Quality Control: a Practical Guide* (ed. D. Kilcast), Woodhead Publishing, Cambridge, pp. 37–48.

IOC (2007a) *Sensory Analysis of Olive Oil – Standard – Sensory Analysis: General Basic Vocabulary*, COI/T.20/Doc. No. 4/Rev. 1, International Olive Council, Madrid.

IOC (2007b) *Sensory Analysis of Olive Oil – Standard – Glass for Oil Tasting*, COI/T.20/Doc. No. 5/Rev. 1, International Olive Council, Madrid.

IOC (2007c) *Sensory Analysis of Olive Oil – Standard – Guide for the Installation of a Test Room*, COI/T.20/Doc. No. 6/Rev. 1, International Olive Council, Madrid.ioc2011a

IOC (2011a) *Sensory Analysis of Olive Oil – Standard – Guide for the Selection, Training and Monitoring of Skilled Virgin Olive Oil Tasters*, COI/T.20/Doc. No. 14/Rev. 3, International Olive Council, Rome.

IOC (2011b) *Sensory Analysis of Olive Oil – Method for the Organoleptic Assessment of Virgin Olive Oil*, COI/T.20/Doc. No. 15/Rev. 4, International Olive Council, Madrid.

IOC (2012a) *World Olive Oil Figures*, International Olive Council, Madrid, http://www.internationaloliveoil.org/estaticos/view/131-world-olive-oil-figures (last accessed 26 May 2013).

IOC (2012b) *Mission Statement*, International Olive Council, Madrid, http://www.internationaloliveoil.org/estaticos/view/100-mission-statement (last accessed 26 May 2013).

IOOC (1996) *Sensory Analysis of Olive Oil – General Methodology for the Organoleptic Assessment of Virgin Olive Oil*, COI/T.20/Doc. No. 13/Rev. 1, International Olive Oil Council, Madrid.

ISO (1985) *Sensory Analysis – Methodology – Flavour Profile Methods*, ISO 6564:1985(E), International Organization for Standardization, Geneva.

ISO (1991) *Sensory Analysis – Methodology – Method of Investigating Sensitivity of Taste*, ISO 3972:1991(E), International Organization for Standardization, Geneva.

ISO (1993) *Sensory Analysis – General Guidance for the Selection, Training and Monitoring of Assessors – Part 1: Selected Assessors*, ISO 8586-1:1993(E), International Organization for Standardization, Geneva.

ISO (2005) *Sensory Analysis – Methodology – General Guidance*, ISO 6658:2005(E), International Organization for Standardization, Geneva.

ISO (2006a) *Sensory Analysis – General Guidance for the Staff of a Sensory Evaluation Laboratory – Part 1: Staff Responsibilities*, ISO 13300-1:2006, International Organization for Standardization, Geneva.

ISO (2006b) *Sensory Analysis – General Guidance for the Staff of a Sensory Evaluation Laboratory – Part 1: Recruitment and Training of Panel Leaders*, ISO 13300-2:2006, International Organization for Standardization, Geneva.

ISO (2006c) *Sensory Analysis – Methodology – Initiation and Training of Assessors in the Detection and Recognition of Odours*, ISO 5496:2006(E), International Organization for Standardization, Geneva.

ISO (2008) *Sensory Analysis – General Guidance for the Selection, Training and Monitoring of Assessors – Part 2: Expert Sensory Assessors*, ISO 8586-2:2008(E), International Organization for Standardization, Geneva.

Kicenik Devarenne, A., and Vossen, P.M. (2010) *Understanding the USDA Olive Oil Standards*, http://cesonoma.ucdavis.edu/files/48283.pdf (last accessed 26 May 2013).

Langstaff, S.A., Aparicio, R., and CSIC–SEXIA Group (2011) *The Defects Wheel for Olive Oil*, http://appliedsensory.com/Defects_Wheels.html (last accessed 26 May 2013).

Lawless, H.T., and Heymann, H. (2010) *Sensory Evaluation of Food: Principles and Practices*, 2nd edn, Springer, New York.

Lesschaeve, I. (2007) Sensory evaluation of wine and commercial realities: review of current practices and perspectives. *American Journal of Enology and Viticulture*, **58**, 252–258.

Mailer, R.J. (2012) *A World Tour of Olive Oil Standards*, PowerPoint presentation at 103rd AOCS Anual Meeting and Expo, Olive Oil Chemistry and Sensory Relationships Short Course, http://aocs.files.cms-plus.com/AnnualMeeting/Presentations/Mailer%20-World-Tour-Olive-Oil-Standards.pdf (last accessed 26 May 2013).

Meilgaard, M., Civille, G.V., and Carr, B.T. (2006) *Sensory Evaluation Techniques*, 4th edn, CRC Press, Boca Raton, FL.

Mueller, T. (2012) *Extra Virginity*, W.W. Norton, New York.

Muñoz, A.M. (2002) Sensory evaluation in quality control: an overview, new developments and future opportunities. *Food Quality and Preference*, **13**, 329–339.

North American Olive Oil Association (2012) *North American Olive Oil Association Introduces Olive Oil Seal Program*, North American Olive Oil Association, Neptune, NJ, http://www.naooa.org/sealprogram (last accessed 26 May 2013).

Olives New Zealand (2012) *Certification*, Olives New Zealand, Hawke's Bay, http://www.olivesnz.org.nz/certification-the-olive-mark (last accessed 26 May 2013).

Prenzler, P., Robards, K., and Bedgood, D. (2007) *Quality Enhancement of Australian Extra Virgin Olive Oils*, Publication No. 06/135, Project No. UCS-33A, Rural Industries Research and Development Corporation, Barton, ACT.

Sivertsen, H.K., and Risvik, E. (1994) A study of sample and assessor variation – a multivariate study of wine profiles. *Journal of Sensory Studies*, **9**, 293–312.

Stone, H.S., and Sidel, J.L. (1993) *Sensory Evaluation Practices*, Academic Press, San Diego, CA.

USDA (2010) *United States Standards for Grades of Olive Oil and Olive-Pomace Oil*, Federal Register, United States Department of Agriculture, Washington, DC, http://www.ams.usda.gov/AMSv1.0/getfile?dDocName=STELDEV3011889 (last accessed 3 August 2013).

Vossen, P.M. (2005) *Olive Oil Processing Technology Influences on Quality*, http://cesonoma.ucanr.edu/files/27187.pdf (last accessed 26 May 2013).

Vossen, P.M. (2010) *International Olive Council (IOC) – Olive Oil Sensory Evaluation Methodology*, http://cesonoma.ucdavis.edu/files/48152.pdf (last accessed 26 May 2013).

Yantis, J.E. (1992) *The Role of Sensory Analysis in Quality Control*, ASTM Manual Series, MNL 14, American Society for Testing and Materials, West Conshohocken, PA.

Zampounis, V. (2006) Olive oil in the world market, in *Olive Oil: Chemistry and Technology*, 2nd edn (ed. D. Boskou), AOCS Press, Champaign, IL, pp. 21–39.

5
Sensory Methods for Optimizing and Adding Value to Extra Virgin Olive Oil

Erminio Monteleone
Dipartimento di Gestione dei Sistemi Agrari, Alimentari e Forestali (GESAAF), University of Florence (Università degli Studi di Firenze), Florence, Italy

5.1 Introduction

It is very well known that the sensory properties of extra virgin olive oil vary widely depending on a number of factors (Servili *et al.*, 2004). Differences in genetic resources, environmental conditions, process specifications, and local know-how induce sensory differences among oils (Caporale *et al.*, 2006). As already pointed out in Chapter 1, the variety of sensory profiles is extraordinarily rich in connection with the biodiversity of the olive and the ability of producers. Hence there is the need for producers and researchers to put emphasis on variety and sensory style for differentiating brands and also on the consistency of the sensory profiles within each brand identity over time.

Most of the attention on the sensory characteristics of olive oil is currently focused on how to evaluate whether a given oil is free of defects and how extra virgin olive oil is qualified. The International Olive Council (IOC) standards for the sensory evaluation of oils represent an effective method to qualify oils in categories such as extra virgin or virgin. These standards consist in evaluating both "positive" and "negative" attributes. The latter is the category of defects that cannot be present in an extra virgin olive oil. Positive

Olive Oil Sensory Science, First Edition. Edited by Erminio Monteleone and Susan Langstaff.

sensory attributes are bitterness, pungency, and fruity notes. It should be emphasized that this necessary evaluation (mandatory in the European Union) is not sufficient to describe the sensory characteristics of extra virgin olive oils (Figure 5.1). The description of the sensory characteristics of oils should be seen as a necessary step to link the world of production with the world of use and consumption. In relation to the production system, the sensory profile of an oil represents a product specification describing characteristics due to cultivar type, climatic conditions, and operational process conditions; it also represents the product specifications necessary to verify the producer's capability to control the critical factors of sensory quality and to guarantee the consistency of the sensory profile of an oil over time.

In relation to the use and consumption of oils, the sensory profile is necessary to improve the communication of the sensory style and the culinary use of a product; it is also necessary to study the sensory functionality of an oil in dish preparations and oil–food pairings and to understand and interpret consumer likes and dislikes. In other words, this information is necessary in order to promote the excellence of an oil.

The aim of this chapter is to present sensory methods and to describe sensory similarities and differences among products based on small-panel data and suitable extra virgin olive oils. Particular relevance is given to the appropriate methods used to obtain so-called perceptual maps, and included are those that are alternatives to conventional descriptive analysis. The chapter introduces dynamic methods of sensory evaluation suitable for olive oils such as Time–Intensity (TI) and Temporal Dominance Of Sensation (TDS) methods. The latter is fully described and discussed in Chapter 7.

The evaluation of sensory defects is not considered here. This evaluation, described in Chapter 4, is a preliminary step in the sensory characterization of oils (see Chapter 1). Furthermore, consumers' data are not considered here.

5.2 Perceptual maps

The use of perceptual maps is common in food and beverage development and optimization. A perceptual map is defined as a "pictorial representation that captures the relationships among a set of products" (Lawless and Heymann, 1998) and it is obtained by submitting sensory data to a variety of multivariate statistical techniques. These techniques can extract from complex data (a matrix with many products and variables) the most important information (how different the products are to each other and how much each variable contributes to the difference among samples) and present the results in a simplified picture or map in two or three dimensions that can be easily understood. Because of its characteristics, perceptual mapping is very attractive but also effective. In Figure 5.2, an example of a perceptual map is illustrated. It shows the differences and similarities among a set of oils: products that are similar to one another are positioned close together.

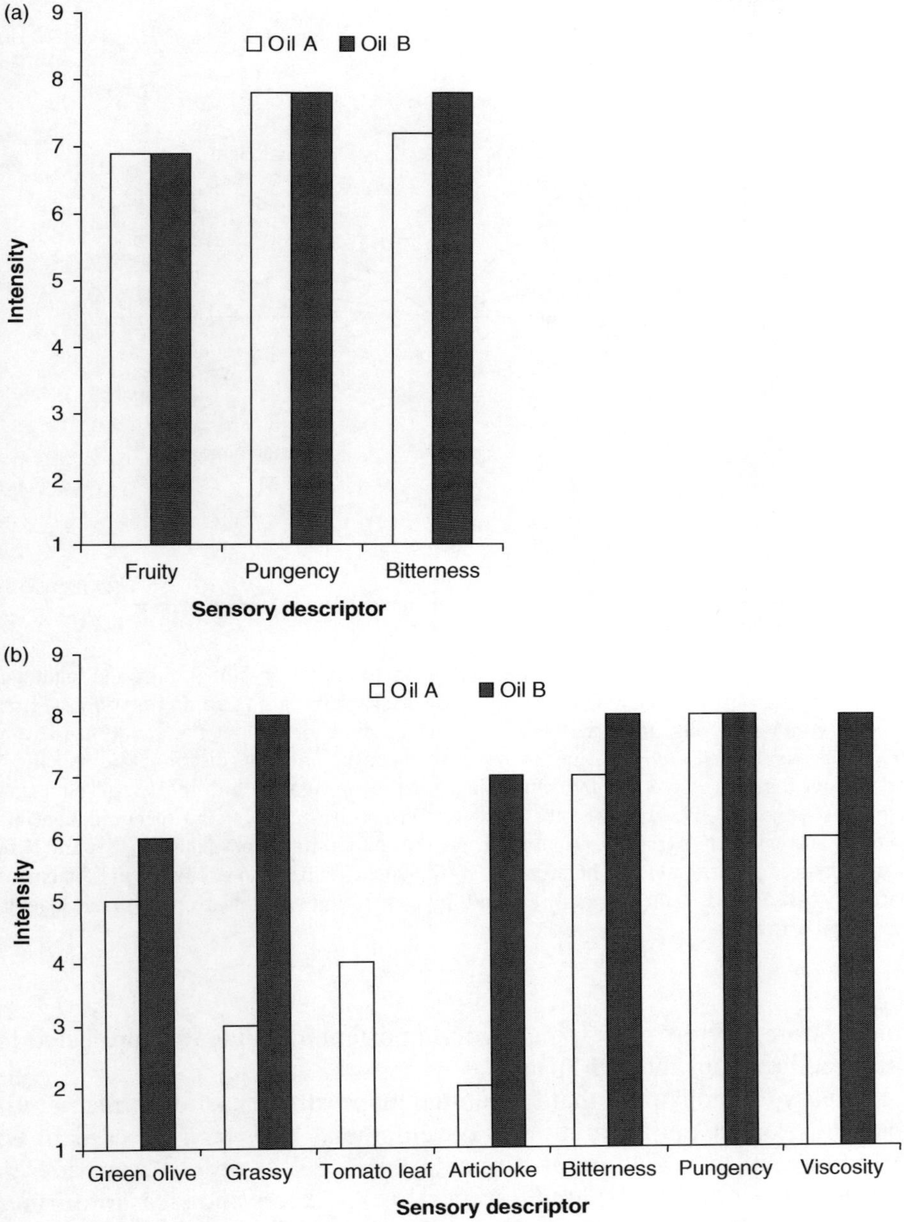

Figure 5.1 Sensory profile of two extra virgin olive oils as described by means (a) the IOC method and (b) conventional descriptive analysis. The bars in (a) indicate that the two oils are very similar whereas those in (b) clearly illustrate the differences between the oils.

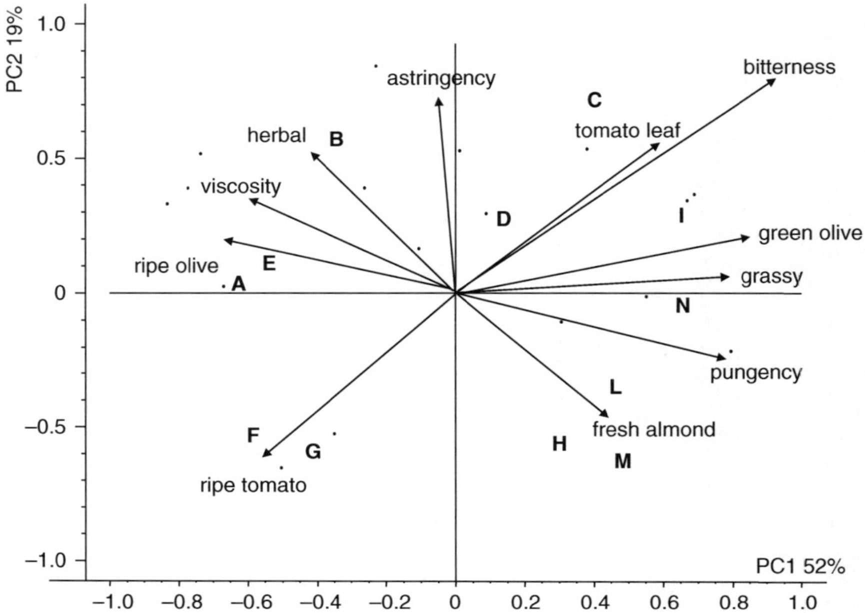

Figure 5.2 Example of a perceptual map. The graph shows the bi-plot (scores and loadings) from principal component analysis (PCA) of sensory descriptive data (11 aroma attributes) from 12 extra virgin olive oils. The first two dimensions account for 71% of the variation (52 and 19%). The first dimension, from left to right of the map, primarily discriminates samples I and N from the rest of the oils by contrasting green olive, grassy, bitterness, and pungency attributes with ripe olive and viscosity attributes. Visual inspection of the second dimension, from the bottom to the top of the map, indicates that oil C differs from samples H, L, and M by contrasting tomato leaf and astringency with fresh almond. Furthermore, the second dimension discriminates samples F and G from A, E, and B by contrasting ripe tomato note to astringency and herbal attributes.

Furthermore, vectors corresponding to important attributes are projected to interpret directions through space.

It is easy to understand that perceptual maps can be useful for many purposes in product optimization and development. They can be used in an early phase of investigation as an explorative tool in order to provide an overview. They can also be used to generate hypotheses and ideas for further experimentations (Næs, Brockhoff, and Tomic, 2010) and to confirm previous hypotheses. Visual inspection of perceptual maps permits the identification and grouping of samples with similar characteristics that at the same time are different from other samples or sample groups. Because of its characteristics, perceptual mapping is very effective in identifying oils with varied sensory styles. The term “style” does not indicate a quality model, but rather the sensory profile that describes an oil (or a group of oils) as different from others.

Table 5.1 Sensory methods applied in laboratory small-panel studies to obtain a perceptual map.

Sensory response	Sensory method	Statistical method
Descriptive data	Descriptive Analysis (conventional profile)	Principal Component Analysis (PCA)
	Temporal Dominance of Sensations	
	Free Choice Profiling	Generalized Procrustes Analysis (GPA)
	Flash Analysis	
Similarity data	Sorting	Multidimensional Scaling (MDS)
	Projective Mapping or "Nappe" method	MDS; GPA; PCA; Multifactor Analysis (MFA)

Perceptual mapping is a necessary step to relate sensory properties effectively to both consumer hedonic responses (or expert quality assessments) by means of preference mapping and the physico-chemical characteristics of samples in multi-product studies.

In detail, by interpreting perceptual maps from small-panel studies data it is possible:

- to compare a product with its competitors;
- to show the effect of modifications on agricultural practices, olive ripening stages, extraction techniques, and oil storage conditions;
- to define desired changes of the sensory properties of a product;
- to group oils in varied sensory styles;
- to select products to study the sensory functionality of oils;
- to select products to study consumer affective responses (e.g., liking, emotions) to oils and oil–food pairings.

Perceptual maps can be obtained using different sensory methods and applying a variety of statistical analyses, as reported in Table 5.1. The table presents methods that can be applied in laboratory small-panel studies to obtain a map. Descriptive analysis is probably the most important tool in sensory analysis in describing products and differences between products; however, alternative ways of capturing sensory differences between products, such as Free Choice Profiling, Flash Analysis, Sorting and Napping, are available for specific aims and needs. In this chapter, these methods are illustrated in consideration of their use in product optimization and development of olive oils with particular reference to extra virgin olive oils.

5.3 Conventional descriptive analysis

Descriptive analysis is a term generally used to describe a sensory method by which identification, quantification, and description of sensory attributes (the so-called sensory profile) of food by human subjects are obtained (Piggott, Simpson, and Williams, 1998). Several descriptive methods are considered: the Flavor Profile (Cairncross and Sjostrom, 1950), the Texture Profile (Szczesniak, 1963), Quantitative Descriptive Analysis (Stone *et al.*, 1974), and Sensory Spectrum (Meilgaard, Civille, and Carr, 1991). However, the most widely used profile technique combines different aspects from the above mentioned methods and is generally named "conventional descriptive analysis" (see ISO, 1994; Lawless and Heymann, 1998, p. 362). Free Choice Profiling (Williams and Langron, 1984), and Flash Analysis (Sieffermann, 2000) are further descriptive techniques that are separately presented in this chapter.

Conventional descriptive analysis has been widely used to provide both qualitative and quantitative measures of food and beverage properties, including extra virgin olive oil (Bertuccioli, 1994; Lyon and Watson, 1994; Mojet and de Jong 1994; Monteleone *et al.*, 1995, 1996, 1997; Caporale *et al.* 2006; Delgado and Guinard, 2011a,b; Dinnella *et al.*, 2012). It is based on independent judgments of panelists (trained subjects) and statistical testing rather that group discussion and consensus procedures. Hence this method is capable of providing a picture of how products differ among themselves, implying a comparison among products. Several products are assessed together, and the descriptive profile of a single product is both placed in and compared with the context of other products. It is a multi-product test that uses a limited number of subjects. It requires a language development process and subjects provide an intensity rating for each descriptive term. The data are averaged across the panel and thus a statistical evaluation of results is obtained.

5.3.1 Subjects

Usually, a descriptive panel involves between 10 and 15 trained assessors, recruited according to their ability to detect differences in important product attributes. Subjects must be qualified prior to their participation. They must be users of the product class to be evaluated. Previous exposure to varied extra virgin olive oils is an important qualification. It is also important that individuals participate in a series of difference tests organized to represent a range of difficulty and to include relevant modalities (olfaction, taste, etc.).

5.3.2 Language development, subject training, and subject reproducibility

Before assessment of products, assessors participate in a series of language sessions managed by a panel leader. Generally, a new panel develops the sensory language itself. This is a consensus-building process aimed at defining the attributes that the panel utilizes to represent their perceptions. Subjects familiarize themselves with the product space and generate attributes that describe the differences among products. The way in which products are presented to the subjects may vary. For example, subjects can be exposed to the entire range of products and then be asked to write down the descriptors that describe perceived differences. The number of training sessions is dependent on the number of the products, but in general 6–10 training sessions of 1 h duration are needed. The initial list of attributes is normally reduced to achieve a list that comprehensively and accurately describes the product space: redundant and/or less cited terms are grouped on a semantic basis and/or eliminated according to the subjects' consensual decisions. To facilitate the consensus and to calibrate the subjects, reference standards are presented to the panel, discussed, and modified if necessary. Refining of descriptive terms, reference standards, and definitions continues until the panelists reach a consensus. Unlike wine or cheese, there are a limited number of scientific papers that have reported attributes generated by descriptive panels to describe the sensory properties of virgin olive oils.

Early in the 1990s, as part of a European Union research program[1], a project was undertaken on the "Sensory and Nutritional Quality of Virgin Olive Oil." A key objective of the project was to establish a standardized terminology for describing virgin olive oils. Sensory characteristics of samples from different countries (Italy, Spain, and Greece), different varieties (Moraiolo, Coratina, Frantoio, Coroneiki, Tzunnati, Picual, and Arbequina), varied olive ripeness levels (unripe, ripe, and over-ripe) and varied extraction technologies (centrifugation, expression, and percolation) were described by three descriptive panels from Italy, the United Kingdom, and The Netherlands by using the descriptive analysis technique. The main results of the study in terms of a sensory vocabulary used to describe virgin olive oils were reported by Lyon and Watson (1994) and Mojet and de Jong (1994). The number of sensory descriptors generated varied considerably, with 26 British attributes, 68 Dutch attributes, and 18 Italian attributes (Table 5.2). Superficially, some of the terms were fairly similar but a number of the terms were panel specific. Differences in language, culture, and food experience affect the way in which samples are described. Nevertheless, the results of the study indicated that oil samples were perceived in a similar way, irrespective of the panel or country that analyzed the oils.

Figure 5.3 reports the perceptual map resulting from a Generalized Procrustes Analysis (GPA) applied on the sensory mean panel data from each panel (Lyon and Watson, 1994). For each sample, the triangles represent the different positioning for each panel. The smaller the triangles, the more similar is the perception of the samples between the countries. It can be noted

[1] EC FLAIR Research Program 1991: The Study of Sensory and Nutritional Quality of Virgin Olive Oil in Relation to Variety, Ripeness, and Extraction Technology.

Table 5.2 Sensory descriptors generated by three panels [Italian (IT), British (UK), and Dutch (NL) evaluating the same extra virgin olive oils.

Sensory dimension	Panel	Descriptors				
Appearance	IT	Yellow	Green			
	UK	Depth	Bright	Yellow	Green	Brown
	NL	Yellow particles	Green syrup	Brown	Glossy	Transparent
Mouthfeel	IT	Astringent				
	UK	Thickness	Throatcatching			
	NL	Velvet	Sticky	Astringent	Cooling	Rough
		Dry	Sharp	Pungent		
Flavor	IT	Tomato leaf	Green olive	Ripe black olive	Grassy	Artichoke
		Apple	Yeast	Bitter	Pungent	
	UK	Grassy	Almond	Banana	Pungent	Tomato
		Hay	Perfume			
	NL	Sea breeze	Prickly	Apple	Twig	Harshy
		Dry wood	Lemon	Orange	Soft fruit	Candies
		Wild flowers	Ferment	Farm	Salad oil	Tallow
		Rancid	Cod liver	Nuts	Medicine	Earthy
		Sweet	Salty	Sour vinegar	Olives	Green leaf
		Grass	Green banana	Herbs	Green pepper	Chilli pepper
		Butter	Rancid	Coconuts	caramel	grotty
		Roasted	Ashtray	Envelope glue	Refinery	Metallic
		Bitter	Green	Fruity	Putty	Dusty

Source: Reproduced from Lyon, D.H., and Watson, M.P. (1994) with kind permission from Grasas Y Aceites.

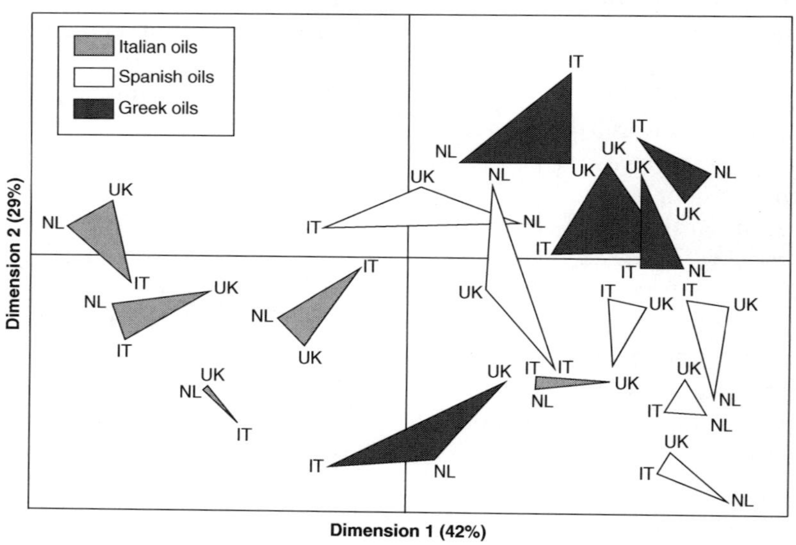

Figure 5.3 Results from the EC FLAIR Research Program 1991: The Study of Sensory and Nutritional Quality of Virgin Olive Oil in Relation to Variety, Ripeness and Extraction Technology". Consensus map resulting from a Generalized Procrustes Analysis (GPA) applied on the sensory mean panel data from the Italian (IT), British (UK) and Dutch (NL) panels. *Source:* Reproduced from Lyon, D.H., and Watson, M.P. (1994) with kind permission from Grasas Y Aceites.

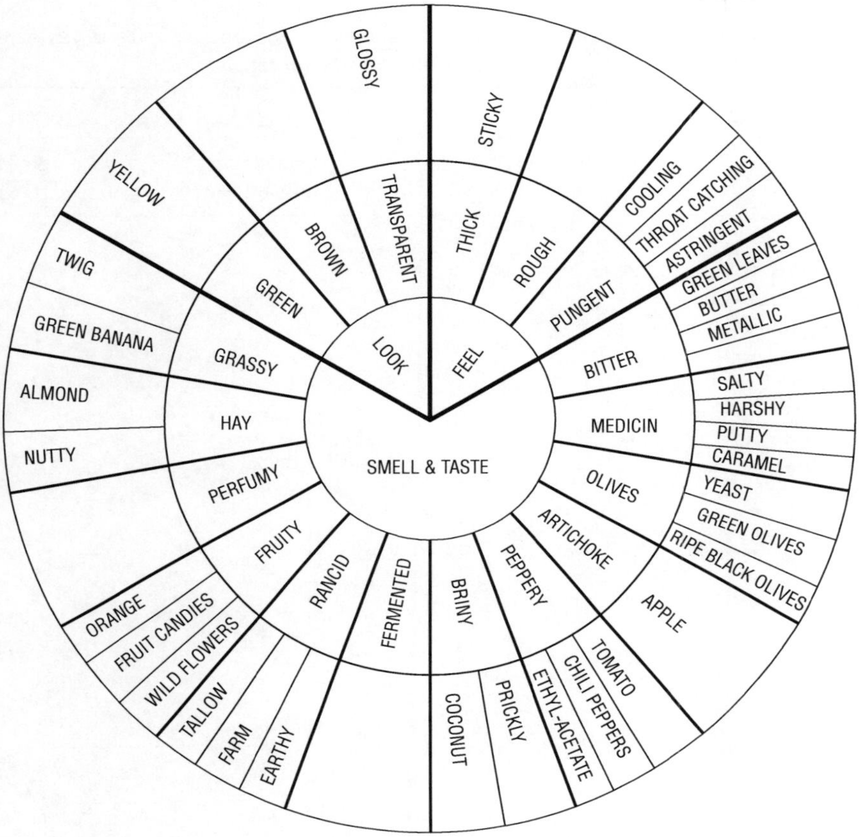

Figure 5.4 The sensory wheel of virgin olive oil. *Source:* Reproduced from Mojet, J., and de Jong, S. (1994) with kind permission from Grasas Y Aceites.

that samples from each country are in a similar area of the plot. Based on this evidence, Mojet and de Jong (1994) studied the correlations between the attributes evaluated by the three panels and generated the first Sensory Wheel of Virgin Olive Oil (Figure 5.4). Using the central attributes from the wheel, they proposed a score form for profiling virgin olive oils.

Further contributions to the sensory description of extra virgin olive can be found in Monteleone *et al.* (1997), Caporale, Policastro, and Monteleone (2004), Gawel (2007), Delagdo and Guinard (2011b), Dinnella *et al.*, (2012) and Recchia, Monteleone, and Tuorila (2012). Table 5.3 reports the terms used to describe the sensory characteristics of extra virgin olive oils according to three key papers: Mojet and de Jong (1994), Delagdo and Guinard (2011b), and Monteleone *et al.* (2012). These papers were selected for two main reasons: (a) descriptive studies were carried out in different countries and in qualified research sensory laboratories and (b) descriptive terms were generated to profile oils with varied origins (country and variety). It can be

Table 5.3 List of sensory descriptors of extra virgin olive oils as reported in three key papers from 1994 to 2012. Descriptors of defects or off-flavors are not reported.

	Ref.		
Descriptor	Mojet and de Jong (1994)	Delagdo and Guinard (2011b)	Monteleone *et al.* (2012)
Grassy	+	+	+
Green fruit (green olives, green banana; green apple)	+	+	+
Ripe fruit (olives, banana, apple)	+	+	+
Tropical fruit	–	+	–
Hay	+	–	–
Tea	–	+	–
Tomato leaf	+	+	+
Tomato fruit	+	–	+
Herbs	–	+	–
Citrus	–	+	+
Floral	+	+	–
Spicy	–	+	–
Nutty	+	+	–
Butter	–	–	–
Mint	–	–	–
Perfume	+	–	–
Almond	+	+	+
Briny	+	–	–
Artichoke	+	–	+
Bitter	+	+	+
Thick	+	+	+
Rough	+	–	–
Pungent	+	+	+
Peppery	+	+	+
Astringent	+	+	+

Source: Adapted from Mojet, J., and de Jong, S. (1994) with kind permission from Grasas Y Aceites; Delgado, C., and Guinard, J. (2011b) with kind permission from Wiley; and Monteleone, E., Bendini, A., Dinnella, C., Gallina Toschi, T., Giomo, A., Migliorini, M., Pagliarini, E., Recchia, A. (2012) with kind permission from Società Italiana di Scienze Sensoriali.

noted that there are several descriptors that are recurrent across the three considered papers. Hence a new panel may adopt an already existing terminology. In that case, if the descriptive vocabulary was developed by another laboratory or in a different country or region, difficulties in understanding and interpreting the terms could arise. This problem could be bypassed by providing full definitions and standards to panelists (Hunter and McEwan, 1998).

Table 5.4 and Table 5.5 report examples of descriptive terms and their relative definitions and standards as reported in Delagdo and Guinard (2011b) and Monteleone *et al.* (2012). During language development, subjects practice scoring products in order to familiarize themselves with products and

Table 5.4 Vocabulary and references used in the descriptive analysis of 23 extra virgin olive oils varying in origin (country and variety).

Descriptive term	Definition	Reference[a]
Grassy	Fresh cut grass	1 drop of *cis*-3-hexen-1-ol (courtesy of Robertet Flavors, Mexico) in 200 ml of Carapelli Extra Light Olive Oil
Green fruit	Unripe fruit (green olive, banana, green, apple, etc.)	Verbal description
Ripe fruit	Ripe fruit (olives, apple, banana, berries, etc.), "sweet aroma"	Verbal description
Tropical fruit	Aroma of pineapple, guava, lychee, mango, etc.	100% Spanish EVOO from Wholefoods
Green tomato	Aroma of green tomato/tomato leaves and tomato seeds	1492 EVOOs
Tea	Aroma of green tea	Bigelow Green Tea (8 bags/5 g per 4 bags) in 100 ml of Carapelli Extra Light Olive Oil
Herbs	Fresh herbs (e.g., rosemary, basil, oregano)	Basil: 10 g of fresh basil in a coffee filter bag in 150 ml of Carapelli Extra Light Olive Oil Rosemary: 10 g of fresh rosemary in a coffee filter bag in 150 ml of Carapelli Extra Light Olive Oil
Citrus	Lemon aroma	Bartolini Lemon Oil EVOO flavored with lemon (product of Italy)
Floral	Flowers aroma	1 g of Miracle by Lancome Solid Fragrance in 100 ml of Carapelli Extra Light Olive Oil
Nutty	Almonds, pecans, walnuts, etc.	Pecans: 28 g of pecans in a coffee filter bag in 150 ml of Carapelli Extra Light Olive Oil Almond: Sweet Almond Oil International Collection
Butter	Aroma of butter	Fresh butter bar, Imperial Butter from Unilever
Spicy	Spices, e.g., pepper/"pungent" smell	Verbal description
Mint	Mint/eucalyptus aroma	1 drop of artificial flavor (courtesy of Robertet Flavors, Mexico) in 100 ml of Carapelli Extra Light Olive Oil
Bitter	Bitter taste	Caffeine (0.7 g/l), Fisher Scientific
Pungent	Trigeminal sensation/chemical irritation	Capsaicin (0.08m g/l), Fisher Scientific
Thick	Thin–thick	Verbal description
Astringent	Dryness in the mouth	Tannic acid (1.1 g/l), Sigma-Aldrich

[a]Fresh ingredients were weighed over a coffee filter bag, sealed, stapled, and dipped into the oil. The standards were prepared at least 3–4 days in advance to let the oil absorb the aromas of the ingredient.
EVOO, extra virgin olive oil.
Source: Reproduced from Delgado, C., and Guinard, J. (2011b) with kind permission from Wiley.

Table 5.5 Vocabulary and references used in descriptive analysis of extra virgin olive oils varying by origin (country and variety).

Descriptive term	Definition	Reference[a]
Green olive	Odor associated with freshly milled green olives	100 g of olive paste from fresh green olives in 100 ml of seed oil. The standard should be presented to the panelists within 60 min of the preparation
Ripe olive	Odor associated with black (ripe) olives malaxed for 30 min	Olive pastes from black (ripe) olives malaxed for 30 min
Grassy	Odor associated with fresh cut grass	14 μl of *cis*-3-hexen-1-ol in 100 ml of seed oil. The standard should be presented to the panelists within 4 h of the preparation
Tomato fruit	Aroma of ripe tomato	100 g of fresh and ripe "pachito" cherries in 100 ml of seed oil. The standard should be presented to the panelists within 60 min of the preparation
Tomato leaf	Aroma of tomato leaves	Fresh tomato leaves
Apple	Aroma of Golden apple	20 g of skin and 20 g of pulp from a ripe Golden apple in 100 ml of seed oil. The standard should be presented to the panelists within 60 min of the preparation
Citrus	Aroma of lemon/orange	2 g of lemon skin and 2 g of orange skin in 100 ml of seed oil. The standard should be presented to the panelists within 60 min from the preparation
Artichoke	Aroma of artichoke	12 g of artichoke heads in 100 ml of seed oil. The standard should be presented to the panelists within 60 min. of the preparation
Astringency	Dryness in the mouth	Aqueous solution of aluminum potassium sulfate (0.3 g/l)
Bitterness	Bitter taste	Water–oil emulsions of quinine dihydrochloride solutions (intensities from weak to strong 50, 100, and 200 ppm). The model oil should be prepared by using an odorless and tasteless food-grade seed oil and a food-grade water–oil emulsifier
Pungency	Leaving a burning sensation in the back of the throat	Verbal description
Viscosity	Thin–thick	Verbal description

[a]Odorless and tasteless food-grade seed oil is used to prepare references in oil.
Source: Adapted from Monteleone, E., Bendini, A., Dinnella, C., Gallina Toschi, T., Giomo, A., Migliorini, M., Pagliarini, E., Recchia, A. (2012) with kind permission from Società Italiana di Scienze Sensoriali.

the scale rating system. Different scaling methods are used by different researchers, and their efficiency seems to be similar. The unstructured, 10 cm linear scale anchored "not perceived" at the left end and "very intense" at the right end (other anchors are frequently used) is very common, but 9-point category and 15-point unlabeled box scales are also used.

In conventional descriptive analysis, it is important to determine assessor reproducibility. Normally, panel performance is evaluated at the end of the training period by having the panel evaluate, in triplicate, a subset of samples to be used for the real study. Data are frequently analyzed for each attribute by means of a two-way (sample and assessor) or three-way (sample, assessor, replication) analysis of variance (ANOVA) to determine whether there are significant assessor–sample interactions. The significant effect of this interaction implies that the panel leader will determine which assessor should be further trained in the use of which attribute. The need for solid and validated information from sensory data in decision-making processes is extremely clear to sensory scientists. It is not the case that user-friendly statistical tools for validating panel performance are now available for free. Panel Check software developed at Nofima (Norway) allows the panel leader to control the quality of sensory profile data using both univariate and multivariate approaches (Næs, Brockhoff and Tomic, 2010). The software provides a series of graphs for quick visual inspection of data:

- mean and standard deviation and box plot for all assessors for each attribute;
- line plot showing the panel average;
- individual scores for all assessor for each product.

The *F*-values of either two or three-way ANOVA associated with sample, assessor, and their two-way interaction effects are plotted in different colors depending on the significance level, allowing for the rapid detection of significant effects. Furthermore, the comparison of assessors' ability to detect differences among products for each attribute can be checked using the p-MSE plot. The plot allows the analyst easily to detect subjects able to discriminate between samples and, at the same time, able to reproduce their own scores reliably. Finally, the software permits an overall assessment of assessor differences using all variables simultaneously by means a methodology referred to as Tucker1. This is a multivariate method (consensus principal component analysis) able to detect assessors who differ from the rest and attributes that are affected by poorly performing assessors.

5.3.3 Experimental design

The sample evaluation in descriptive analysis is run after defining an appropriate experimental design. This term indicates a series of experimental procedures that have been developed to provide as much information as possible in the most efficient way. A design can be defined only after the aim and the unique properties of the experiment have been stated. In descriptive analysis, the objective is to collect a sensory profile of a heterogeneous group of treatments (products) using a defined number of assessors and replicates. In general, the design should take into account assessor variation, presentation order effect, first-order and carryover effects, and any specific limitation associated with samples and assessors (Piggott, Simpson, and Williams, 1998).

It is possible to say that when planning a profile study there are two important aspects to consider: the first is the initial choice of products (Næs, Brockhoff and Tomic, 2010). It is important to select products carefully. When the objective is to compare a product with similar products on the market, products that span the entire variation in the study must be included. Some sensory methods applied to obtain perceptual maps (e.g., Free Choice Profile, Flash Analysis, and Projective Map) can be very helpful for this purpose.

As already stated, before running a descriptive analysis, samples should be evaluated to test the absence of defects. The second important aspect of an experimental design is how to present the samples to panelists. In sensory olive oil research, the use of a complete design with replicates involving three factors, sample, replicate, and assessors, is extremely frequent. Randomization is a key principle in experimental design. A proper randomization ensures that the effect of extraneous factors is averaged out in the long run. The presentation order represents a source of variation of sensory data in themselves and a balanced design for first-order and carryover effects is needed. This can be obtained by adopting modified Latin-square designs reported by MacFie *et al.* (1989). Software for collecting sensory data and running tests normally allow the experimenter easily to design the presentation order with respect to this important requirement. In descriptive analysis, each assessor is asked to replicate the evaluation of samples. The number of replicates depends on the size of the differences that the experimenter is required to detect. Small expected differences require a higher number of replicates. However, the number of replicates in olive oil sensory studies is frequently in the range 2–4 with 10–12 panelists involved in the test.

5.3.4 Sensory procedure

Bitterness, astringency, pungency, and peppery are common descriptors of extra virgin olive oils due to their phenol content and profile. These sensations tend to persist for a fairly long time after swallowing, showing a clear after-effect that can vary strongly among olive oils in intensity and duration and might affect consumer acceptance (Esti *et al.* 2009; Caporale *et al.*, 2006). Hence they are important sensory characteristics of oils. In a study that explored the dynamic perception of bitterness and pungency by time–intensity measurements (Sinesio, Moneta, and Esti, 2005), the attribute variation over time showed that each sensation acts according to a regular temporal sequence. The difference between the two attribute maxima, of ~10 s (in the order bitterness and pungency) is independent of the intensity. In agreement with these observations, Dinnella *et al.* (2012) showed that the dynamic changes of sensory dominances when tasting oils for 90 s follow the temporal sequence bitterness, pungency, astringency.

A good procedure to describe the sensory characteristics of olive oils should consider the main sensory properties of oils in relation to the following points:

- Conditions of constant stimulation determine a decrease in responsiveness to bitterness (adaptation).
- Astringency is a tactile sensation perceived as a diffuse stimulus in the mouth and commonly described as a puckering, roughening, and drying of the oral surface (Lee and Lawless, 1991). The perceived intensity of an astringent stimulus increases with repeated ingestion (Lyman and Green, 1990). Because of this well-known carryover effect, the evaluation of astringent products such as olive oil with a very high phenolic content cannot be made using a typical side-by-side comparison (Lesschave and Noble, 2005).
- Pungency and peppery are burning sensations and in general have a long-lasting nature. They are defined as chemesthetic sensations (chemical responsiveness mediated by trigeminal nerves). When the rest period between the evaluation of samples is omitted (or it is too short), the perceived strength of these sensations continues to build to higher levels.

Considering all these aspects, it seems appropriate to suggest the sensory procedure for profiling extra virgin oils described by Monteleone *et al.* (2012). Panelists are presented with up to four samples per session (served monadically). Each sample, identified by a three-digit code, is placed in a 100 ml amber-glass vessel containing 30 ml of oil, covered with a plastic Petri dish. The presentation order of samples should be balanced for first-order and carryover effects. Following the presentation order, subjects are asked to smell a sample and score the intensity of aroma (odor by nose) descriptors. Then they are asked to pour part of the sample into a teaspoon (around 3.5 ml), take it into their mouth and rate the perceived viscosity. Panelists are instructed to hold the sample in their mouth for up to 8 s. spit it out, and, after a further 12 s, rate the perceived intensity of bitterness, pungency, and astringency. Finally, subjects are asked again to pour the sample into a teaspoon, take it into the mouth, and rate the intensity of odors perceived retronasally. Specific rinsing procedures between the evaluation of two samples are required to control possible carryover effects. For this purpose, after each sample, subjects can be instructed to rinse their mouths with distilled water for 30 s, chew some plain crackers (or plain unsalted white bread) for 30 s, and finally rinse their mouths with water for a further 30 s. Tests should be conducted in isolated booths, under red light (in order to limit visual bias). Scores are frequently recorded directly on a computer system using dedicated software. When more than four samples are evaluated, it is possible to run more than two sessions per day. However, a break of at least 1 h between each session is recommended. In these cases, the presentation order of samples should be balanced within each replicate rather than each session.

Other general rules are recommended. Before the evaluation, oils should be kept at a temperature ranging from 14 to 15 °C in containers of made inert materials and impermeable to light, and closed tightly. The presence of air in the headspace of storage containers should be avoided. Oils should be presented at room temperature (around 25 °C). Oils should be evaluated within 20 min of the sample preparation.

5.3.5 Data analysis

ANOVA is the most common statistical method used to test the significance of effects included in descriptive analysis experimental designs. As reported above, a three-way ANOVA, involving sample, assessor, and repetition as fixed factors and all first-order interactions, is frequently computed for each attribute in order to evaluate panel performance. When replicate and interaction effects are not significant for the tested attributes, the panel is assumed to have performed well and the data are analyzed in order to study similarities and differences between products. Spider-plots are used to represent the sensory profile of each sample using the mean (over assessors and replicates) intensity score.

In product development studies, mixed ANOVA models are often computed before submitting sensory data to a multivariate analysis in order to obtain a perceptual map. Two-way mixed models are used to test the importance of sensory attributes (sample effect) considering assessors at random. In these cases, Næs, Brockhoff and Tomic (2010) suggested checking for interaction effects in order to obtain a complete overview of differences between products. Nonsignificant sample effects identify unimportant attributes in discriminating between products. These attributes are not used in perceptual mapping.

Principal component analysis (PCA) is a very well-known multivariate statistical method and probably the most commonly applied approach to obtain a perceptual map from descriptive data (Martens and Martens, 2001). For this purpose, PCA models are computed on the product-by-attribute matrix, after having averaged out both replicates and assessors. The method is based on the computation of the most interesting directions of variability, called principal components. The main results are presented graphically in a score plot, which describes the relations between the products, and a loading plot, which describes the relations between the sensory attributes and the principal components. Martens and Martens (2001) proposed a modification of the loading plot that can facilitate the interpretation of the importance of variables in discriminating between products, particularly when the PCA model relies on nonstandardized sensory data, namely a correlation loading plot. This is a two-dimensional scatter plot of correlations between the principal components and the variables themselves. The advantage of this plot is that the researcher can directly obtain information about how much each variable is explained by each component. Moreover, the correlation plot also provides the possibility of drawing circles in the plot corresponding to various degrees of explained variances. Typically, circles for 100% explained variance and for 50% explained variance for the two components are drawn. Furthermore, in order to improve the visual interpretation of the correlation plot, samples can be included as dummy (down-weighted) variables in the data matrix (Martens and Martens, 2001). The graphical interpretation of the plots follows simple rules (Næs, Brockhoff and Tomic, 2010):

- Products which are close to each other have similar overall properties and samples which are far apart are very different.
- Attributes which are close are strongly and positively correlated, whereas those on opposite sides of the origin have a negative correlation.
- Products to the left of the score plot are characterized by attributes to the left in the loading plot, and products to the top of the score plot are characterized by attributes to the top in the loading plot, etc.
- The higher the explained variance, the more valid is the information obtained from the perceptual map.

In PCA models, each component describes as much variance as possible and each new component is orthogonal to the previous one. The explained variance is largest for the first component, next largest for the second component, and so on. The problem is, how many components can be looked at safely? Often, two or three components are computed in order to explain a substantial amount of variance in the data, but even for a small number of components there is the interest in evaluating how reliable the components are. Cross-validation is frequently used to test the goodness of PCA models. The procedure computes an explained "validation" variance for each component. The explained validation variance can be plotted as a function of the number of components. The point where the curve flattens out and becomes stable is where one should stop interpreting components.

Figure 5.5 presents examples of all the plots mentioned above.

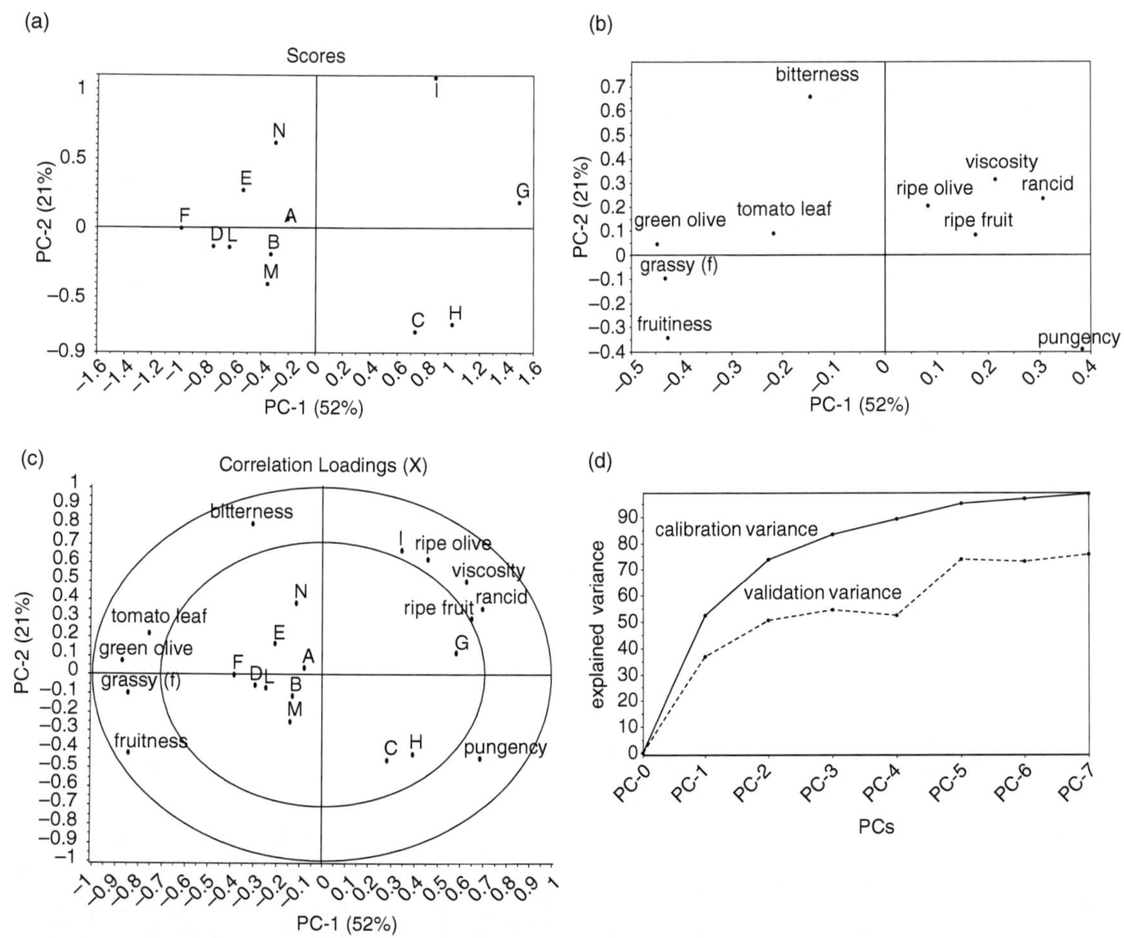

Figure 5.5 PCA of descriptive sensory data: 11 virgin olive oils and 10 attributes. (a) Score plot, which gives information about patterns in the samples; (b) loading plot. which shows which variables are important and which variables correlate; (c) correlation loading plot, where the importance of individual variables is visualized more clearly than in the standard loading plot; (d) explained variance plot, which gives an indication of how much

5.4 Alternative descriptive methods to conventional descriptive analysis

Alternative and complementary methods to conventional descriptive analysis are often used to evaluate a set of products, providing quick access to a relative sensory positioning of a set of products. The best use of these tools is for rapid analyses. Free Choice Profiling (FCP) and Flash Analysis (FA) are the two methods reported here. These methods differ from conventional descriptive analysis in the following characteristics:

- Each panelist creates their own list of descriptive terms: panelists are not extensively trained to create a consensus vocabulary for the product.
- The data set obtained from these methods requires a specific statistical procedure to be analyzed in order to obtain a perceptual map: the Procrustes Analysis.

In product development, these methods can be seen as convenient sensory mapping tools for conducting the preliminary phases of sensory studies. In fact, they may provide a general overview of the sensory differences between products, allowing the researcher to select samples that span the entire sensory space. Selected samples can be further studied by mean conventional descriptive analysis and in consumer studies. FCP and FA are also particularly useful when only a one-shot evaluation is possible. They can also be used for conducting the preliminary phases of the language development step in conventional descriptive analysis.

5.4.1 Free Choice Profiling

FCP (Williams and Langron, 1984) is a descriptive technique in which each assessor produces individual profiles of the products, using his or her own terms for describing products. The spatial configurations derived from individual profiles are analyzed by Generalized Procrustes Analysis (GPA), resulting in a consensus configuration (a perceptual map) revealing the interrelationships between the samples for the panel as a whole. In sensory food studies, the method has been largely used with product experts, sensory evaluation experts who have previously participated in several descriptive evaluations, and consumers (Jack and Piggott, 1991–1992). In olive oil studies, FCP has barely been applied. Guerrero, Romero, and Tous (2011) proposed the method in order to explore the information contained in the generic descriptors (such as fruity) of the official sensory profile of virgin olive oil. However, its use can be appropriate in many steps of olive oil product optimization and development. In presenting this method, we do not consider its consumer application.

In FCP, panelists are allowed to term the sensations they perceive and rate them on a scale. Some panelists may use very few descriptors and others may use many. This technique can be applied for many purposes of perceptual mapping, particularly for explorative purposes when a representation of differences between products is required in a short amount of time. For instance, it can be used to check immediately how much and why a new product differs from its competitors or regular products. In this case, its application is suggested with panels composed of either product experts or subjects who have previously participated in several descriptive analyses of the olive oils. It should be recalled that FCP shows the main differences among samples and makes it possible to interpret them, but it does not indicate the stable differences among products that are often very important to product developers (Lawless and Heymann, 1998).

Data from FCP are treated by means GPA (Gower, 1975). This procedure allows one to obtain a consensus space from individual spatial configurations (one for each panelist for each replication) by means of three steps: translation, scaling, and rotation. These steps produce a perceptual space for each panelist which is matched as closely as possible with other panelists (Piggott and Watson, 1992). Then, the consensus map is calculated as an average configuration for all panelists. This is usually simplified as a reduced dimensional plot by PCA. Thus, the consensus configuration is interpreted similarly by the PCA map. The residual errors, which mean the distances between the panelists' individual configurations and the consensus, can be used to identify outliers or groups.

Figure 5.6 shows the consensus map describing differences among 11 oils. It can be noted that oils are represented by an ellipse which is an empirical picture of the variation in the positions of the oils. If ellipses are superimposed,

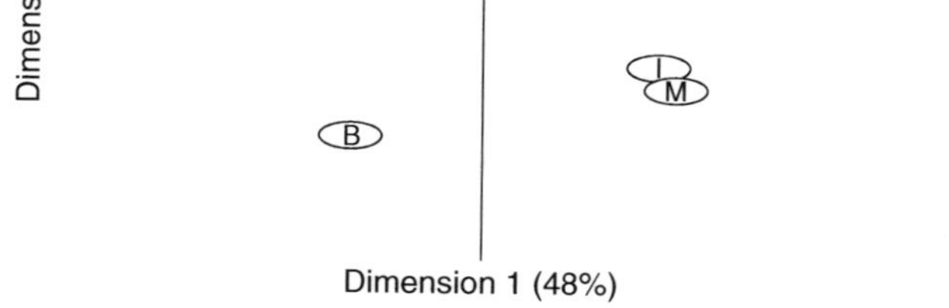

Figure 5.6 Generalized Procrustes Analysis: example of consensus plot.

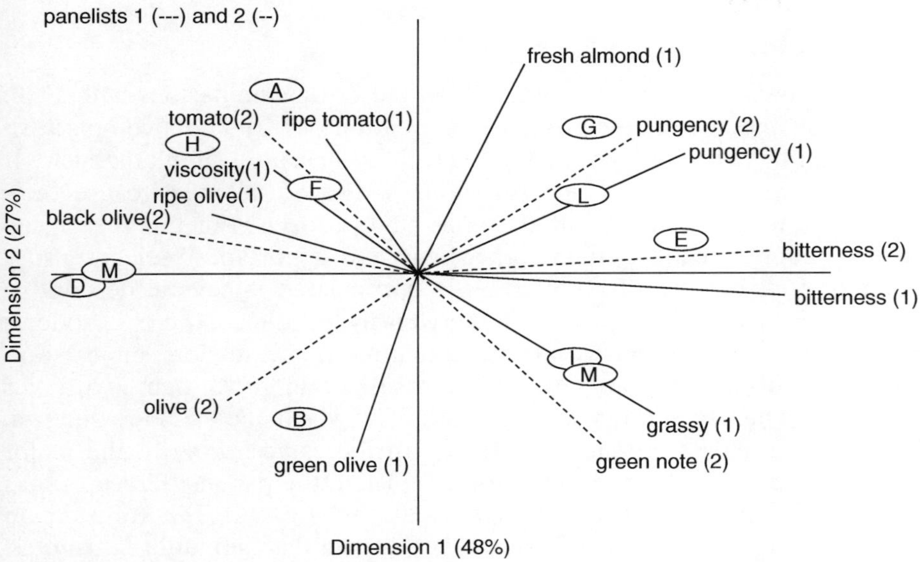

Figure 5.7 Generalized Procrustes Analysis: attribute vectors for subjects 1 and 2.

they are not significantly different. It is possible to plot the descriptors used by the panelists into the consensus space and interpret them in the same fashion as the descriptors on a PCA plot (Figure 5.7). The consensus configuration may be interpreted in terms of each individual panelist's vocabulary and the researcher can evaluate how different terms used by different subjects are related to each other.

The sensory procedure in FCP does not differ from those reported for the conventional profile. Two or three replicates should be run to verify both panelists and panel consistency. Replicate samples should plot together. When it is not possible to run replicates of the entire set of samples, duplicate samples should be included in the test and they should plot together in the consensus map. A permutation test on the data is suggested in order to estimate the statistical validity of the dimensionality of a consensus model. This procedure rearranges at random each individual's rating scores and produces new permuted data sets in which the scores no longer correspond to the products to which they were originally given. The GPA is carried out on a number of permuted data sets (e.g., 50) and the percentage of variance accounted for is computed for each permuted data set. The distribution of these values reflects the probability that a variance accounted for occurs by chance alone. A variance accounted for in the real data higher than the 95th percentile of this distribution indicates a probability of less than 5% that the consensus generated is arisen by chance.

5.4.2 Flash Analysis

Flash Analysis (FA) (Sieffermann, 2000) has strong similarities with FCP. In fact, this method is based on the combination of FCP and a comparative evaluation of the whole product set, which is presented simultaneously. It is applied with a small number of subjects who are allowed to use their own list of attributes to profile the presented products. For this reason, no panel training is needed. On the other hand, it is important to select sensory evaluation experts who have previously participated in several descriptive evaluations, even if they are not necessarily product experts. Products are presented anonymously. At the beginning of the session, subjects are provided with all the samples. They are asked to rank them (with ties allowed) along the sensory characteristics for which the samples differed. In fact, each expert produces their own attributes according to the major differences perceived among the products. Then, the panelist directly ranks products on the chosen descriptors, attribute by attribute. This forces them to focus on the perceived differences and to use discriminating attributes. The sessions are individual and could last from 40–75 min to about half a day. This is because no limitations are given regarding the number of sensory characteristics that they should evaluate, the number of retastings, or the breaks needed. As a repeatability probe, a product can be replicated in the sample set, or the whole evaluation can be replicated. All sessions can be carried out individually, on separate days, and at separate locations, provided that all samples are available. GPA is frequently applied to the data from the FA profile to assess the consensus between assessors' sensory maps (Delarue and Sieffermann, 2004). Considering the main sensory properties of extra virgin olive oils (bitterness, pungency, astringency) and that samples are directly compared with each other, it is strongly suggested that the FA method should be applied when a limited number of samples (4–5) are tested and when well-experienced assessors are recruited. The method could be of some utility when sensory differences between a few samples have to be quickly evaluated.

5.5 Perceptual maps from similarity data

Sorting and projective mapping (or "Nappe" method) are methods frequently applied to obtain representations of the differences amongst products. Because of their immediateness and ease of application, they are assumed to be useful tools in product development (Buck, 2007).

In the sorting method, subjects are asked to sort items into two or more groups: items that have something in common are placed in the same group, whereas items that differ from one another should be placed in different groups. In this method, similarity is a group-derived estimate; in fact, similarity is inferred from the number of times two items are sorted into the same group across a panel of participants. Data from the sorting task for each sample are converted into a similarity matrix by summing over all participants the number of times each pair of oils is sorted into the same group. This matrix is analyzed with multidimensional scaling to obtain a map. After completing their sorting task, subjects can be asked to give a few words or descriptors that provided the criteria on which they had based their sorting. This information is used to interpret the differences among products.

Sorting tasks have rarely been applied in olive oils studies. In a consumer study, Recchia, Monteleone, and Tuorila (2012) asked subjects to sort four samples. Research studies (unpublished data) conducted at the Sensory Laboratory of the University of Florence allowed us to suggest the use of sorting to categorize up to eight oils when working either with trained assessors or with product experts. Sorting can be usefully applied to study differences among products and to verify the correspondence of some prototypes to specific profiles. For instance, the method might be used with product experts to verify the correspondence of an oil blended from different production batches to a specific sensory style. In a sorting task, product experts use a personal criterion based on their technical knowledge. Thus results from sorting tasks depend mainly on previous knowledge of participants, that is, the level of expertise influences the categorization. It should be clear to the reader that a consensus map from a sorting test performed by experts can be very different, in terms of relative positioning of samples, from a perceptual map obtained from descriptive data. Product experts tend to discriminate samples mainly on their quality (e.g., absence of defects, balance, harmony), rather than on their specific sensory properties. Oils with different sensory profiles categorized as having similar qualities are closely positioned on the consensus map. In contrast, when the sorting task is applied with assessors specifically trained to profile oils, the resulting perceptual map describes the main differences and similarities among the products in relation to specific sensory properties. Therefore, depending on the objective of the test, assessors should be qualified on the basis of clearly defined criteria. A sorting task is usually simple and easy to perform, but there are limitations. As observed by Nestrud and Lawless (2008), one of the limitations of sorting is that it provides only grouped or aggregate data on similarity, which means that it only puts similar products into nominal level categories. Furthermore, the stability of models needs to be checked. From a practical point of view, the simplest way to check the reliability of results is to insert a blind duplicate of one of the products to see whether they plot together on the map.

Projective mapping may provide more graded information than sorting, because it is based on the individualization of similarities and differences using a graphic representation and not a nominal categorization (Pagès, 2005; Nestrud and Lawless, 2008). In fact, this is a simple user-friendly technique that allows subjects (naive consumers, trained subjects, and experts) to express perceptual similarities and groupings among a set of products by placing them on a two-dimensional surface. Subjects are simply required to place products on the surface according to the similarities and differences they perceive, so that similar products appear near to one another and different products further apart. Placing the items on a two-dimensional surface provides a potentially richer amount of information than the simple categorical information of sorting tasks: measuring the distance on the surface amongst products for each person, it is possible to calculate the greater and the smaller similarities (and differences) among products.

This method was originally developed by Risvik and colleagues as a possible alternative to profiling and (dis)similarity scaling methods to obtain perceptual maps (Risvik *et al.*, 1994). From its first application, this method has been viewed as a useful technique for the study of consumer perception. In further work by this group, the mapping dimensions from naive consumers were compared with those from the profile data. The results showed that "the best similarity was found when comparing the first dimension, thus suggesting good agreement on the obvious aspects of the product" (Risvik, McEwan, and Rodbotten, 1997).

In earlier studies, more than one multidimensional method was applied to analyze projective map data, such as MDS, GPA, and PCA. More recently, Pagès (2003, 2005) renewed the projective mapping technique, terming it the "Nappe" method (or napping). This method consists in collecting the perceived differences among products by positioning the products on a sheet of blank paper (in French the word "nappe" means "tablecloth"). The data obtained (the coordinates for each product) are treated by a multiple factor analysis (MFA).

In a further study (Morand and Pagès, 2006), results from GPA and MFA, applied to the same "nappe" data set, were compared, showing that the two approaches give very similar configurations. These findings were subsequently confirmed by Nestrud and Lawless (2008).

With "napping," it is possible to get a representation of the products that integrates the relative importance for the subjects of the characteristics of the products; however, this does not characterize the product itself. Hence the "nappe" method is presented as a complementary method and not as an alternative to descriptive analysis. As any other method applied in food sensory science, the "nappe" procedure shows some weakness, as reported in most of the papers cited above. One important issue is certainly the reliability of the results from this method: results are strongly dependent on the initial question, which needs to be attentively and consciously determined. Another issue is the real contribution of the second dimension of the maps in discriminating among samples; actually, subjects seem to discriminate on the basis of only one dimension.

When applying projective mapping, the following rules are suggested:

- Define the objective of the test, select an appropriate set of samples, and recruit qualified subjects.

- Give instructions about the principle and procedure of the method to the subjects, as outlined by Pagès (2005) and Nestrud and Lawless (2008):
  - *Principle*
    You are asked to evaluate the similarities (or dissimilarities) between several oils. You have to do this according to your own criteria, those that are significant to you. You do not have to indicate your criteria. There is no good or bad answer.
  - *Procedure*
    First, it is important to taste each sample and make any notes or comments about the sample according to your own criteria in the space provided below and on the back of the sheet. Be sure to save enough oil for three or four tastes. Cleanse your palate as needed with water and unsalted crackers. Then, after tasting all the samples, you have to position the oils on the paper in such a way that two oils are very near if they seem identical to you and that two oils are distant to one another if they seem different to you. This must be done according to your own criteria. Do not hesitate to express strongly the differences you perceive by using most of the sheet. When the operation is finished, write down on the sheet the number of the oil in the place it occupies.
  - Notes or comments about the samples can be directly plotted in the consensus map by using MFA (Pagès, 2005). Alternatively, they can be used to arrange a follow-up session to retaste the samples and rate the generated attributes (Nestrud and Lawless, 2008; Perrin *et al.* 2008).
- Run the test with a number of subjects from 15 to 20.
- Run a pilot test to check the number of products to taste. It is important to add duplicate samples in the group of examined products; it is suggested that the number of oils to be tested should be limited to eight.
- Before submitting data to multidimensional methods (GPA or MFA), a visual inspection of data is strongly suggested in order to remove unreliable subjects from the data set.

5.6 Temporal aspects of sensory characteristics of olive oils: Time–Intensity (TI) and Temporal Dominance of Sensations (TDS)

TI is a very well-known method to evaluate the temporality of the sensory perception in food products. In olive oil studies, there are few example of the application of this method. Caporale, Policastro, and Monteleone (2004) used TI to investigate the extent to which the green aroma affects the perception of bitterness in virgin olive oil and described a clear bitterness enhancement induced by a cut grass odorant. Sinesio, Moneta, and Esti (2005) applied the TI method to characterize temporal differences of bitterness and pungency perceptions in stored virgin olive oils. The aim of the study was to determine TI bitterness and pungency evolution in mono-varietal oils derived from autochthonous olive varieties and to follow their evolution in ideal and stressed storage conditions. The same group (Esti *et al.*, 2009) expanded the results from the initial study on temporal intensity variation of bitterness and pungency perception in extra virgin olive oils by exploring the relationship between bitterness and pungency intensity and its phenolic compound contents and their changes during storage due to low and stressed temperature conditions. It was shown that the TI curves for pungency showed a slower rise and decline than those of bitterness for all oil samples.

In a TI measurement, a single sensory characteristic (or occasionally two characteristics) is tracked as it changes over a period of time (Piggott, Hunter, and Margomenou, 2000). The intensity of the same sensation is measured continuously, using a computerized data collection system, and is registered at short time intervals (repeated measures data); the measurement instruments are sensory assessors. For both credibility and precision, many (eight or more) assessors are required in addition to replications. Prior to their participation in the experiment, subjects are trained in the recognition of the sensation to be tested (e.g., bitterness) through standard samples. No more than four samples per session should be evaluated. In order to minimize position bias, the presentation order of the samples within each session should be balanced among assessors. Oils, labeled with a three-digit code, are served in colored tasting glasses to mask color differences, thus eliminating the visual factor. Evaluation should take place in individual sensory testing booths under red light. A horizontal linear scale is displayed on a computer screen where 0 represents no taste and 100 denotes a very strong flavor. Assessors are instructed to take a defined amount of oil (3–5 ml) in their mouth and start the data collection. They are instructed to start rating the intensity of sensation as soon as the sample is in their mouth by moving the rating line away from the 0 position. After 10 s, the subjects are prompted by a screen signal to spit out (or swallow) the sample and to continue assessing the intensity of the evaluated sensation. The computer continues to collect data either until it registers the 0 value again or for a maximum time (around 2 min for oils). The assessors' ratings are recorded by computer every 0.5 s for 1 min. Normally assessors are asked to rinse their mouths with distilled water, eat plain crackers, and rinse again with water after each sample. A 2 min interval between each sample is suggested (Caporale, Policastro, and Monteleone, 2004). Individual TI curves from each assessor can be obtained by plotting collected intensity data as function of time.

It is also possible to obtain average curves (across assessors and replications) for each sample. Owing to the large differences between assessors, the computation of average curves is more complex than it might seem. According to Sinesio, Moneta, and Esti (2005), the application of a non-centered PCA on the row data matrix instead of arithmetic averaging over the individual allows for the representation of all the individual curves and retained information regarding the level and variability in the analysis (it does not

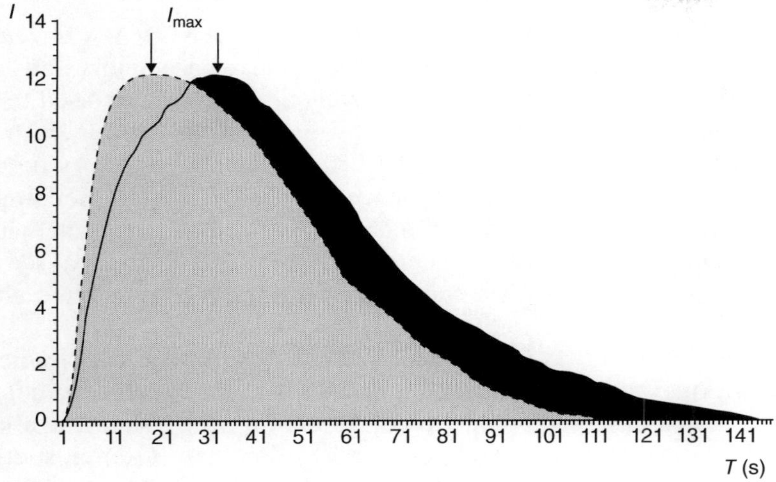

Figure 5.8 Scheme of time–intensity bitterness (- - -) and pungency (—) curves in extra virgin olive oils. *Source:* Reprinted from Esti, M., Contini, M., Moneta, E., and Sinesio, F. (2009) with kind permission from Elsevier.

change the shape of the average curve). An example of an average TI curve referred to olive oil bitterness is shown in Figure 5.8. The most common approach in comparing TI curves between samples is the computation for each sample of a number of curve parameters. In their study on olive oils, Sinesio, Moneta, and Esti (2005) considered the following: maximum perceived intensity (I_{max}) during the time of measurement; reaction time to the stimulus, when it is first perceived after initial exposure (T_{onset}) (s); time to reach maximum intensity of the sensation (T_{max}) (s); plateau time, or duration of the maximum intensity ($T_{plateau}$) (s); extinction time, when the sensation returns to zero (T_{ext}) (s); area under the curve before maximum (A_{before}); area under the plateau ($A_{plateau}$); area under the curve after maximum (A_{after}); rate of intensity increase before Imax (slope) (V_1); and rate of intensity decrease after I_{max} (slope) (V_2). These parameters from each individual curve can be submitted to several ANOVA models considering sample, repetition, and assessor effects (Lawless and Heymann, 1998; Piggott, Hunter, and Margomenou, 2000). For instance, Sinesio, Moneta, and Esti (2005) computed a three-way ANOVA model on the above-mentioned parameters of bitterness and pungency curves to evaluate the effects of panelists, samples, and test replicates and all fixed effects. They reported that panelists were a significant source of variation for a few TI parameters. However, they underlined that this is to be expected and common in TI measurements because individuals can differ in sensitivity and use of scale, thus providing different patterns of response. However, test replicates and interactions between panelists were not significant. ANOVA with repeated measure models are frequently used to treat TI curve parameters statistically.

Although the method has been extensively and successfully used to study the temporal aspects of the perception of a given sensory attribute in a product, when it is used for several attributes it quickly becomes very time consuming because one run is needed for each attribute. This limitation was clear to Pangborn (1989), who predicted increasing sensory research activities in analyzing temporal aspects of perception. Temporal dominance of sensation (TDS) (Pineau *et al.*, 2009) is certainly the most innovative method alternative to TI.

TDS was developed as of 1999 at the Centre Européen des Sciences du Goût, in Dijon, France. It consists in presenting to the panelist the complete list of attributes on a computer screen. Thereafter, the panelist is asked to assess which of the attributes is perceived as dominant (i.e., most striking perception at a given time). During the testing of one product, the panelist is free to select an attribute several times. Conversely, another attribute may not be selected at all. In the course of the evaluation, when the panelist considers that the dominant attribute has changed, they have to select the new dominant attribute, and so on, until the perception ends. For each run, this method allows the collection of a sequence of sensory attributes quoted at different times during the tasting.

The TDS method is applied with a trained panel, which means that, as for descriptive panels, qualified subjects develop a vocabulary to describe the sensations they perceive considering the temporal evolution of the sensations. They also establish a consensual definition of each attribute and are trained in the method in itself. For the training sessions, a dominant attribute is defined as the attribute associated with the sensation catching the attention at a given time. Thus the dominant attribute is not necessarily the one with the highest intensity.

The size of a TDS panel is similar to that of descriptive panels. Sample presentation follows the same rules of descriptive analysis. In order to avoid any order effect of the list of attributes, the order of descriptors should be balanced among subjects using Williams Latin squares. The application of the method requires dedicated software for data acquisition, now commercially available. This software also allows for the analysis of the data and the creation of TDS curves for each product. The computation of TDS curves was well explained by Pineau *et al.* (2009). The procedure considers each attribute separately. For each point in time, the proportion of runs (subject $\times$ replication) for which the given attribute is assessed as dominant is computed. These proportions, smoothed using appropriate procedures, are plotted against time and are called TDS curves (Figure 5.9). On the plot it is possible to draw a line corresponding to the "chance level," which means the dominance rate that an attribute can obtain by chance. Its value is equal to $1/p$, where p is the number of attributes. It is also possible to drawn a line corresponding to the 95% significance level.

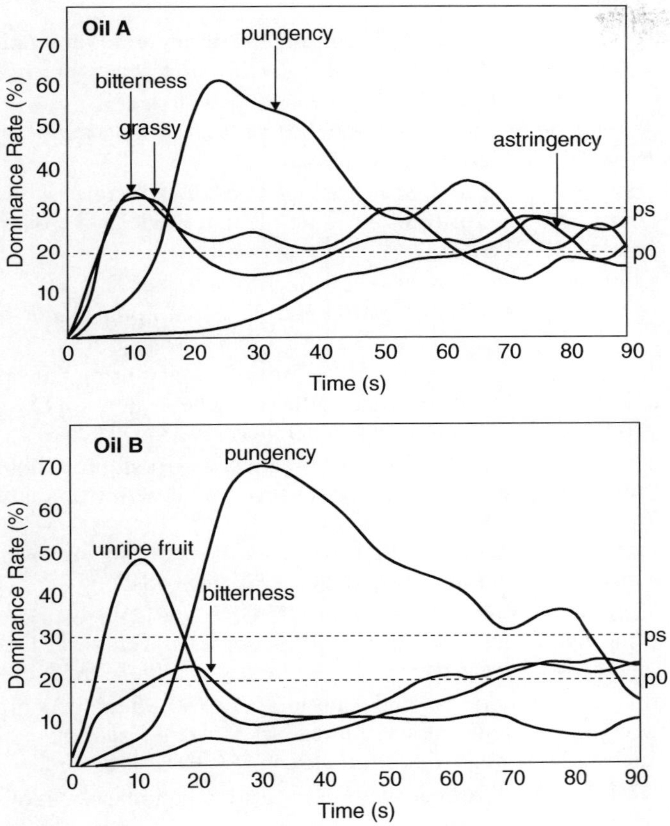

Figure 5.9 TDS curves of olive oil samples: p0 represents the chance level and ps the 95% significance level. *Source:* Reprinted from Dinnella, C., Masi, C., Zoboli, G., and Monteleone, E. (2012) with kind permission from Elsevier.

TDS seems to be more appropriate than descriptive analysis to study sensory interactions occurring when tasting food (such as suppression, release from suppression, and enhancement). For this reason, TDS seems to be a promising method also for developing new approaches in studying oil–food pairing, as demonstrated by Dinnella *et al.* (2012).

More details on TDS application and data validation are reported in Chapter 7.

References

Bertuccioli, M. (1994) A study of sensory and nutritional quality of virgin olive oil in relation to variety, ripeness and extraction technology. Overview of three year study and conclusion. *Grasas y Aceites*, **45**, 55–59.

Buck, D. (2007) Methods to understand consumer attitudes and motivations in food product development, in *Consumer-Led Food Product Development* (ed. H.J.H. MacFie), Woodhead Publishing, Cambridge, pp. 141–157.

Cairncross, S.E., and Sjostrom, L.B. (1950) Flavor profiles – a new approach to flavor problems. *Food Technology*, **4** (8), 308–311.

Caporale, G., Policastro, S., and Monteleone, E. (2004) Bitterness enhancement induced by cut grass odorant (*cis*-3-hexen-1-ol) in a model olive oil. *Food Quality and Preference*, **15**, 219–227.

Caporale, G., Policastro, S., Carlucci, A., and Monteleone, E. (2006) Consumers' expectations for sensory properties in virgin olive oils. *Food Quality and Preference*, **17**, 116–125.

Delarue, J., and Sieffermann, J.-M. (2004) Sensory mapping using flash profile comparison with a conventional descriptive method for the evaluation of the flavour of fruit dairy products. *Food Quality and Preference*, **15**, 383–392.

Delgado, C., and Guinard, J. (2011a) How do consumer hedonic ratings for extra-virgin olive oil relate to quality ratings by experts and descriptive analysis ratings? *Food Quality and Preference*, **22**, 213–225.

Delgado, C., and Guinard, J. (2011b) Sensory properties of Californian and imported extra virgin olive oils. *Journal of Food Science*, **76**, S170–S176.

Dinnella, C., Masi, C., Zoboli, G., and Monteleone, E. (2012) Sensory functionality of extra-virgin olive oil in vegetable foods assessed by Temporal Dominance of Sensations and Descriptive Analysis. *Food Quality and Preference*, **26**, 141–150.

Esti, M., Contini, M., Moneta, E., and Sinesio, F. (2009) Phenolic compounds and temporal perception of bitterness and pungency in extra-virgin olive oils: changes occurring throughout storage. *Food Chemistry*, **113**, 1095–1100.

Gawel, R., (2007) *Olive Oil Tasting Wheel*, www.aromadictionary.com/oliveoilwheel.html (last accessed 28 May 2013).

Gower, J. C. (1975) Generalized procustes analysis. *Psychometrika*, **20**, 33–51.

Guerrero, L., Romero, A., and Tous, J. (2001) Importance of Generalised Procrustes Analysis in sensory characterisation of virgin olive oil. *Food Quality and Preference*, **12**, 515–520.

ISO (1994) ISO 11035. *Sensory Analysis. Identification and Selection of Descriptors for Establishing a Sensory Profile by a Multidimensional Approach.* International Standards Organization, Geneva.

Jack, F.R., and Piggott, J.R. (1991–1992) Free choice profiling in consumer research. *Food Quality and Preference*, **3** (3), 129–134.

Lawless, H.T., and Heymann, H. (1998) *Sensory Evaluation of Foods: Principles and Practices*, Chapman and Hall, New York; 2nd edn, 1999, Springer, New York.

Lee, C.B., and Lawless, H.T. (1991) Time-course of astringent materials. *Chemical Senses*, **16**, 225–238.

Lesschaeve, I., and Noble, A.C. (2005) Polyphenols: factors influencing their sensory properties and their effects on food and beverage preference. *American Journal of Clinical Nutrition*, **81**, 330–335.

Lyman, B.J., and Green, B.G. (1990) Oral astringency: effects of repeated exposure and interactions with sweeteners. *Chemical Senses*, **15**, 151–164.

Lyon, D.H., and Watson, M.P. (1994) Sensory profiling: a method for describing the sensory characteristics of virgin olive oil. *Grasas y Aceites*, **45**, 20–25.

MacFie, H.J., Bratchell, N., Greenhoff, K., and Vallis, L.V. (1989) Designs to balance the effect of order of presentation and first-order carry-over effects in Hall tests. *Journal of Sensory Studies*, **4** (2), 129–148.

Martens, H., and Martens, M. (2001), *Multivariate Analysis of Quality: an Introduction*, John Wiley & Sons, Ltd, Chichester.

Meilgaard, M., Civille, G.V., and Carr, B.T. (1991) *Sensory Evaluation Techniques*, CRC Press, Boca Raton, FL [2nd edn, 2000].

Mojet, J., and de Jong, S. (1994) The sensory wheel of virgin olive oil. *Grasas y Aceites*, **45**, 42–47.

Monteleone, E., Caporale, G., Lencioni, L., Favati, F., and Bertuccioli, M. (1995) Optimization of virgin olive oil quality in relation to fruit ripening and storage, in *Food Flavours: Generation, Analysis and Process Influence. Developments in Food Science*, vol. **37**A (ed. G. Charalambous), Elsevier, Amsterdam, pp. 397–418.

Monteleone, E., Caporale, G., Carlucci, A., and Bertuccioli, M. (1996) Prediction of virgin olive oil sensory profile. *Industrie Alimentari*, **35**, 1066–1072.

Monteleone, E., Carlucci, A., Caporale, G., and Wakeling, I. (1997) Consumer preference of extra virgin olive oil. *La Rivista Italiana delle Sostanze Grasse*, **74**, 415–421.

Monteleone, E., Bendini, A., Dinnella, C., Gallina Toschi, T., Giomo, A., Migliorini, M., Pagliarini, E., Recchia, A. (2012) L'olio extra vergine di oliva, in *Atlante Sensoriale dei Prodotti Alimentari*, Società Italiana di Scienze Sensoriali, Tecniche Nuove, Milan, pp. 114–129.

Morand, E., and Pagès, J. (2006) Procrustes multiple factor analysis to analyse the overall perception of food products. *Food Quality and Preference*, **17**, 36–42.

Næs, T., Brockhoff, P.B., and Tomic, O. (2010) *Statistics for Sensory and Consumer Science*, John Wiley & Sons, Ltd, Chichester.

Nestrud, M.A., and Lawless, H.T. (2008) Perceptual mapping of citrus juices using projective mapping and profiling data from culinary professionals and consumers. *Food Quality and Preference*, **19**, 431–438.

Pagès, J. (2003) Recueil direct de distances sensorielles: application à l'évaluation de dix vins blancs de Val de Loire. *Sciences des Aliments*, **23**, 679–888.

Pagès, J. (2005) Collection and analysis of perceived product inter-distances using multiple factor analysis: application to the study of 10 white wines from the Loire Valley. *Food Quality and Preference*, **16**, 642–649.

Pangborn, R.M. (1989) The evolution of sensory science and its interaction with IFT. *Food Technology*, **43** (9), 248–256, 307.

Perrin, L., Symoneaux, R., Maitre, I., Asselin, C., Jourjon, F., and Pagès, J. (2008) Comparison of three sensory methods for use with the Napping^{TM} procedure: case of ten wines from the Loire valley. *Food Quality and Preference*, **19**, 1–11.

Piggott, J.R. and Watson, M.P. (1992). A comparison of free choice profiling and the repertory grid method in the flavour profiling of cider. *Journal of Sensory Studies*, **7**, 133–145.

Piggott, J.R., Simpson, S.J., and Williams, S.A.R. (1998) Sensory analysis. *International Journal of Food Science and Technology*, **33**, 7–18.

Piggott, J.R., Hunter, A.E., and Margomenou, L. (2000) Comparison of methods of analysis of time–intensity data: application to Scotch malt whisky. *Food Chemistry*, **71**, 319–326.

Pineau, N., Schlich, P., Cordelle, S., Mathonnière, C., Issanchou, S., Imbert, A., Rogeaux, M., Etiévant, P., and Köster, E. (2009) Temporal dominance of sensations: construction of the TDS curves and comparison with time–intensity. *Food Quality and Preference*, **20** (6), 450–455.

Recchia, A., Monteleone, E., and Tuorila, H. (2012) Responses to extra virgin olive oils in consumers with varying commitment to oils. *Food Quality and Preference*, **24**, 153–161.

Risvik, E., McEwan, J.A., Colwill, J.S., Rogers, R., and Lyon, D.H. (1994) Projective mapping: a tool for sensory analysis and consumer research. *Food Quality and Preference*, **5**, 263–269.

Risvik, E., McEwan, J.A., and Rodbotten, M. (1997) Evaluation of sensory profiling and projective mapping data. *Food Quality and Preference*, **8**, 63–71.

Servili, M., Selvaggini, R., Esposto, S., Taticchi, A., Montedoro, G., and Morozzi, G. (2004) Health and sensory properties of virgin olive oil hydrophilic phenols: agronomic and technological aspects of production that affect their occurrence in the oil. *Journal of Chromatography A*, **1054**, 113–127.

Sieffermann, J.M. (2000) Le profil flash – un outil rapide et innovant d'évaluation sensorielle descriptive, in *AGORAL 2000*, XIIèmes Rencontres "l'Innovation: de l'Idée au Succès," Montpellier, France, pp. 335–340.

Sinesio, F., Moneta, E., and Esti, M. (2005) The dynamic sensory evaluation of bitterness and pungency in virgin olive oil. *Food Quality and Preference*, **16**, 557–564

Stone, H., Sidel, J., Oliver, S., Woolsey, A., and Singleto, R.C. (1974) Sensory evaluation by quantitative descriptive analysis. *Food Technology*, **28** (11), 24–34.

Szczesniak, A.S. (1963) Objective measurements of food texture. *Journal of Food Science*, **28**, 410–420.

Williams, A.A., and Langron, S.P. (1984) The use of free-choice profiling for the evaluation of commercial ports. *Journal of the Science of Food and Agriculture*, **35**, 558–568.

6 Consumer Research on Olive Oil

Claudia Delgado,[1,2] Metta Santosa,[1,3] Aurora Gómez-Rico,[1,4] and Jean-Xavier Guinard[1]

[1] *Department of Food Science and Technology, University of California, Davis, CA, USA*
[2] *Sensory and Statistics, Gustatec Division, Chromocell Corporation, North Brunswick, NJ, USA*
[3] *Sensory Scientist and Consultant, Hayward, CA, USA*
[4] *AINIA, Parque Tecnologico de Valencia, Paterna Valencia, Spain*

6.1 Introduction

For decades, understanding how consumers behave has been critical for the development of new products, the acceptability of existing ones, and also the profitability of companies. Olive oil as a product has been recognized as a staple food for the countries in the Mediterranean area. According to data from the International Olive Council in 2009, the cultivation of olives is mainly located in the countries of the Mediterranean basin, which account for 98% of world olive cultivation. Most of the world olive oil production is also located in the Mediterranean countries, with Spain, Italy, and Greece being the major producers. In recent years, however, interest in olives and olive oil has extended olive cultivation to other regions, such as the United States, South America, South Africa, and Australia.

Traditionally, the major producing countries are also the major consumers of olive oil. For example Greece has the highest per capita consumption (25 kg per capita), followed by Spain (12.6 kg per capita) and Italy (12.3 kg per capita); these three countries account for approximately 60% of the world's olive oil consumption (Zampounis, 2006). Although consumption outside the Mediterranean area is still relatively low, popularity has been increasing owing to the evolution of consumer preferences towards healthier,

Olive Oil Sensory Science, First Edition. Edited by Erminio Monteleone and Susan Langstaff.

more nutritious, and more natural foods (Bagnoli, 2007). In this context, olive oil fulfills consumer expectations because of its nutritional content (high content of oleic acid and balance between monounsaturated and polyunsaturated fatty acids) and the presence of minor compounds that have antioxidant properties (polyphenols and tocopherols), in addition to the volatile compounds that are responsible for its particular flavor.

Our group has conducted a number of sensory and consumer studies to document the sensory properties of olive oil and to understand the perceptions of and preferences for olive oil, and the knowledge and attitudes of consumers towards olive oil, with the overarching goal of increasing extra virgin olive oil (EVOO) appreciation and consumption in the United States (Santosa, Abdi, and Guinard, 2010; Santosa and Guinard, 2011; Delgado and Guinard, 2011a, 2011b, 2012). This chapter describes our approach to consumer research and how we have applied it to the study of consumer behavior as it pertains to olive oil.

6.1.1 Consumer research model

The various attempts to develop food choice behavioral models confirm that food consumption behaviors are complex. Hence there is a need to develop a consumer research model that allows operational investigation of consumer behavior without any constraint from particularly predefined sets of factors. Meiselman (2003) suggested that food quality perceptions and acceptance research should investigate three main factors: the food itself, who is eating the food and the characteristics, and the environment in which the food is consumed. Some studies have shown that blind sensory or consumer acceptance tests did not provide similar results when conducted non-blind, for example, when other information such as brand, price, nutritional information, manufacturing process, and so on was available to subjects (Solheim and Lawless, 1996; Guinard, Uotami, and Schlich, 2001; Caporale and Monteleone, 2004; Di Monaco *et al.*, 2004). Often, favorable results from consumer acceptance tests did not translate into purchase or consumption of the new products, or product success in the market place (Garber *et al.*, 2003). Grunert (2003) proposed that food choice analysis needed to include both consumption and purchase aspects for better predictions of product success.

A consumer research model that extends the three-factor research approach proposed by Meiselman (2003) to include broad consumption behaviors of interest beyond the traditional food quality perceptions and acceptance was developed by our group. This consumer research model relates consumer behavior to three types of variables: product variables, consumer variables, and context variables (Figure 6.1). The model provides a simple framework to select appropriate methodologies to investigate consumption behaviors. Product variables are all the characteristics of the product, such as price, brand image, package, label information, and sensory properties, among others. Consumer variables are mostly demographics and

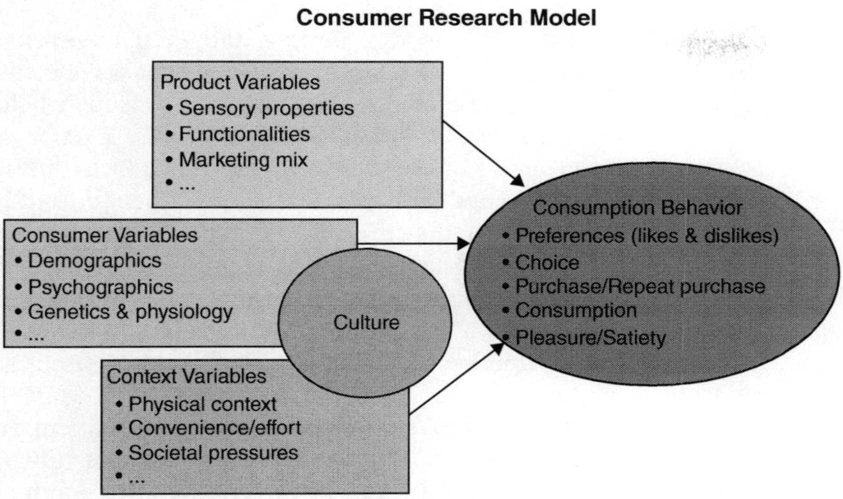

Figure 6.1 Consumer research model for studying consumer behavior related to consumption.

psychographics, but also include anatomy, physiology, and genetics. Context variables can summarily be defined as all those variables that are not product or consumer variables. In practice, they include the physical environment of consumption but also the nonphysical context. Examples of context variables are the location of consumption, the presence or absence of others, the political, economic, and social climate, and societal trends and pressures. The consumption behavior measures that can be modeled with this model are variables such as preferences, liking, purchase intent, repeat purchase, and satisfaction, among others. The model also considers culture as a set of variables that overlaps the consumer variables and the context variables. This model has a number of implications for consumer research.

First, the model is a multivariate model. It recognizes that a multitude of factors affect consumer behavior, and that each variable may be weighted differently in its influence on behavior. Because it is a multivariate model, multivariate statistics are required to analyze the effect of the different variables on the behavioral measure whenever quantitative methods are used to investigate the behavior. In turn, because multivariate statistics are used, trends and patterns are uncovered rather than quantifiable cause–effect relationships.

Second, the traditional scientific method that investigates the effect of one or more independent variables on a dependent variable of interest is no longer applicable here, or, if it is, it provides only partial and somewhat limited explanations for consumer behavior. For example, one can study the effect of sugar and fat content on liking for ice cream, or the effects of consumer age and gender on liking for ice cream, but the outcomes of such experiments will only begin to explain the behavior of ice cream consumption.

Third, the way to investigate the effects of the multiple independent variables on the dependent variable is not to manipulate all the independent variables according to some experimental design, and then measure the dependent variable, but rather to cast a very broad net over a wide range of levels for all the variables in the model and monitor the consumer behavior of interest across these combined variables and levels. In practice, this means measuring the behavior of interest for a range of products differing in many product variables (i.e., finding and/or creating variance in product variables), among a range of consumers differing in many consumer variables (i.e., finding variance in consumer variables), and in a range of contexts differing in many context variables (i.e., finding and/or creating variance in context variables). Because of the sheer number of potential variables in the model, however, it may not be possible to effect the data collection entirely with quantitative means.

Fourth, the model combines discrete and nondiscrete independent variables. This adds to the challenge of analyzing the effects of these independent variables on the dependent variable(s), because it requires combining parametric and nonparametric statistics. Furthermore, some of the variables in the model are difficult to quantify, hence limiting the exclusive application of quantitative methods.

Therefore, and fifth, investigating the effects of the product, consumer, and context variables in the model on consumer behavior requires a combination of qualitative and quantitative approaches. Indeed, the trends that we seek to uncover and the relative contributions of the variables that we want to elucidate can be resolved either through multivariate analysis of quantitative data or by deriving key learnings from qualitative approaches.

6.1.2 Consumer research methodology

It is well known in the sensory and consumer literature that both sensory properties and nonsensory factors are important determinants of food choice (Meiselman and MacFie, 1996; Cardello, 2003; Jaeger, 2006; Köster, 2009). In order to understand how these factors affect consumer preferences for olive oil, it is important to be familiar with the fundamentals of consumer research methodologies. There are two main approaches of consumer research methodologies – quantitative and qualitative methods. This chapter illustrates some examples of quantitative and qualitative research methods as applied to olive oil.

6.1.2.1 Quantitative versus qualitative Quantitative methods in consumer research utilize systematic experimental and sampling designs that produce numerical measurement results from which statistical analysis would be employed to analyze the collected data, and to deduce theories or models for the phenomenon or questions under study.

In quantitative research methods, consumers indicate their response to a product or topic using numerical parameters; these numbers are translated into means, percentages, or other statistics of importance for the researcher.

Survey questionnaires are the most common form of quantitative methods (McCarthy and Perreault, 2005). A survey is a method of gathering and collecting data from a particular sample in order to generate quantitative parameters. There are several methods of data collection: ballot, mail, telephone, and Internet. The selection of the collection method depends on the purpose of the research and also the impact of the selected method on survey costs and errors (Groves, 2004; Dillman, 2007).

Qualitative research methods, in contrast, employ a series of techniques that do not involve the generation of numerical parameters. for example. one-to-one interviews, focus groups, and observation/ethnography studies. Qualitative methods instead provide specific and in-depth details of how consumers perceive or use a product/service. Qualitative data involve words and visual images of products and objects, all of which can be subjected to analysis. The downside is that the statistical analysis is limited; some researchers may argue against the use of qualitative methods because of the lack of statistical validity and the fact that the consumer sample may not be representative of the target population; however, it is important to keep in mind that the researcher should pay more attention to the objective of the test in order to select the method that best fits those objectives because the methods can be overused or used inappropriately. A good recommendation to follow is that one consumer methodology is not a substitute for another.

6.1.2.2 Statistics Statistical methods are needed in consumer research to understand and draw conclusions about the numbers generated through quantitative studies. The following are important definitions with which to be familiar:

- Descriptive statistics: refers to the description of the numbers, for example, mean, standard deviation, variance, range.
- Inferential statistics: refers to the use of statistics to infer, from sample facts, about the population from which it came. In this regard, a parameter is a fact concerning the population whereas a statistic is a fact concerning a sample.

In sensory and consumer research, the tests are divided into parametric and nonparametric tests; the former refer to tests used to analyze data from interval or ratio scales (continuously distributed, following a normal distribution), whereas the latter are designed to handle ordinal data (ranks) and nominal data (categories).

The types of information collected in quantitative measurements can be categorized into nominal, ordinal, interval, and ratio data (O'Mahony, 1986):

- Nominal data are categorical in nature and without any ordering information.
- Ordinal data are ordered rank data without any information indicating the size of difference or spacing between the objects.

- Interval data assume equal space between the data objects without a true zero.
- Ratio data have equal space between the data objects with a true zero.

Numerical measures can be derived from either ranking or rating methods. In the case of rating methods, various scales, depending on the purpose of the information gathered such as the nine-point hedonic scale (Peryam and Pilgrim, 1957), labeled affective magnitude scale, agreement disagreement rating scale, and so on (Solheim and Lawless, 1996), can be used to collect numbers representing the intensity or the magnitude of differences among objects studied (Lawless and Heymann, 2010). Different methods of statistical analysis need to be selected to analyze the information collected depending on the type of data.

The data generated by the different steps of this consumer research on olive oil (survey: attitudes towards olive oil, likes and dislikes, etc.) was evaluated through traditional univariate and multivariate statistical analyses. Table 6.1 summarizes the methods applied in our consumer research. Excellent sources are available to gain a deeper understanding of the statistical methods applied to consumer research (Gacula and Singh, 1984; Lawless and Heymann, 2010; Meullenet, Xiong, and Findlay, 2007; Næs, Brockhoff, and Tomic, 2010; O'Mahony, 1986).

Univariate analysis consists of a set of statistical techniques that have the objective of evaluating the effect of a single variable. Analysis of variance, correlation, and Student's *t*-test are common univariate methods applied to consumer research.

Statistical analysis for categorical data normally involves testing the associations between the variables or modeling the dependent categorical variable as a function of one independent categorical variable or several of them (Agresti, 2007). Examples of statistics for categorical data analysis are logistic regression and log–linear modeling. Logistic regression is used to model a binary variable output in terms of some categorical variables. Log–linear analysis models frequencies for multiple categorical variables to find the simplest, best-fitting model explaining the observed data.

Multivariate analysis (MVA) consists of a set of statistical techniques that can be utilized to analyze several variables at once (Abdi, 2003). MVA allows one to discover which variables are different when considered jointly or which variables are linearly correlated to explain some latent phenomena (Næs and Risvik, 1996). Part of the MVA outputs is graphs for exploration of the relationships among the variables and among the objects.

To analyze a set of interval or ratio data, principal component analysis (PCA) is utilized to decompose data sets with correlated measurements into orthogonal, uncorrelated variables. Usually only a few of the new principal components are needed to explain most of the variance found in the original variables. To analyze a set of nominal or ordinal data, correspondence

Table 6.1 Summary of statistical methods applied in our consumer research model.

Method	Reason to use
ANOVA	To determine panel performance (reproducibility, agreement, discrimination ability) Real means of the EVOO samples in each attribute Model: Attribute = Product + Judge + Replication + Product × Judge + Judge × Replication + Product × Replication
Fisher's LSD	Comparison of each EVOO versus the rest of the samples
Pearson's correlation coefficient	Determination of some possible relationship among the attributes
PCA	To obtain the image representation of the EVOO by the sensory properties given by the descriptive panel Relate reproducibility of the panel Compare quality ratings and liking scores of consumers
GPA/MFA	Determination of panel performance and agreement for the descriptive analysis Characterization of the products Establish relationships among different data set, in this case consumer, descriptive, experts, instrumental data Understand consumer responses in survey research
MANOVA	Explore the relationship among all the attributes and their significance in the descriptive analysis
CVA	Characterization and differentiation of the different oils Group the similar oils Panelists' performance
PLS	Determining relationships among different kinds of data bases, e.g., consumers versus descriptive, consumers versus instrumental, consumers versus experts, instrumental versus descriptive
CA	Graphical visualization of categorical variables in the survey research Measure of associations between variables
MCA	Multivariate correspondence analysis to analyze more than one categorical variable

analysis (CA) or multiple correspondence analysis (MCA) is used to detect significant associations between two variables or several variables at once, respectively. CA may produce either a symmetrical map, which plots the principal coordinates of the rows and columns, or an asymmetric map, which shows "describing variables" (usually the rows) in standard coordinates and the "variable being described" (the columns) in principal coordinates (Blasius and Greenacre, 2006). The measure of variance is closely related to the chi-squared statistic. The objective of CA is to represent the maximum possible variance in a map of few dimensions. The map is interpreted in how the categories lie relative to one another and how the other categories are spread out relative to the former categories. The origin coincides with the average for the data set at hand. Therefore, the categories are judged in terms of how much they deviate from the average. MCA investigates relationships within a set of variables by analyzing the indicator matrix or Burt matrix. There is no distinction between the describing variables and variables to be described in MCA: all variables have the same status (Blasius and Greenacre, 2006).

For data measuring similarity or distance, multidimensional scaling (MDS), additive trees, or cluster analysis could be used to represent graphically the original similarity or distances found in the data (Abdi, 2003). To analyze two or more data sets at the same time, multiple factor analysis (MFA) allows the analysis of several data tables at once. MFA can be used to analyze a mixed table containing both metric and categorical data (Zarraga and Goitisolo, 2006). The analysis contains two parts. The first part involves performing PCA on the metric data and MCA on the categorical data. The first eigenvalues of each analysis are stored for the second part of the analysis, which involves performing weighted PCA analysis on the stored data. The results highlight the relationship among the tables using RV coefficients.

6.2 Applications to olive oil

6.2.1 What and how consumers think, know, and feel about olive oil

6.2.1.1 Focus groups Focus groups are a qualitative research methodology that involves a small number of targeted consumers, usually 3–4 groups of 8–12 consumers, answering questions about and discussing their attitudes, perceptions, beliefs, and /or opinions regarding a product, issue, or service. Consumers interact among themselves and there is a moderator who follows a carefully planned discussion guide covering the topics of interest (Calder, 1977; Casey and Krueger, 1994). The application of focus groups includes obtaining background information, identifying and exploring concepts, identifying beliefs, opinions, attitudes, and motivations, generating hypotheses, identifying new opportunities, and pretesting questionnaires for comprehension, among others.

It is recommended to have a professional moderator as sometimes consumers tend to change the direction of the discussion. The moderator is also in charge of encouraging shy participants to express their opinions and controlling strong personalities within the group. Usually the sessions are recorded and the audiotape records are translated into verbatim reports prior to being content analyzed (Krueger, 1998).

Focus groups were carried out with 8–10 consumers in each of four groups for a total of 35 consumers. Four individual sessions were conducted on two consecutive days in November 2008, with one session in the morning and another in the afternoon. Each session lasted 1.5–2 h. A focus group discussion guideline was developed to help the moderator lead the sessions. The sessions included a projective technique, a series of open-ended questions, and an informed tasting of four EVOO products. The word association projective technique was meant to uncover objects, beliefs, or concepts associated closely with olive oil. Open-ended questions were utilized during the discussion to allow focus group participants to determine the direction of the response and provide opportunities for the participants to answer from a variety of dimensions (Krueger, 1988). The informed tasting involved a preference ranking test with four EVOO samples differing in sensory characteristics and packaging appearance and was meant to access how consumers were affected by sensory and nonsensory characteristics of the oils, and to elicit consumer language related to olive oil. At the end of the sessions, consumers completed an exit survey covering basic socio-demographic information and received a $15 gift certificate.

The script developed for the groups, the interaction among the consumers in the groups, the probing by the moderator on specific issues, the word association exercise, and the tasting of EVOO products combined to produce valuable insights into consumer perceptions of olive oil and their consumption patterns. Specifically, this experimental approach yielded information regarding what concepts consumers associate with olive oil, the range of experiences they have had with olive oil, their olive oil purchasing and consumption habits, what they know about EVOO, the language they use to describe olive oil, and their awareness of a number of emerging issues and concerns associated with olive oil production and marketing. This information confirmed our hypotheses that EVOO consumption is partially driven by some of the current food innovation themes, especially "foods for health" and "local foods," and that existing knowledge and consumption patterns influence future adoption of California-produced EVOO to varying extents among the Northern California consumers who participated in this study. The modified focus group sessions enabled us to accomplish our research goals of investigating consumer perceptions and behavior towards olive oil.

Overall it was found that:

1. Consumers have different concepts associated with olive oil. When asked to share what things first came to their mind when consumers heard the terms "olive oil," 10 thematic categories emerged from the projective technique. Those categories and the words associated with them are shown in Table 6.2.
2. Cultural tradition or heritage (e.g., Italian descendant) plays a key role in the exposure of consumers to olive oils. However, for those consumers

Table 6.2 Words and phrases elicited in the projective technique during focus groups sessions.

Natural or living objects	Beautiful trees; rich variety (of olives); orchard
Physical characteristics	Exotic labels; bottles
Sensory characteristics	Good taste; salad taste; gold and rich, lovely, fresh fruity taste; buttery taste, rich gold and green color; stimulating pungency of the olive taste
Quality	Purity
Food items or dishes	Salad; pesto; bread; pasta; fish
Usages	Curing of the olives; cooking; good cooking; my mom's cooking; how I use more than a liter a month
Emotion or nostalgia	Family; good memories; very soothing, calming aspects of the fat
Health benefits	Most healthful oil; health benefits; healthy; good for you; good for me – the kind of oil that helps deal with some of the bad fats in your body
Geographical locations	Italy; France: Mediterranean
Fictional character	Popeye

with no previous exposure during childhood or family cooking, the following events were mentioned as factors that help them to be aware of olive oil as a fat choice in the diet or for cooking: (i) restaurant dining, (ii) married life, (iii) college life, (iv) having a trip overseas or living abroad, and (v) media exposure.

3. Consumers indicated they were motivated to buy olive oil because of its sensory characteristics, the health benefits, its uses, cultural tradition, and the geographic/region of origin of the oil.
4. When buying olive oil, consumers indicated that they consider the price, value, quality, place of origin, flavor/taste, whether it is organic or not, the volume or size, the purpose of use, reliability, familiarity, appearance, and packaging features. However cost/price was one of the most cited factors involved in the purchase decision.
5. Olive oil taste/flavor and health benefits are the major motivations for consuming olive oils; however, consumers showed limited knowledge about actual health benefits (monounsaturated fatty acids, antioxidants, etc.).
6. Consumers are willing to set at least two different price points for buying EVOO, depending on the other consumption factors (location of purchase, purpose of use, etc.).
7. Media exposure; for example, Tom Mueller's 2007 article on olive oil fraud in *The New Yorker* made some consumers aware of olive oil adulteration and quality issues.
8. Consumers feel that they know little or nothing about what EVOO is or what "extra virgin" means. However, most consumers know about the regions where olive oil is produced (e.g., Italy, California, Mediterranean countries).

9. The nonblind tasting of four EVOOs produced consumers' language mostly related to sensory characteristics of olive oil rather than nonsensory aspects.
   a. Positive attributes: fruitiness.
   b. Attributes with bipolar connotations: butter or buttery, bitter flavor or aftertaste, pungency, pepper aftertaste/finish or peppery flavor, grass or grassy flavor.

Exposure to the sensory properties (potentially bitterness and pungency) of California-made EVOOs would be beneficial to increase the consumption within the United States. Consumers must be educated or at least informed about the nature, properties, and styles of olive oil that are "extra virgin" to establish regional identities for California (Santosa, 2010). In any emergent market, education and exposure are critical factors to increase consumption.

According to Santosa (2010), in order for consumers to increase their knowledge of the health benefits of olive oil, producers should include specific information on the relevant bioactives on their product labels, within regulatory constraints. The label also should indicate information relevant to the shelf life of the product (e.g., harvest date, date bottled, best-consumed-before date). This study showed that there are opportunities to develop EVOO products based on the several factors that seem to motivate purchase decision and consumption.

6.2.1.2 Means–end chain analysis (MECA) MECA is a market research technique that has been widely and successfully used to connect consumer values and product-feature preferences (Pieters, Baumgartner, and Allen, 1995). The model assumes a hierarchy of goal levels at which a product's attributes provide positive consequences and in turn satisfy values that are desired by consumers (Reynolds and Guttman, 1988).The technique involves an individual interview called a laddering interview, which asks the consumer a series of "why" questions, to discover the important consequences associated with the identified attributes and ultimately uncover the important values associated with the identified consequences that consumers seek when consuming the products. A hierarchical value map (HVM) is usually constructed in the form of a tree diagram to represent the associations between the product attributes, the consequences, and the values.

We applied this methodology to understand why consumers consume olive oil and what motivates them to purchase olive oil (Santosa and Guinard, 2011). The objectives of the study were to investigate the hierarchical links among the olive oil product attributes, the consequences associated with those product attributes, and the personal values associated with olive oil consumption. We recruited a total of 63 consumers (40 female, 23 male, ages 20–86 years) living in Northern California. Consumers participated in one-to-one interviews using the soft laddering technique. The authors chose that technique because consumers may use unwanted strategic processes, and the degree of knowledge about the product category is either low or high (Grunert and Grunert, 1995). Consumers were asked two main questions: (1) Why do you consume olive oil? (2) Why did you buy the olive oil(s) that you currently have at home? An example of the process was proportionated to each consumer before starting the interview questions, and each of them completed an exit survey regarding some basic demographics using an electronic survey method. Consumers received a $10 gift certificate at the completion of the session. The interviews were conducted between January and March 2009, each lasting 30–40 min.

We analyzed the interview data following the guidelines described by Reynolds and Guttman (1988). Three clusters based on the origin of the olive oil consumed at home were uncovered: (1) consumers of olive oils from America or California, (2) consumers of imported olive oils, and (3) consumers of both imported and local olive oils. Three HVMs were built from the results, one for each cluster (Figure 6.2, Figure 6.3, and Figure 6.4).

We found that consumers who purchased locally had a less complex HVM than consumers who purchased imported oils. Flavor, healthy oil, and monounsaturated fat were reported as important product characteristics for the three clusters; those product characteristics were motivators to consume olive oil. The differences were the values associated with each characteristic; for example, consumers who consumed imported oil associated these characteristics with health consciousness, quality of life, happiness, and exciting life;

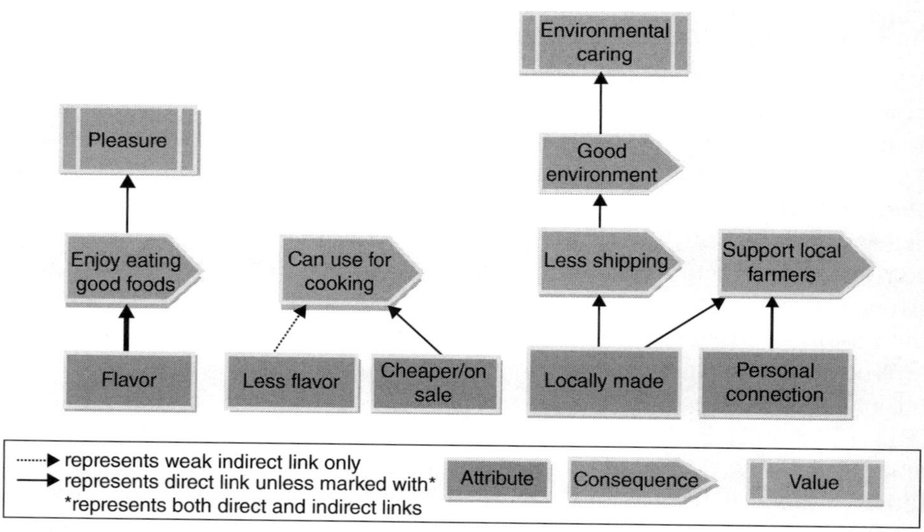

Figure 6.2 Hierarchical value map for olive oil purchase motivations among consumers who consumed mostly locally made olive oil. Source: Reprinted from Santosa, M., and Guinard, J.-X. (2011) with kind permission from Elsevier. For a color version of this figure, see the color plates section.

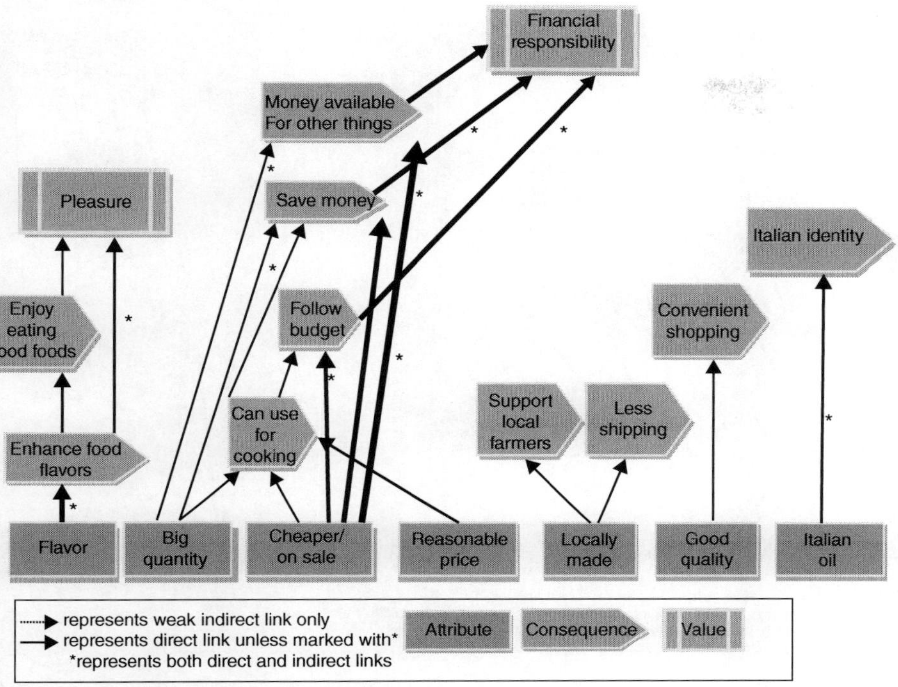

Figure 6.3 Hierarchical value map for olive oil purchase motivations among consumers who consumed mostly imported olive oil. Source: Reprinted from Santosa, M., and Guinard, J.-X. (2011) with kind permission from Elsevier. For a color version of this figure, see the color plates section.

consumers who consumed local oils associated them with feeling good and social company; and consumers who consumed both local and imported oils associated those properties with feeling good and quality of life.

Consumers associated the flavor of olive oil with the value of enjoyment when eating foods. In the case of consumers who consumed imported oils and those who consumed both imported and local oils, the flavor of olive oil was seen as a way to enjoy cooking with the consequence of excitement in their lives.

In terms of the health benefits, the three groups concurred that healthier oil and monounsaturated fat were associated closely with health benefits that would translate into consumers maintaining health in order to function better or enjoy life.

The laddering interview and MECA, undertaken to investigate Northern California consumers' motivations to consume and purchase EVOO, successfully identified attributes, consequences, and values that are relevant to this consumer behavior. This marketing research method focused on understanding the individual consumer as a person. The findings indicated that despite the different levels of abstraction, the main purpose of uncovering the

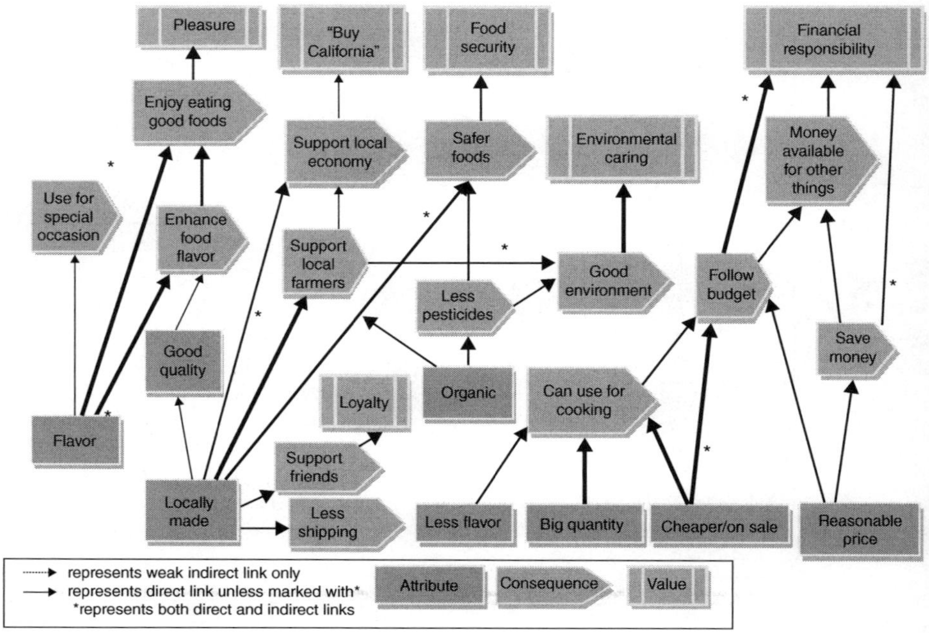

Figure 6.4 Hierarchical value map for olive oil purchasing motivations among consumers who consumed both imported and locally made olive oil. Source: Reprinted from Santosa, M., and Guinard, J.-X. (2011) with kind permission from Elsevier. For a color version of this figure, see the color plates section.

motivations underlying consumer behavior related to consumption and purchase could still be fulfilled.

6.2.1.3 Sorting A sorting task is a simple method for collecting similarity data that has commonly been used in psychology since the 1970s (Healy and Miller, 1970). Subjects are normally asked to sort together stimuli based on their perceived similarity. The task is very simple and less time consuming than other methods such as pairwise similarity evaluation.

Sorting is especially advantageous when the number of stimuli to investigate is large. Sorting tasks have been found to have a good correlation coefficient with pairwise similarity scaling results (Rao and Katz, 1971). It also produced similar results to triadic similarity methods (MacRae, Howgate, and Geelhoed, 1990). More recently, the sorting task has been shown to produce data similar to quantitative ratings coming from descriptive trained panelists (Cartier *et al.*, 2006), but it was also noted that there were some perceptual differences between the product map derived from the sorting method and that derived from traditional descriptive analysis, with slightly less product discriminations obtained with sorting.

The analysis of similarity data gathered from sorting tasks can be done with multidimensional scaling (MDS). MDS produces an average product map and individual data are lost during the production of this product map. A newer MVA technique called DISTATIS was developed to circumvent this problem (Abdi *et al.*, 2007). DISTATIS produces two maps, one for the subjects and the other for the products. The consumer map shows how much the subjects agree or disagree from one another whereas the product map shows a compromised product positioning.

We used a sorting method to explore the effect of product variables in our consumer research model (Santosa, Abdi, and Guinard, 2010). We modified the simple sorting task into two stages and incorporated a verbal description task at the end of the sorting in order to allow a finer product perceptual discrimination and understanding of fundamental perceptual dimensions obtained from the DISTATIS analysis, respectively. In this study, 25 EVOOs were sorted in two stages based on visual assessment of the bottles. Nine of the oils were imported products, with eight of them (products 1–8) bought at local supermarkets (Nugget, Safeway, and Davis Co-Op) in Davis, CA, and one (product 9) donated by an olive oil producer in Spain. Sixteen of the oils (products 10–25) were commercial oils donated by olive oil producers in California and were made from olives grown in California, except for one product that also included olives grown in other countries. Only a few of the donated olive oils can be found in the supermarkets that consumers typically go to for grocery shopping. Social representation theory was utilized to explain why consumers had higher discriminations for imported EVOO products than for California-made EVOO products. This research also highlighted the fact that Northern California consumers were not very familiar with California-produced EVOO products. Consumer language elicited through the verbal description task could be used in consumer education to ensure effective communication messages.

Figure 6.5 shows the results of the two-stage sorting method. We found a greater discrimination of imported oils (products 1–9) than of California oils (products 10–25). Factors related to the product such as those mentioned during the focus groups, namely sensory factors, flavor, price, size or volume, region of origin, and factors related to the applicability of the product such as the use of olive oil, were also revealed by the modified sorting task. The degree of quality of the oil was associated with the shopping place, the nature of the olives, and the production scale and size.

Additionally, the study helped us understand consumer perceptions of the packaging of olive oil. Bottles that were shorter, smaller, with square or round shapes were associated with cooking oil. In contrast, thinner bottles were perceived as hard to handle but stylish, and to be given as a gift.

Some consumers used wine terminology to describe some of the California oils, particularly when the bottle shape resembled that for wine or the product label included the harvest year and the region of origin. Social

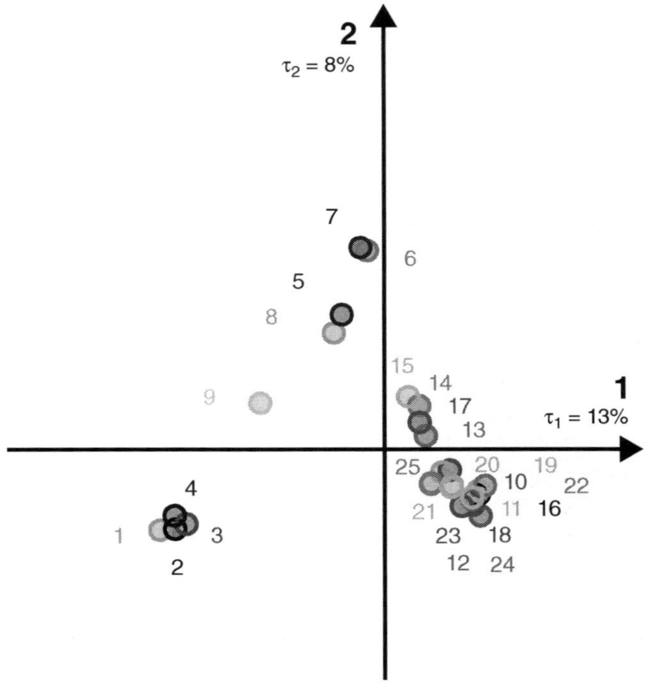

Figure 6.5 Positioning of 25 EVOOs and the associated confidence intervals from the second stage of the sorting task. For a color version of this figure, see the color plates section.

representation theory (Agostinous, 2001) and consumer learning by analogy theory (Gregan-Paxton and John, 1997) could be used to explain why consumers often referred to wine characteristics to define their perceptions for the set of EVOO samples during the modified sorting task, and why the California-produced oils were not differentiated much. Social representation (SR) theory (Moscovici, 2000) assumes that for new perceptions to form with regard to the unfamiliar, unnamed, or noncategorized objects, a linkage to existing knowledge structures needs to be made so that objects become more familiar and better understood

6.2.1.4 Survey work A survey questionnaire is a quantitative instrument designed to obtain information about a particular topic of interest (i.e., products, services, issues, etc.). Depending on the objectives, surveys can be administered through the telephone, mail, or Internet.

Surveys need to be carefully designed, otherwise the researchers run the risk of collecting inaccurate information. According to Tourangeau, Rips, and Rasinski (2000), answering a survey questionnaire involves four processes, namely comprehension, retrieval, judgment, and response. Consumers need to understand the question, and in this regard the language used to formulate the questions needs to be appropriate for the target population. Clear, not vague, and not ambiguous is the rule when designing questions. Then, once the consumer understands the question, the next step is to estimate which of the retrieved information is relevant to answer that question. Memory plays a key role in this process, so attention must be paid to the time frame of the question. During the judgment phase, consumers decide how much information they want to share, and in this step they consider sensitive information, the social desirability of the answer, and the probable accuracy of the answer, and questions related to sensitive information such as income or family status should be carefully phrased in order to avoid inaccurate answers due to consumers feeling an invasion to their privacy. Usually, potentially sensitive information such as demographics is included at the end of the questionnaire. The last step is giving the answer to the question.

A total of 178 consumers completed our survey on olive oil in 2009 (Santosa, 2010). The survey design included several question formats, including multiple-choice, yes/no, check all that apply, and open-ended questions. The main topics covered in the survey were oil consumption and purchase behaviors, demographics, and psychographics.

In terms of purchase behavior as it related to the origin of the oil, the study revealed three consumer segments: (a) imported oils buyers (39.9%), (b) locally made oils buyers (20.2%), and (c) both imported and locally made oil buyers (39.9%). These segments were used to understand further the relationship between consumption and purchasing habits. Given the categorical nature of the data collected, CA was applied in order to obtain a graphic representation of the variables that allowed for an understanding of patterns and trends associated with the variables. In the case of no clear trends for each of the consumer segments, the data were pooled and MCA was used instead.

Overall, the results of the survey research agreed with those from the focus groups; for example, the majority of the consumers surveyed indicated that one of the main reasons to start consuming olive oil was because of their awareness of olive oil's health benefits. Those consumers were not previously exposed to olive oil when still living at home or because of cultural traditions. In the case of those consumers previously exposed to olive oil in early stages of their life or from cultural tradition, they continued to use olive oil as part of their daily diet. Triggers such as "starting to cook," "their spouse uses olive oil to cook," "a recipe calls for olive oil," "tasting it at a fair or event," and "having it as a condiment for bread dipping at a restaurant" were the main reasons for a great number of consumers to start consuming olive oil. Very few indicated that they started consuming olive oil after living overseas or taking a trip overseas. Receiving olive oil as a gift was not a main reason for starting to consume olive oil for the majority of consumers.

The frequencies of consumption of olive oil differed from those of the other oils and fats. Figure 6.6 shows how the particular oil consumption

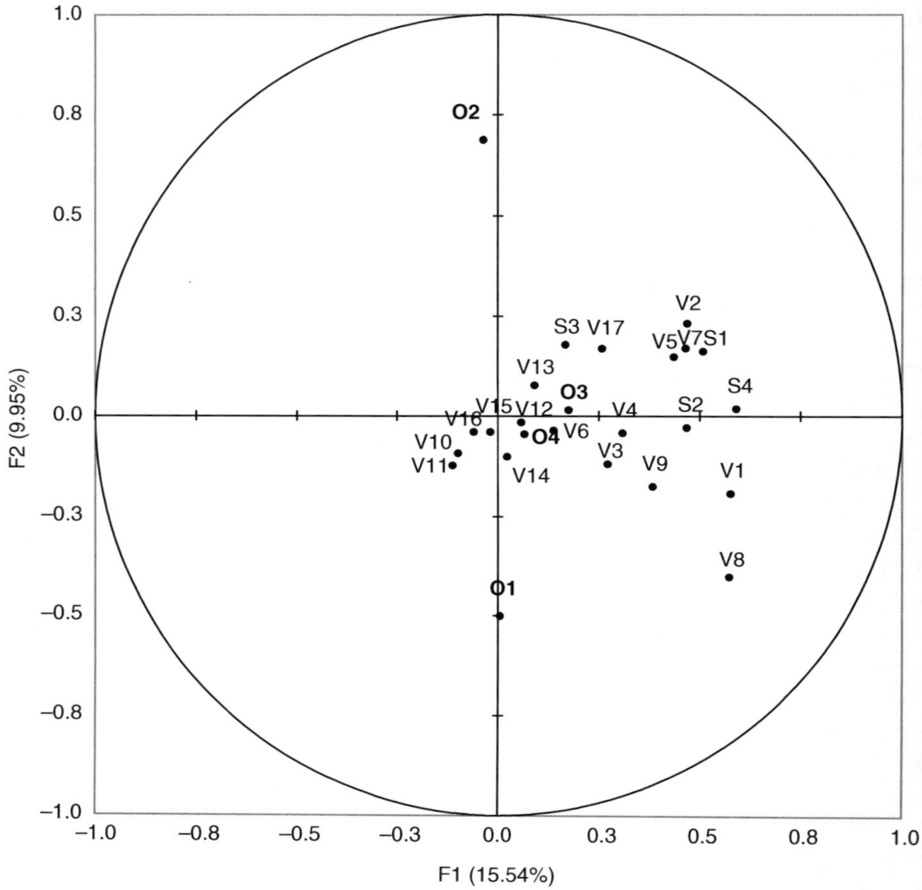

Figure 6.6 Multiple factor analysis (MFA) results: relationships of vegetable oils, specialty oils, spray oils, and olive oil consumption.

frequencies were associated with other oils' consumptions. Higher frequencies of EVOO consumption (O1) were highly correlated with frequencies of consumption of canola oil (V1), sesame oil (V8), olive oil spray (S4), and to a slightly lesser extent peanut oil (V9) and canola oil spray (S2). Furthermore, consumption frequencies of non-EVOO such as pure olive oil and light tasting olive oil (O2) was highly correlated with consumption frequencies of corn oil (V2), vegetable oil (V5), and blend of vegetable oils (V7) and butter spray (S1), and to a lesser extent with canola oil spray (S2) and olive oil spray (S4). Consumption frequencies of flavored olive oil (O3) and also other vegetable oils, specialty oils, and cooking spray with flour (S3), for instance, was much more variable or was not very high among the consumers. Consumption of EVOO (O1) was associated with consumption of coconut oil (V10), grape seed oil (V11), and walnut oil (V14) among other oils, albeit at less frequent rates.

In terms of where consumers bought olive oil and their relationship with the brands they usually bought, consumers who typically bought store brands were more likely to shop at discount retailers (i.e., warehouse clubs such as Sam's and Costco); these consumers also indicated buying their olive oil at the supermarket. Consumers who only bought local brands reported obtaining their oils from specialty/gourmet stores, during olive oil tasting fairs or food festivals, and directly from olive oil farmers or producers. Consumers who bought imported oils bought them online and from wineries. Overall, we found that consumers who usually bought store-brand olive oil tended to pay \$5–7.99, whereas consumers purchasing local brands or both local brands and store brands paid \$16–20.99 and \$21–25.99, respectively, and for those purchasing imported brands, they paid even more, \$26–30.99 or \$31 and above. The three groups also shared the middle price ranges (\$8–9.99 and \$10–15.99), as shown in Figure 6.7.

Figure 6.8 shows the correspondence analysis of the reasons why consumers consumed olive oil and consumers segmented based on origin of olive oil consumed. For those consumers consuming both locally and imported oils, they did so because the oils were locally produced and because olive oil

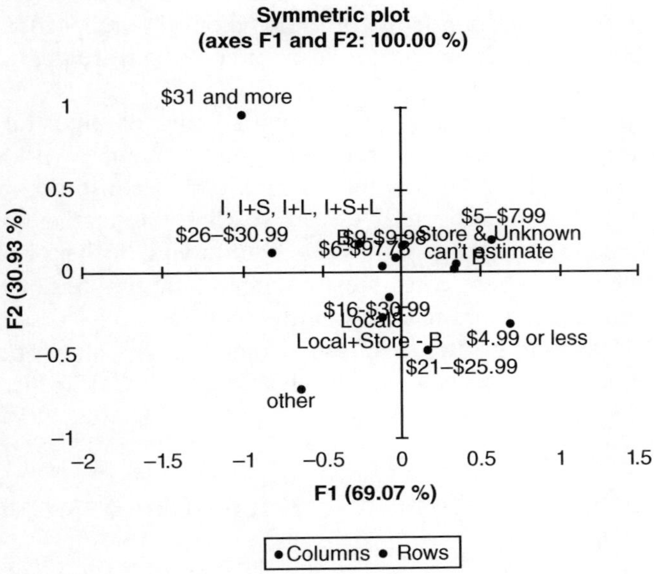

Figure 6.7 Correspondence analysis of typical price ranges paid for olive oils bought for own consumption and use at home and consumer clusters based on olive oil brands purchased.

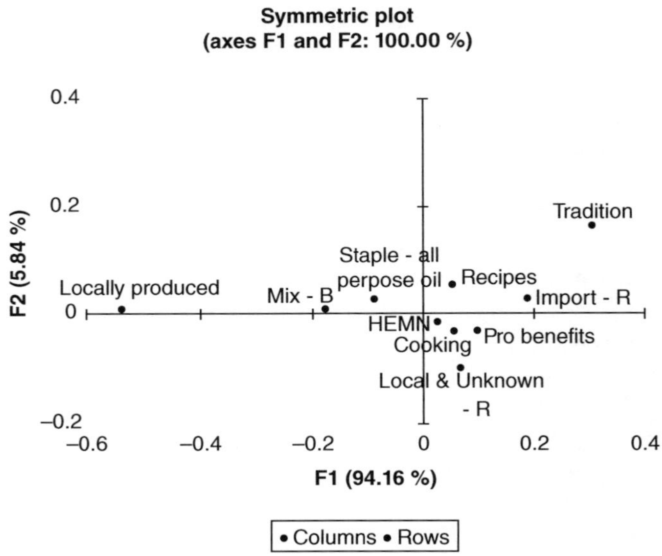

Figure 6.8 Correspondence analysis of the reasons why consumers consumed olive oil and consumers segmented based on origin of olive oil consumed.

was the main oil in the household or a staple/cooking oil. Those consuming imported oil kept doing it because of tradition and family heritage, or because a recipe called for it. Independently of the kind of oil purchased, all segments agreed that flavor, health benefits, and use for cooking were the main reasons to consume olive oil.

There were some differences based on the kind of olive oil purchased (brand and region of origin) in terms of the frequency with which consumers read the labels; for example, the majority of consumers who generally bought store brands reported "never" reading the package label. Those who "rarely" read the package labels reported buying both store brands and locally produced oils. Those consumers buying the three types of brands indicated "sometimes" reading the labels on the package.

Similarly to the focus groups results, consumers in this survey believed that Italy was the best producing region for EVOO, with California mentioned second most often, and Spain third.

6.2.2 Preference mapping – which olive oils consumers like or dislike

Traditional sensory research with consumers involves mostly quantitative methods to measure consumer hedonic/liking ratings for a particular set of products with certain sensory properties in order to identify what the consumer likes or dislikes.

Sensory scientists and marketers typically obtain the most reliable information when sensory data (such as the results of a descriptive analysis) are combined with consumer data – not just the hedonic scores, but also demographic and psychographic information.

6.2.2.1 Preference mapping Preference mapping is frequently the preferred method to uncover market segmentation and find drivers of liking in products for a given consumer population or segments and also to identify the most preferred products (Meullenet, Xiong, and Findlay, 2007; Næs, Brockhoff, and Tomic, 2010). Applications of preference mapping are well documented in the literature, and their uses extend to the food and beverage fields (Elmore *et al.*, 1999; Guinard, Uotani, and Schlich, 2001; van Kleef, van Trijp, and Luning, 2006). Moreover, preference mapping techniques are usually selected as the "benchmark" methods with which to compare alternative approaches (McEwan and Thomson, 1989; ten Kleij and Musters, 2003; Faye *et al.*, 2006; Ares *et al.*, 2010; Dooley, Lee, and Meullenet, 2010).

Lawless and Heymann (2010) define preference mapping techniques as follows:

Internal preference mapping refers usually to a PCA with the products as samples (rows) and the consumer hedonic scores as the variables (columns). This map is based only on consumer data.

External preference mapping creates a spatial representation for the descriptive data, and then the individual consumer's hedonic responses are projected onto the product space by regressing each consumer's response onto the spatial dimensions of the products.

The primary goal of our research (Delgado-Gutierrez 2010; Delgado and Guinard, 2011a; 2011b) was to determine sensory quality factors in EVOO and their contribution to consumer preferences, through the development of statistical models using novel and traditional methods such as preference mapping. Twenty-two commercial EVOOs were used for these studies.

In order to understand consumer preferences for olive oil, the first part of the project had the objective of developing a specific descriptive analysis methodology for EVOO, in order to characterize the sensory properties of a set of California EVOOs (labeled U1 to U11) and imported EVOOs (labeled with the corresponding letter of the country of origin: Spain, S1 to S2; Italy, I1 to I4; Australia, A1; and Chile, C1 to C2, and generic brands from Span and

Italy as GS and GI, respectively). To achieve this goal, a generic descriptive analysis was conducted according to the methodology described by Lawless and Heymann (2010).

The process started with the recruitment of the panelists based on their interest in olive oil and their availability; after selection, the group was trained with the objective of selecting the terminology and vocabulary to describe the olive oil samples, and to achieve concept alignment of those attributes, each panelist completed 10 training sessions. FIZZ software (Biosystemes) was used to build an automated session. After training, the performance of the panel was assessed. Five samples were randomly selected from the total sample set and evaluated in duplicate. In cases of disagreement among panelists or misunderstanding of some attributes, more training was provided in either group or individual sessions. Once the panel was deemed ready, they proceeded to evaluate the whole set of samples. A total of 22 attributes were defined by the panel, with 20 differing significantly among the oils. EVOOs were evaluated in triplicate, with five samples evaluated per session. The order of presentation of the EVOOs was randomized using a Latin-square design provided by the FIZZ software. Additional details about the study can be found in Delgado and Guinard (2011b). The results were analyzed by univariate analysis [three-way analysis of variance (ANOVA) and Fisher's LSD multiple comparisons] and multivariate analysis, particularly MANOVA and canonical variate analysis (CVA).

Figure 6.9 shows the results of the descriptive analysis. California EVOOs were mainly characterized by the following attributes: green fruit, grassy, bitter, nutty, ripe fruit, and pungent. The oil from Australia and the two from Chile were similar to those from California. The two generic brands showed a certain degree of rancidity and other defects, and the oils from Spain were defined by tropical, herbal, and floral notes. The Italian oils used in this study were nutty and buttery but some of them exhibited some degree of rancidity.

We then investigated consumer preferences and drivers of liking for the 22 EVOOs (Delgado and Guinard, 2011a). Similarly to the survey study, the questionnaire design used different approaches to understand the acceptability of the oils among consumers. An exit survey was included that measured behavioral factors influencing olive oil consumption and some attitudes and beliefs regarding olive oil.

A total of 110 consumers participated. Hedonic scores were obtained for each sample using the nine-point hedonic scale (Peryam and Pilgrim, 1957). External preference mapping was used to understand the relationship between the sensory properties and the hedonic responses (Figure 6.10). As mentioned before, the descriptive analysis ratings were analyzed through a CVA, then the consumers' hedonic scores were regressed onto the CVA space derived from descriptive analysis to understand how the hedonic data related to the EVOOs' characteristics. Cluster analysis (Ward's method, Euclidean distance) was applied to the hedonic scores to understand further

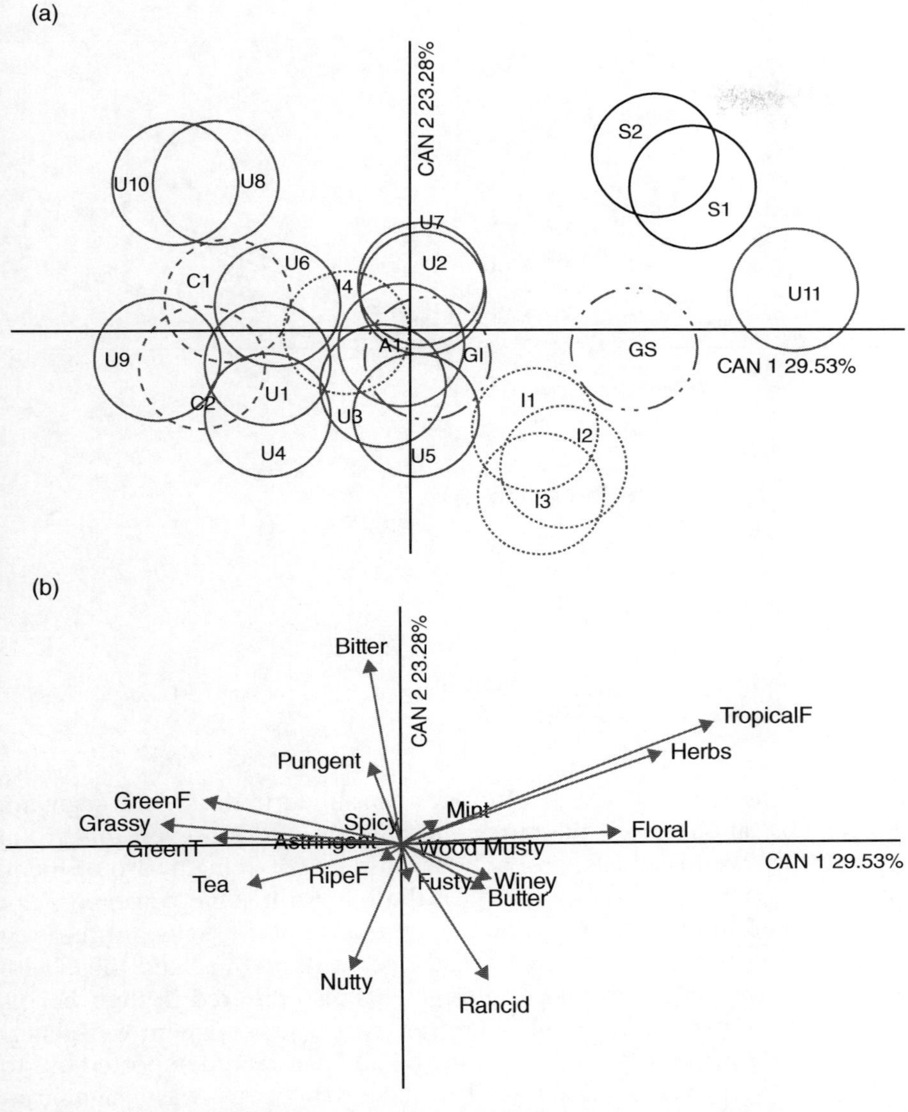

Figure 6.9 Canonical variate analysis (CVA) for 22 products in two dimensions. (a) Distribution of the products; (b) attributes. For a color version of (a) see the color plates section.

the preferences for this consumer population; this analysis uncovered three main segments, as shown in Figure 6.10.

The three segments had different demographics and preferences based on the sensory characteristics of the oils. Overall, and with the exception of a few consumers, it was found that the three segments were consistent in their dislike of bitterness and pungency, and did not see them as positive

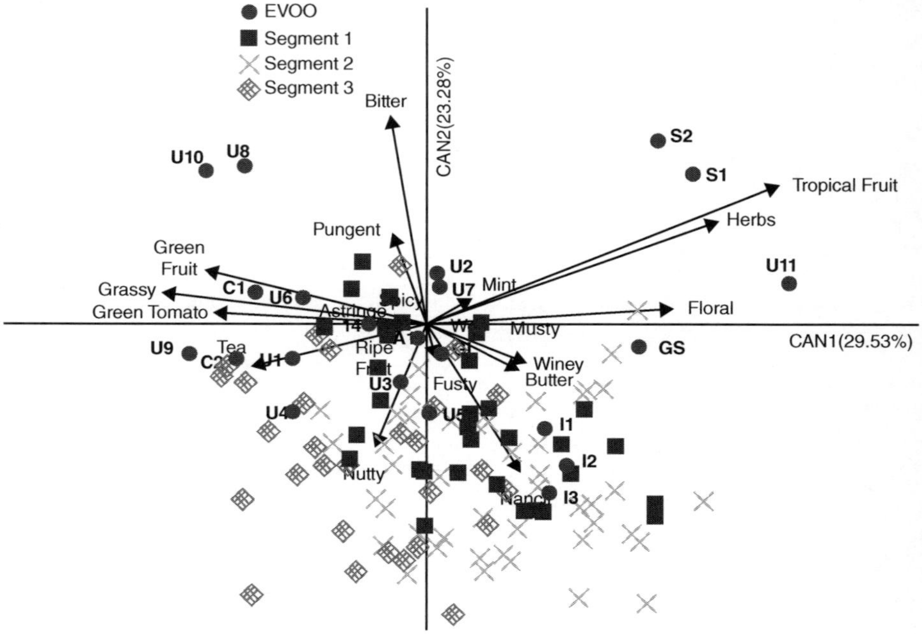

Figure 6.10 Preference mapping of EVOOs with a group of US consumers. For a color version of this figure, see the color plates section.

qualities. Those attributes were critical for consumers in the second segment, who categorically rejected the bitter and pungent oils; however, this group was not sensitive to defects, probably because of not being aware of them. The first segment liked the majority of the oils, with some consumers not being opposed to the bitterness and pungency. The third segment preferred oils that were nutty, with green tea and ripe fruit, and disliked those with tropical fruit or herbal aromas. The segments also differed in their beliefs and attitudes regarding olive oil. Interestingly, the first segment was generally concerned with the reputation of the oil and considered imported oils to be of better quality. For consumers in segment two, price was a significant factor in their purchase decision, and they saw EVOO as a healthy product; however, they did not find differences in its taste. Consumers in segment three reported the information on the label of the product as being important, and indicated that they used EVOO in salad and saw the use of EVOO for cooking as expensive.

6.2.2.2 Consumer perceptions of the packaged product Another quantitative consumer test was conducted with 102 consumers to evaluate the effect of the EVOO packaging (bottle, label, etc.) on consumer acceptance of the

Table 6.3 Pearson's correlation coefficients among overall degree of liking and degree of liking ratings for the packaging–labeling variables evaluated. Values in bold are significantly different from zero with a significance level $\alpha = 0.05$.

	Overall_liking	Size	Shape	Material	Label_design	Pict_label	Color_label
Overall_liking	**1**						
Size	0.296	**1**					
Shape	**0.765**	**0.503**	**1**				
Material	**0.741**	0.442	**0.839**	**1**			
Label_design	**0.845**	0.062	0.433	0.331	**1**		
Pict_label	**0.719**	0.004	0.290	0.165	**0.961**	**1**	
Color_label	**0.827**	0.003	0.371	0.320	**0.947**	**0.899**	**1**

product. The study consisted of one session in which consumers evaluated 18 EVOO bottles presented in a randomized order across consumers. The 18 EVOOs were selected from the 22 oils evaluated in the preference mapping study, but they were tested by a different population of consumers. For each EVOO bottle, consumers indicated their overall degree of liking on the nine-point hedonic scale (Peryam and Pilgrim, 1957), and also their liking of bottle size, bottle shape, container material, label design, pictures or illustrations on the label and colors on the label.

Overall degree of liking of the EVOO bottles was significantly correlated with liking for the other packaging–labeling variables evaluated in the hedonic test (shape, container material, design of the labels, pictures on the labels, and colors of the bottle and labels), except for the size of the EVOO bottles, as shown in Table 6.3.

Figure 6.11 shows the internal preference mapping for the 18 EVOO bottles. EVOO bottles from California were generally liked the most, except U4, whereas imported EVOO bottles GB1, GB5, and S2 were liked the least. Furthermore, the figure shows some spread of the consumers along the top and bottom right quarters of the biplot; this segmentation was confirmed by preference clustering analysis (Ward's method, Euclidean distance), which uncovered two main clusters. Consumers from Group 1 ($N = 59$) are shown in red and consumers from Group 2 ($N = 43$) are shown in blue in Figure 6.11. However, it is important to remark that both groups were located on the right side of the biplot, so these two preference clusters were actually quite close in their preference patterns.

It is possible to confirm which EVOO bottles each group liked and disliked by comparing the biplot below, with the consumers identified by color (left plot), and the biplot that shows the products (right side). It would appear that the consumers in Group 1 liked California EVOO bottles, except U4, whereas consumers from Group 2 liked the EVOO bottles I1, I2, I3, and C1 and disliked U4, S2, GB1, and GB5.

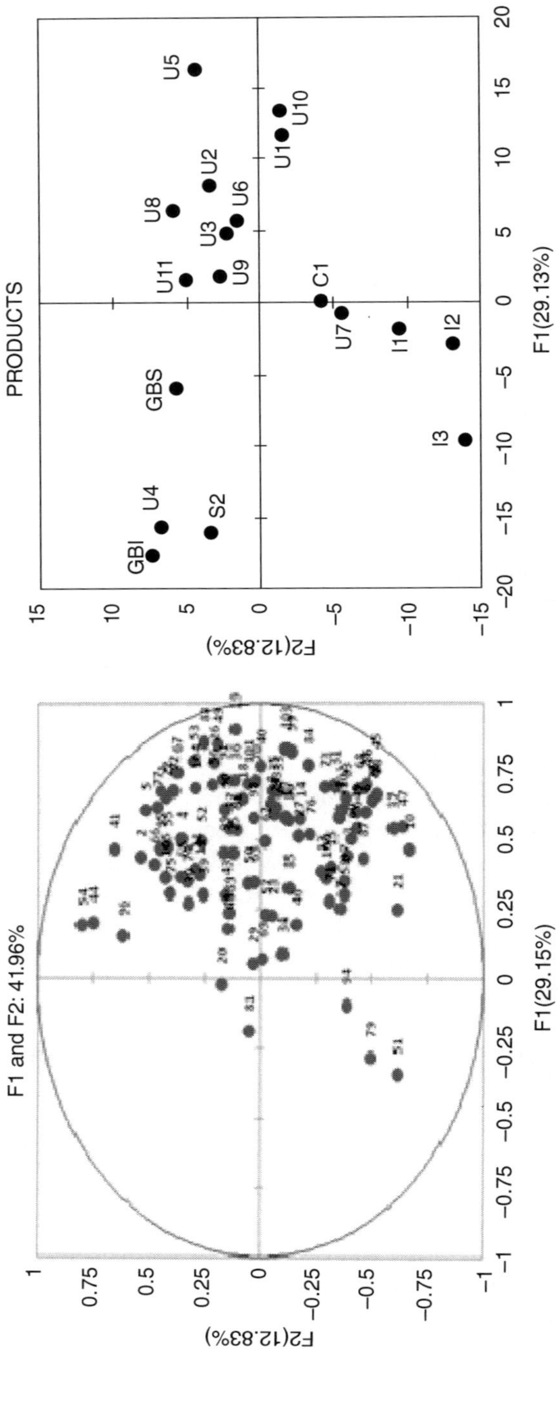

Figure 6.11 Internal preference mapping for overall liking of the 18 EVOO bottles. For a color version of this figure, see the color plates section.

6.3 Conclusion

This chapter presented a detailed overview of the application of our consumer research model to the study of US consumer perceptions of olive oil, sensory and otherwise. It combined quantitative and qualitative approaches and used a range of multivariate statistical techniques.

Our research clearly shows enormous potential for olive oil consumption to grow further in the United States, as consumer knowledge about and appreciation of EVOO and its sensory and health properties expand.

References

Abdi, H. (2003) Multivariate analysis, in *The Sage Encyclopedia of Social Sciences Research Methods* (eds M.S. Lewis-Beck, A. Bryman, and T.F. Liao), Sage, Thousand Oaks, CA.

Abdi, H., Valentin, D., Chollet, S., and Chrea, C. (2007) Analyzing assessors and products in sorting tasks: DISTATIS, theory and applications. *Food Quality and Preference*, **18**, 627–640.

Agresti, A. (2007) *An Introduction to Categorical Data Analysis*, John Wiley & Sons, Inc., Hoboken, NJ.

Ares, G., Giménez, A., Barreiro, C., and Gámbaro, A. (2010) Use of an open-ended question to identify drivers of liking of milk desserts. Comparison with preference mapping techniques. *Food Quality and Preference*, **21** (3), 286–294.

Bagnoli, C. (2007) Consumer trends: the future is now, presented at the 7th Pangborn Sensory Science Symposium; 12–16 August 2007, Minneapolis, MN.

Blasius, J., and Greenacre, M. (2006) Correspondence analysis and related method in practice, in *Multiple Correspondence Analysis and Related Methods* (eds M. Greenacre and J. Blasius), Chapman & Hall/CRC, Boca Raton, FL, pp. 3–40.

Calder, B.J. (1977) Focus groups and the nature of qualitative marketing research. *Journal of Marketing Research*, **14**, 353–364.

Caporale, G., and Monteleone, E. (2004) Influence of information about manufacturing process on beer acceptability. *Food Quality and Preference*, **15** (3), 271–278.

Cardello, A.V. (2003) Consumer concerns and expectations about novel food processing technologies: effects on product liking. *Appetite*, **40** (3), 217–233.

Cartier, R., Rytz, A., Lecomte, A., Poblete, E., Krystlik, J., Belin, E., and Martin, N. (2006) Sorting procedure as an alternative to quantitative descriptive analysis to obtain a product sensory map. *Food Quality and Preference*, **17**, 562–571.

Casey, M.A., and Krueger, R.A. (1994) Focus group interviewing, in *Measurement of Food Preferences*, (eds H.J.H. MacFie and D.M.H. Thomson), Blackie, London, pp. 77–96.

Delgado, C., and Guinard, J.-X. (2011a) How do consumer hedonic ratings for extra virgin olive oil relate to quality ratings by experts and descriptive analysis ratings? *Food Quality and Preference*, **22** (2), 213–225.

Delgado, C., and Guinard, J.-X. (2011b) Sensory properties of Californian and imported extra virgin olive oils. *Journal of Food Science*, **76** (3), S170–S176.

Delgado, C., and Guinard, J.-X. (2012) Internal and external quality mapping as a new approach to the evaluation of sensory quality – a case study with olive oil. *Journal of Sensory Studies*, **27** (5), 332–343.

Delgado-Gutierrez, C. (2010) Sensory quality in extra virgin olive oil, PhD dissertation, University of California, Davis.

Di Monaco, R., Cavella, S., Di Marzo, S., and Masi, P. (2004) The effects of expectations generated by brand name on the acceptability of dried semolina pasta. *Food Quality and Preference*, **15** (5), 429–437.

Dillman, D.A. (2007) *Mail and Internet Surveys: the Tailored Design Method*, John Wiley & Sons, Inc., Hoboken, NJ.

Dooley, L., Lee, Y.-S., and Meullenet, J.F. (2010) The application of check-all-that-apply (CATA) consumer profiling to preference mapping of vanilla ice cream and its comparison to classical external preference mapping. *Food Quality and Preference*, **21**, 394–401.

Elmore, J.R., Heyman, H., Johnson, J., and Hewett, J.E. (1999) Preference mapping: relating acceptance of "creaminess" to a descriptive sensory map of a semi-solid. *Food Quality and Preference*, **10**, 465–475.

Faye, P., Brémaud, D., Teillet, E., Courcoux, P., Giboreau, A., and Nicod, H. (2006) An alternative to external preference mapping based on consumer perceptive mapping. *Food Quality and Preference*, **17** (7–8), 604–614.

Gacula, M.C., and Singh, J. (1984) *Statistical Methods in Food and Consumer Research*, Academic Press, Orlando, FL.

Garber, L.L., Jr, Hyatt, E.M., and Starr, R.G., Jr. (2003) Measuring consumer responses to food products. *Food Quality and Preference*, **14** (1), 3–15.

Gregan-Paxton, J., and John, D.R. (1997) Consumer learning by analogy: a model of internal knowledge transfer. *The Journal of Consumer Research*, **24** (3), 266–284.

Groves, R.M. (2004) *Survey Methodology*, John Wiley & Sons, Inc., Hoboken, NJ.

Grunert, K.G. (2003) Purchase and consumption: the interdisciplinary nature of analyzing food choice. *Food Quality and Preference*, **14** (1), 39–40.

Grunert, K.G., and Grunert, S.C. (1995) Measuring subjective meaning structures by the laddering method: theoretical considerations and methodological problems. *International Journal of Research in Marketing*, **12**, 209–225.

Guinard, J.-X., Uotani, B., and Schlich, P. (2001) Internal and external mapping of preferences for commercial lager beers: comparison of hedonic ratings by consumers blind versus with knowledge of brand and price. *Food Quality and Preference*, **12** (4), 243–255.

Healy, A., and Miller, G.A. (1970) The verb as the main determinant of the sentence meaning. *Psychonomic Science*, **20**, 372.

Jaeger, S.R. (2006) Non-sensory factors in sensory science research. *Food Quality and Preference*, **17** (1–2), 132–144.

Köster, E.P. (2009) Diversity in the determinants of food choice: a psychological perspective. *Food Quality and Preference*, **20** (2), 70–82.

Krueger, R.A. (1988) *Focus Groups: a Practical Guide for Applied Research*, Sage, Newbury Park, CA.

Krueger, R.A. (1998) *Analyzing and Reporting Focus Group Results*, Sage, Thousand Oaks, CA.

Lawless, H.T., and Heymann, H. (2010) *Sensory Evaluation of Food*, Springer, New York.

MacRae, A.W., Howgate, P., and Geelhoed, E.N. (1990) Assessing the similarity of odours by sorting and triadic comparison. *Chemical Senses*, **15**, 661–699.

McCarthy, E.J., and Perreault, W.D. (2005) *Basic Marketing: a Global Managerial Approach*, 15th edn, McGraw-Hill/Irwin, Homewood, IL.

McEwan, J.A., and Thomson, D.M.H. (1989) The repertory grid method and preference mapping in market research: a case study on chocolate confectionery. *Food Quality and Preference*, **1** (2), 59–68.

Meiselman, H.L. (2003) A three-factor approach to understanding food quality: the product, the person and the environment. *Food Service Technology*, **3**, 99–105.

Meiselman, H.L., and MacFie, H.L.H. (1996) *Food Choice, Acceptance and Consumption*, Blackie, London.

Meullenet, J.-F., Xiong, R., and Findlay, C.J. (2007) *Multivariate and Probabilistic Analyses of Sensory Science Problems*, Blackwell, Ames, IA.

Moscovici, S. (2000) *Social Representations: Explorations in Social Psychology*, Polity, Cambridge.

Næs, T., and Risvik, E. (1996) *Multivariate Analysis of Data in Sensory Science*, Elsevier, New York.

Næs, T., Brockhoff, P.B., and Tomic, O. (2010) *Statistics for Sensory and Consumer Science*, John Wiley & Sons, Ltd, Chichester.

O'Mahony, M. (1986) *Sensory Evaluation of Foods: Statistics Methods and Procedures*, Marcel Dekker, New York.

Peryam, D.R., and Pilgrim, F.J. (1957) Hedonic scale method of measuring food preferences. *Food Technology*, **11** (September), 9–14.

Pieters, R., Baumgartner, H., and Allen, D. (1995) A means–end chain approach to consumer goal structures. *International Journal of Research in Marketing*, **12**, 227–244.

Rao, V.R., and Katz, R. (1971) Alternative multidimensional scaling methods for large stimulus sets. *Journal of Marketing Research*, **8**, 488–494.

Reynolds, T.J., and Guttman, J. (1988) Laddering theory, method, analysis, and interpretation. *Journal of Advertising Research*, **28** (1), 11–31.

Santosa, M. (2010) Analysis of sensory and non-sensory factors mitigating consumer behavior: a case study with extra virgin olive oil, PhD dissertation, University of California, Davis.

Santosa, M., and Guinard, J.-X. (2011) Means–end chains analysis of extra virgin olive oil purchase and consumption behavior. *Food Quality and Preference*, **22** (3), 304–316.

Santosa, M., Abdi, H., and Guinard, J.-X. (2010) A modified sorting task to investigate consumer perceptions of extra virgin olive oils. *Food Quality and Preference*, **21** (7), 881–892.

Solheim, R., and Lawless, H.T. (1996) Consumer purchase probability affected by attitude towards low-fat foods, liking, private body consciousness and information on fat and price. *Food Quality and Preference*, **7** (2), 137–143.

ten Kleij, F., and Musters, P.A.D. (2003) Text analysis of open-ended survey responses: a complementary method to preference mapping. *Food Quality and Preference*, **14** (1), 43–52.

Tourangeau, R., Rips, L.J., and Rasinski, K.A. (2000) *The Psychology of Survey Response*, Cambridge University Press, Cambridge.

Van Kleef, E., van Trijp, H., and Luning, P. (2006) Internal versus external preference analysis: an exploratory study on end-user evaluation. *Food Quality and Preference*, **17** (6), 387–399.

Zampounis, V. (2006) Olive oil in the world market, in *Olive Oil Chemistry and Technology* (ed. D. Boskou), AOCS Press, Urbana, IL, pp. 21–39.

Zarraga, A., and Goitisolo, B. (2006) Multiple factor analysis of mixed tables of metric and categorical data, in *Multiple Correspondence Analysis and Related Methods* (eds M. Greenacre and J. Blasius), Chapman & Hall/CRC, Boca Raton, FL, pp. 352–367.

7 Sensory Functionality of Extra Virgin Olive Oil

Caterina Dinnella
Dipartimento di Gestione dei Sistemi Agrari, Alimentari e Forestali – GESAAF, University of Florence, Florence, Italy

7.1 Introduction

The wide variation of the micro-component fraction accounts for the extraordinary sensory diversity of extra virgin olive oils as a function of cultivars, pedoclimatic conditions, agricultural practices, olive ripening stage, and extraction techniques (Servili *et al.*, 2004). Thus, extra virgin olive oils show a broad range of completely different sensory profiles.

Harmony/complexity (defined as the balance of all positive attributes: fruitiness, bitterness, pungency, aroma compounds), persistence (lasting perception of positive attributes), and balance (relative differences among intensity of positive attributes) are often used to define sensory quality categories in extra virgin olive oil competitions (Bongartz and Oberg, 2011). Furthermore, extra virgin olive oil styles are defined as mild, medium, and robust depending on whether they are perceived to have a low, medium, or high combined level of positive attributes (mainly bitterness and pungency) (Gawel and Rogers, 2009).

Extra virgin olive oils ranked as excellent/premium by experts often do not meet consumer hedonic expectations (Delgado and Guinard, 2011), which can be only partially explained by the low consumer exposure to bitter, astringent, and pungent oils. Current methods used to classify extra virgin olive oils into sensory quality categories and to test consumer preferences involve evaluations of oils on their own and therefore do not consider the conditions that consumers experience when consuming olive oil in real life (i.e., mixed

Olive Oil Sensory Science, First Edition. Edited by Erminio Monteleone and Susan Langstaff.

with other foods). Therefore, the question arises of whether the extra virgin olive oil sensory diversity can actually be experienced by consumers. This is an important aspect for extra virgin olive oil valorization and market positioning. In fact, flavor, taste, and culinary use represent factors motivating olive oil purchase and consumption for both experienced and emergent market consumers (Krystallis and Ness, 2003; Caporale *et al.*, 2006; Delgado and Guinard, 2011). Hence there is a need for information about the best conditions of use that allow an oil to express its sensory potential. The "harmonic" pairing (intensely bitter and pungent oils suited for strongly flavored foods and those low in bitterness and pungency suited for delicately flavored foods) (Cerretani *et al.*, 2007) only marginally depict the wide spectrum of potential effect of extra virgin olive oil on the sensory properties of a dish. The relationship between olive oil profile strength/complexity and its capacity for modifying sensory properties of a dish is not obvious.

A dish can be considered as an item combination, where these items may be different ingredients and also stimuli of different sensory modalities, which interact extensively with each other, leading to the overall perceptual experience (Nestrud, Ennis, and Lawless, 2012). Item interactions occurring when a food is eaten can be grouped into three main categories: chemical interactions inducing structural, composition, or physical modification of sensory active molecules; physiological interactions related to the potential for one compound to interfere with taste receptor cells or taste transduction mechanisms associated with another compound; and cognitive interactions related to the central processing of signals induced by different sensory active compounds (Keast and Breslin, 2002). These interactions might lead to perceptual phenomena described as mixture suppression (an individual sensory stimulus is perceived as less intense in a blend than when experienced alone), adaptation (lowering of sensory system responsiveness after exposure to a constant stimulus), and release from suppression (following adaptation to one stimulus in a mixture, other stimuli are less suppressed and perceived at increased intensity) (Lawless, 2000). Moreover, different sensory modalities can interact with each other and the stimulation of one sensory modality gives rise to a perceptual experience in another sensory modality (Robertson and Sagiv, 2005). Interactions among different sensory modalities are responsible for enhancement/suppression of taste intensity induced by mixing congruous/incongruous odor and taste (Stevenson, Prescott, and Boakes, 1999; Caporale, Policastro, and Monteleone, 2004, Pfeiffer *et al.*, 2006).

Another important aspect to consider is the dynamic nature of food perception defined as a temporal series of events (Piggott, 2000). The perception of sensory properties changes during consumption of a food product according to the sequence of events that develop from first bite to swallow. In fact, the food destructuring needed for bolus formation induces sequential texture changes and the release of sensory active molecules (sapid, odorous, and irritant stimuli) in the mouth and in the retro-nasal region, thus

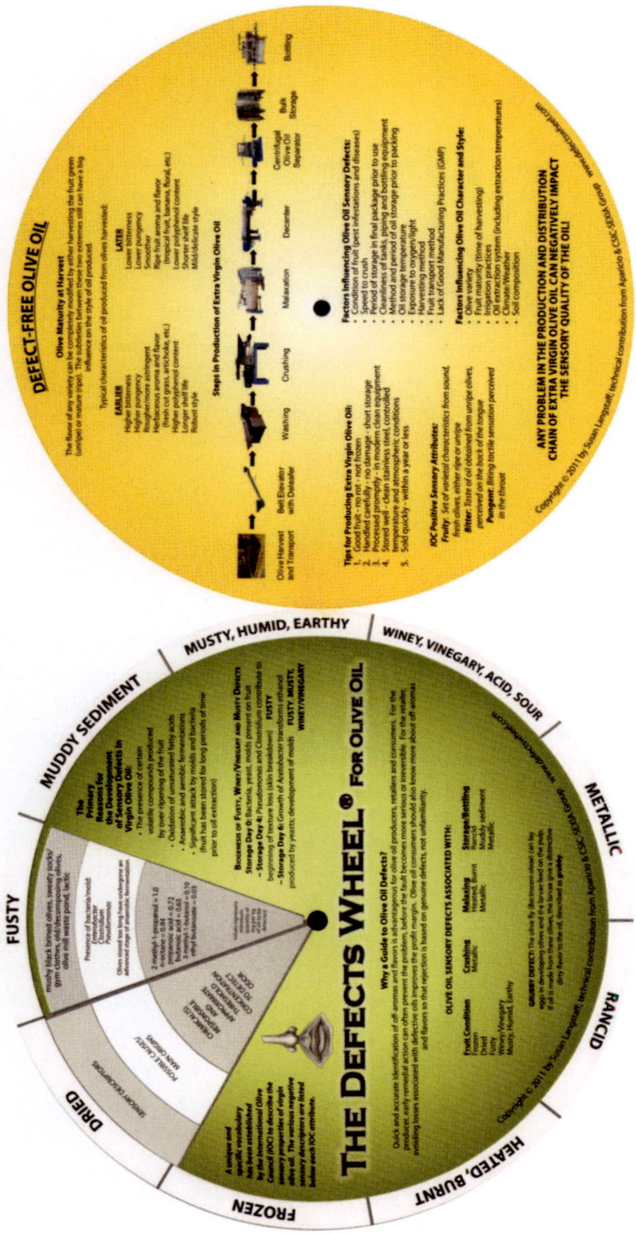

Plate 4.2 The Defects Wheel for Olive Oil.

Olive Oil Sensory Science, First Edition. Edited by Erminio Monteleone and Susan Langstaff.

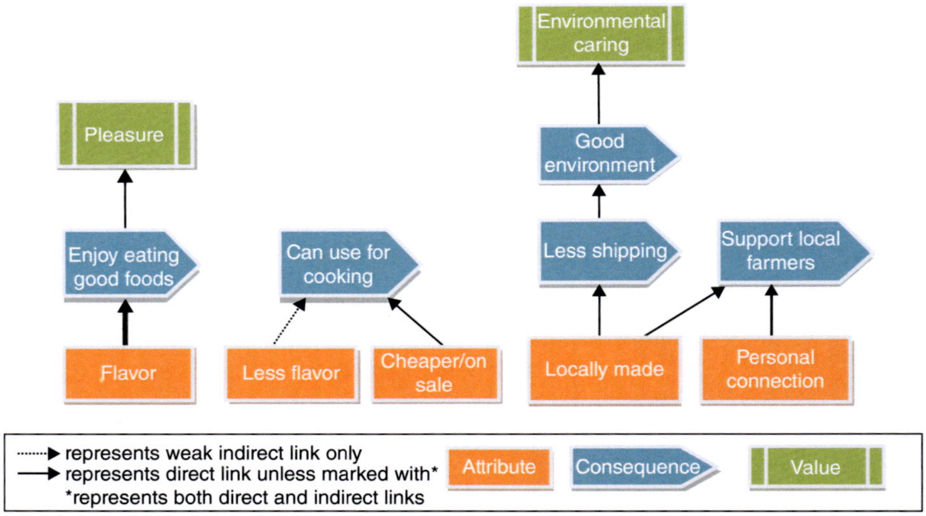

Plate 6.2 Hierarchical value map for olive oil purchase motivations among consumers who consumed mostly locally made olive oil. *Source:* Reprinted from Santosa, M., and Guinard, J.-X. (2011) with kind permission from Elsevier.

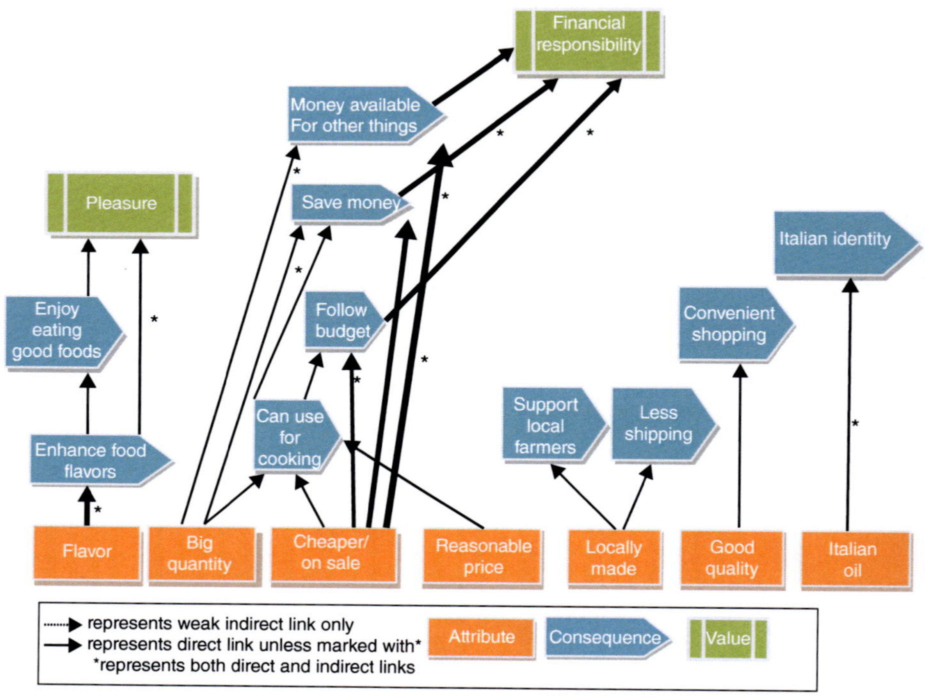

Plate 6.3 Hierarchical value map for olive oil purchase motivations among consumers who consumed mostly imported olive oil. *Source:* Reprinted from Santosa, M., and Guinard, J.-X. (2011) with kind permission from Elsevier.

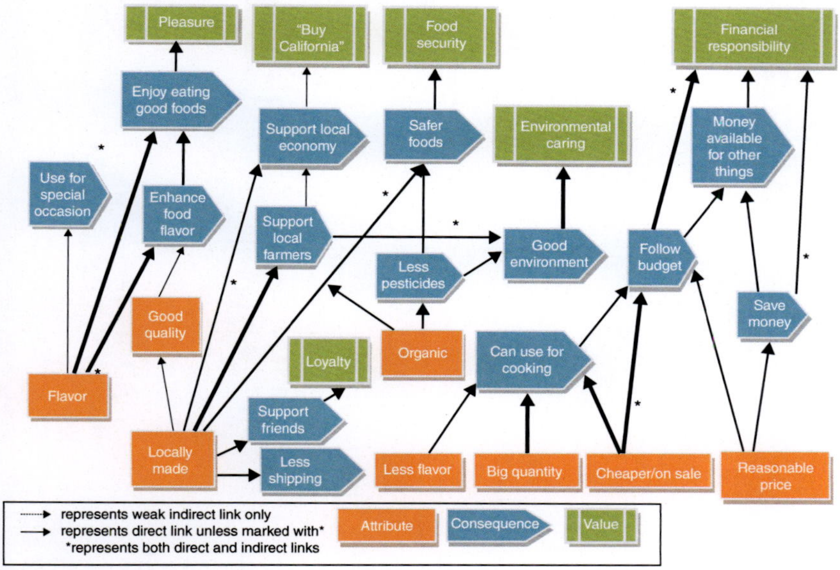

Plate 6.4 Hierarchical value map for olive oil purchasing motivations among consumers who consumed both imported and locally made olive oil. *Source:* Reprinted from Santosa, M., and Guinard, J.-X. (2011) with kind permission from Elsevier.

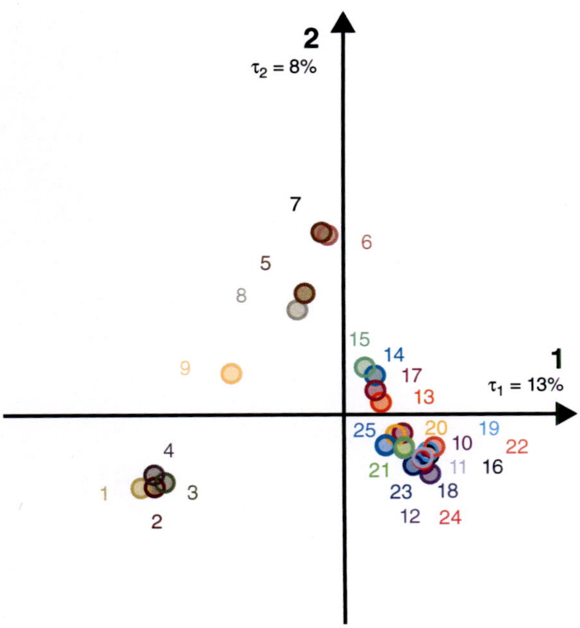

Plate 6.5 Positioning of 25 EVOOs and the associated confidence intervals from the second stage of the sorting task.

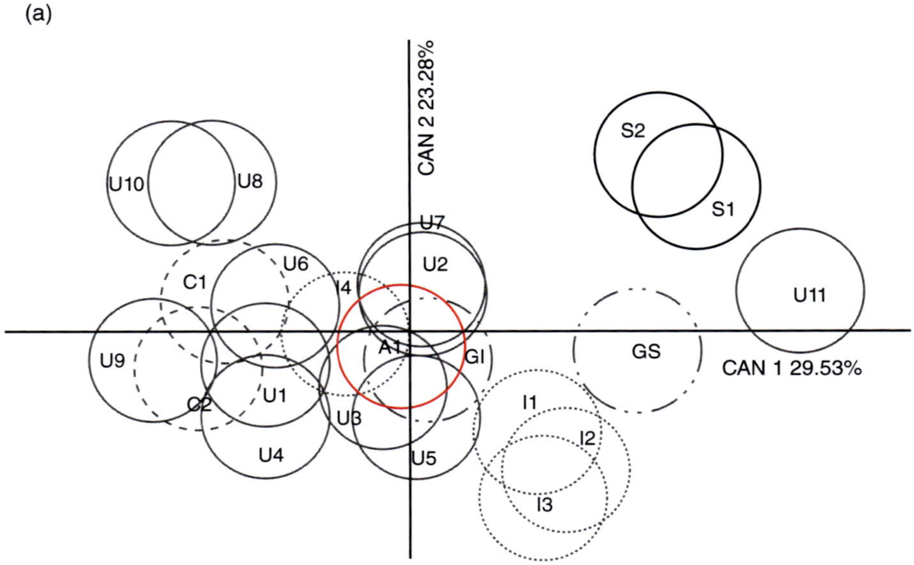

Plate 6.9 Canonical variate analysis (CVA) for 22 products in two dimensions. (a) Distribution of the products.

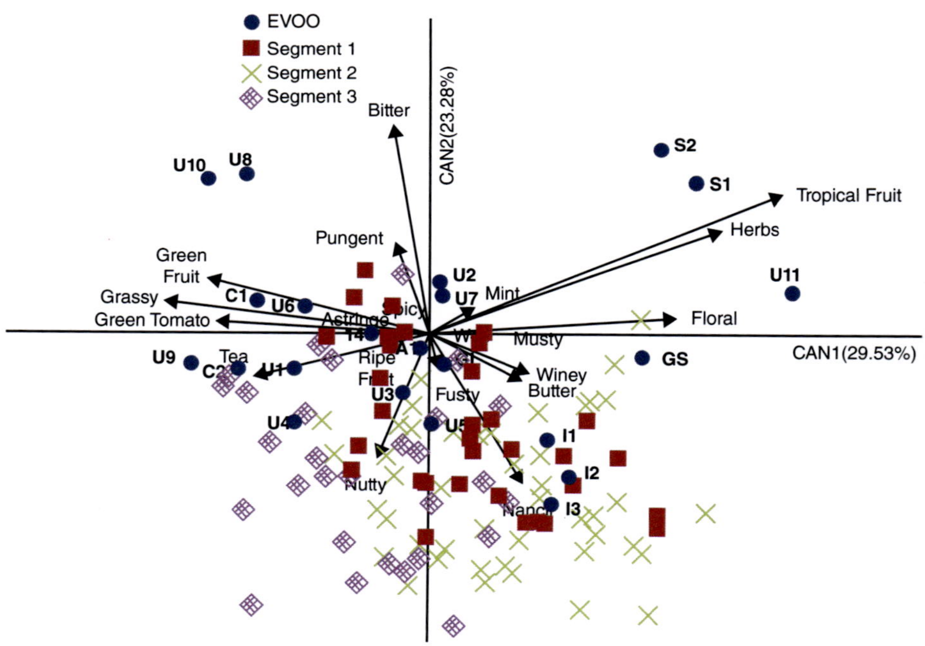

Plate 6.10 Preference mapping of EVOOs with a group of US consumers.

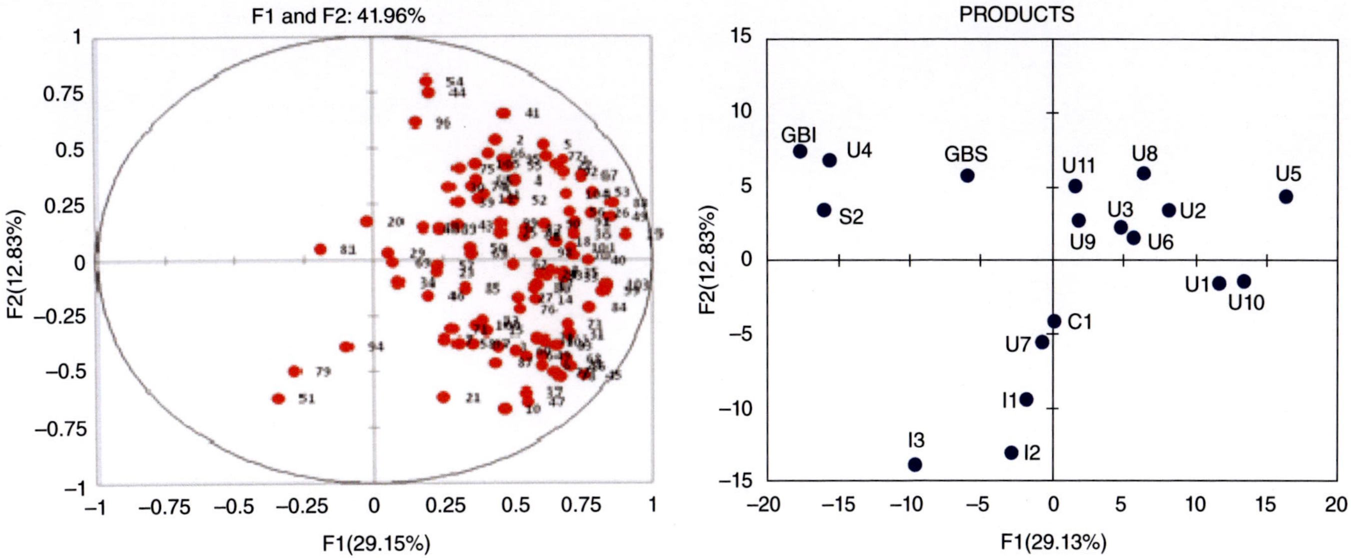

Plate 6.11 Internal preference mapping for overall liking of the 18 EVOO bottles.

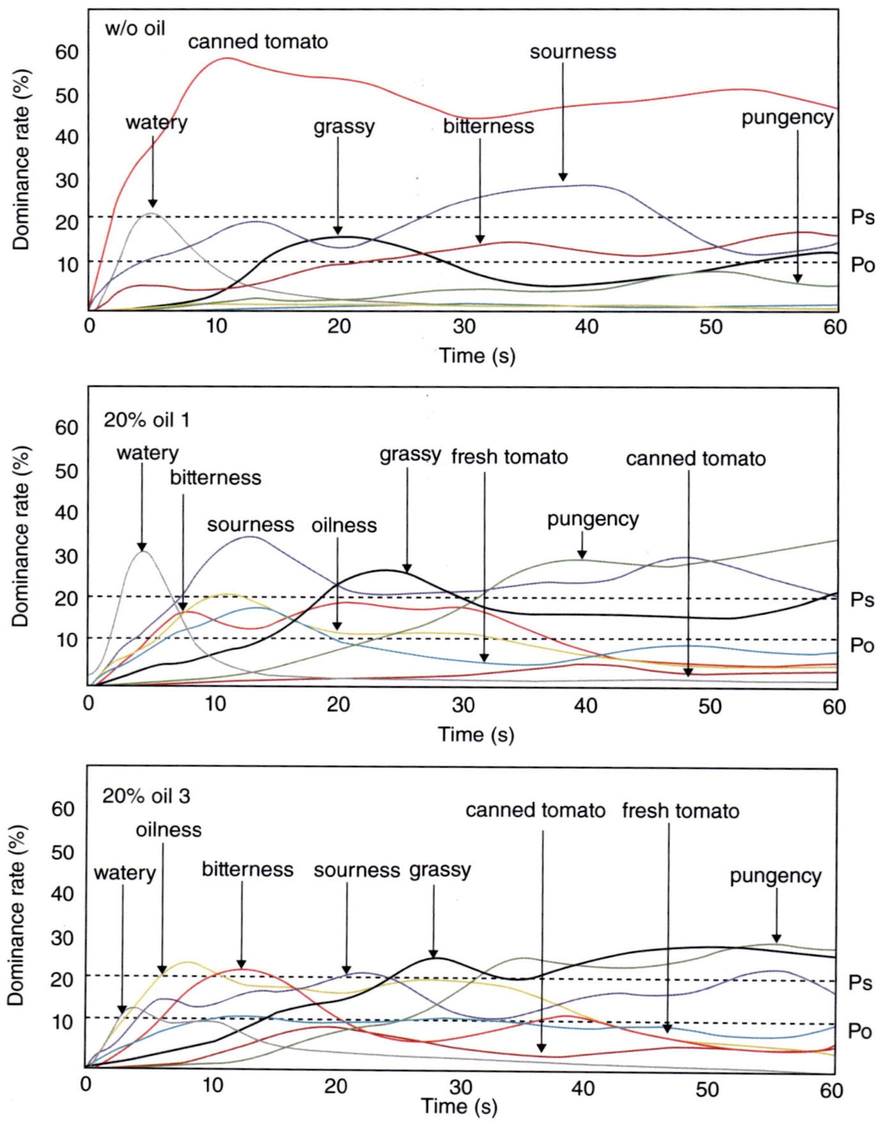

Plate 7.9 TDS curves for a tomato sauce evaluated without oil and in combination with 20% of oil 1 and oil 3. Evaluations were performed by a trained panel of 13 subjects. A list of eight attributes was used: sourness, bitterness, pungency, watery, oiliness, grassy, canned tomato, and fresh tomato. p0 represents the chance level and ps represents the 95% significance level.

Plate 9.1 Main geographic characteristics of Spain. *Source:* own elaboration (base map from IGN, 2012).

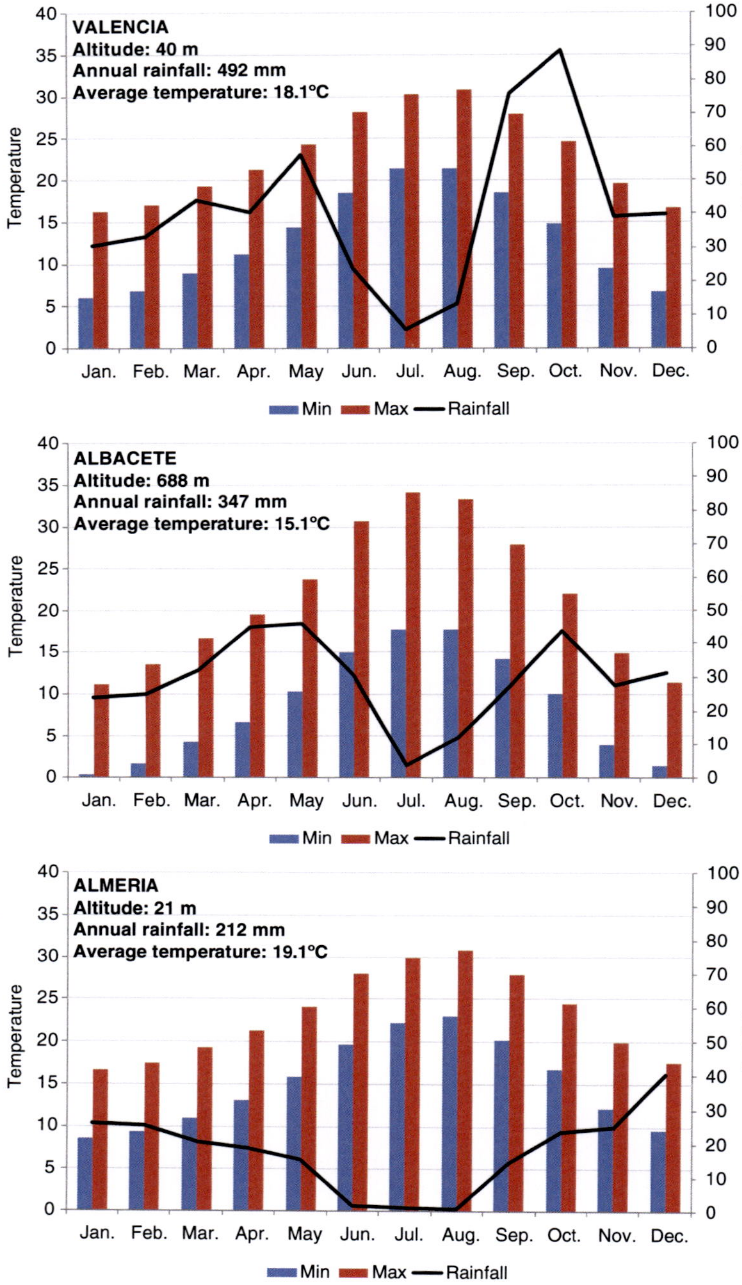

Plate 9.2 Climate diagrams of a typically Mediterranean climate (Valencia), a Mediterranean climate with cold winters (Albacete) and a dry Mediterranean climate (Almeria): average values between 2000 and 2011; temperatures in °C and rainfall in mm. *Source:* own elaboration (climate data obtained from AEMET, 2012).

Plate 9.3 Main Spanish cultivars.

Plate 9.4 Registered Protected Designations of Origin in Spain.

Plate 9.5 Examples of some Spanish cultivars.

Plate 10.1 Areas of production of Italian single cultivar oils, presented during the 9th National Fair of the Single Cultivar Oils promoted and organized by ASSAM – Agenzia Servizi Settore Agroalimentare delle Marche (Office of Services for the Agro-Food Sector in the Marche region).

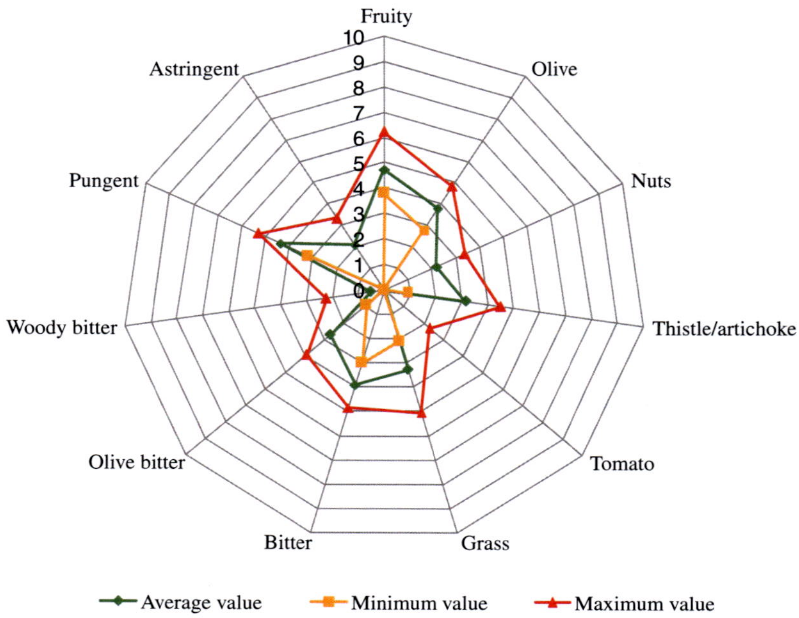

Plate 10.10 Typical radar chart of an extra virgin olive oil produced in the Tuscany region (data processing by Regione Toscana, 2012).

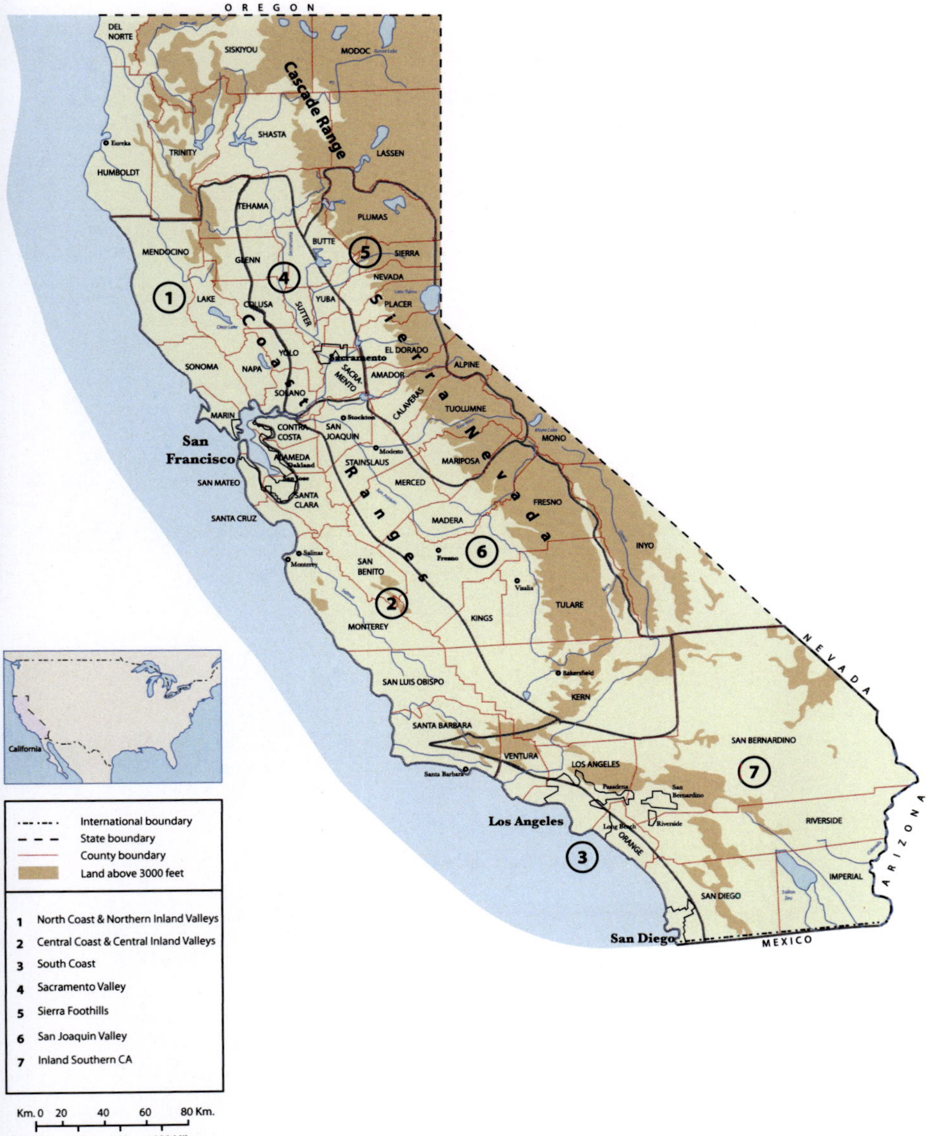

Plate 12.1 Olive-growing regions in California.

Plate 13.1 Mechanical harvesting of olives in Australia.

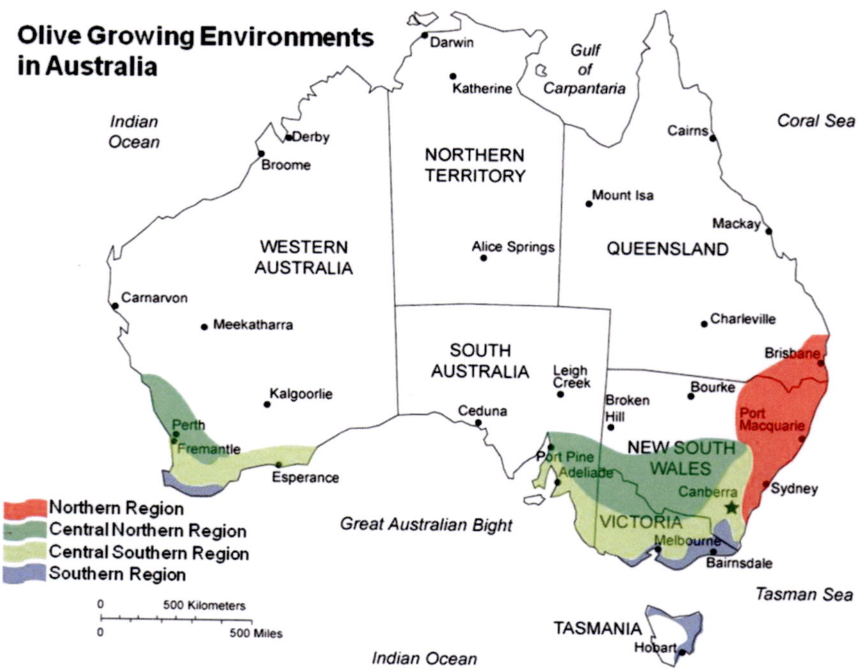

Plate 13.2 Olive-growing environments in Australia.

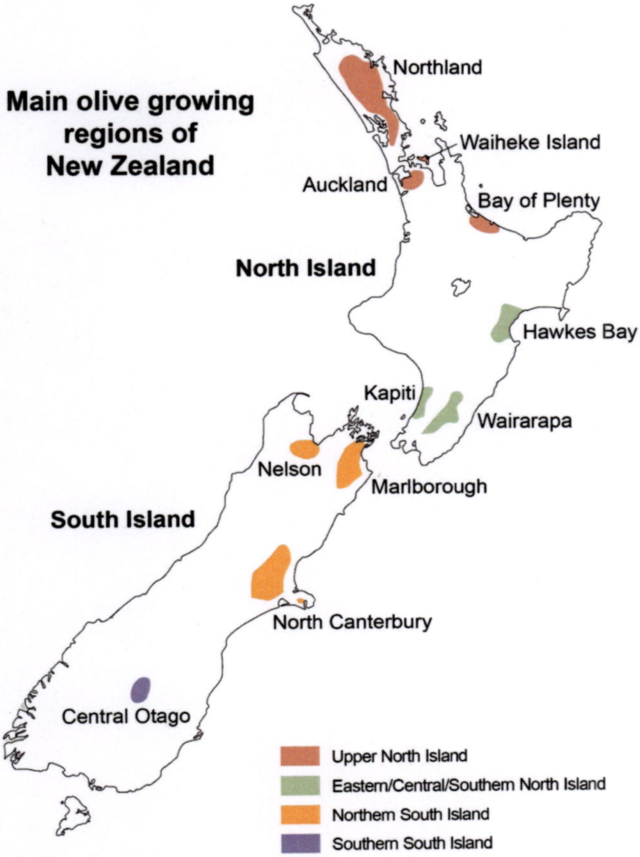

Plate 13.3 Main olive growing regions of New Zealand.

Plate 14.1 Olive distribution in South America.

leading to the dynamic overall perception (Lenfant *et al.*, 2009; Déléris *et al.*, 2011 Saint-Eve *et al.*, 2011). In addition, the perception resulting from the in-mouth transformation partially determines the hedonic response associated with food consumption (Sudre *et al.*, 2012).

The complex network of interactions underlying the perceptual experience accounts for the weak relations found between the performance of a combination and that of its individual components, either in terms of overall liking (Moskowitz and Krieger, 1995) or sensory profile (Aaslyng and Frøst, 2008, 2010).

Assessing the sensory functionality, defined as the ability to impact the sensory profile and the acceptability of the food matrix with which it is combined (Dinnella *et al.*, 2012a), seems to be an appropriate approach for evaluating the sensory performance of extra virgin olive oils. The systematic collection of information relevant to the effects of adding oils with different sensory profiles to various food categories needs sensory methodologies capable of taking into account the complex network of phenomena underlying food perception.

The Temporal Dominance of Sensations (TDS) method describes the evolution of the dominant sensory perceptions during the tasting of food, usually from the point at which the food is put in the mouth until several seconds after swallowing for after-taste evaluation. The perception pattern is represented by curves reporting the frequency with which sensations, reported in a list of several attributes, are considered as dominant by a trained panel during food consumption (Figure 7.1). This multi-attribute descriptive method has been proven to be less time consuming and more effective than other dynamic sensory tests when the whole range of sensations induced by a given food is under investigation (Pineau *et al.*, 2009). The TDS method not only allows the investigation of qualitative changes perceived during eating, but also explicitly considers sensory interactions taking place during food consumption. In fact, subjects are asked to characterize the main temporal aspects of the product simultaneously, thus integrating all the perceived sensations. Comparative studies highlighted the additional information provided by TDS in respect to static profiling for differentiating either products with complex and long-lasting sensations (Labbe *et al.*, 2009) or those only showing subtle sensory differences (Meillon *et al.*, 2010). TDS profiling also provided a better understanding of the food texture oral change (Lenfant *et al.*, 2009) and of its role on aroma and taste perception with respect to static profiling tests or other dynamic sensory procedures (Déléris *et al.*, 2011; Saint-Eve *et al.*, 2011). The dynamic and multisensory aspects of food perception play an important role in processes used by consumers to assess acceptability and sensory properties of food products (Blake, 2004; Sudre *et al.*, 2012). TDS was successfully applied in studies on sensory dimensions that participate in defining multi-attribute drivers of consumer food preference (Meillon *et al.*, 2010). Recently, the method was proved to be more appropriate than descriptive analysis

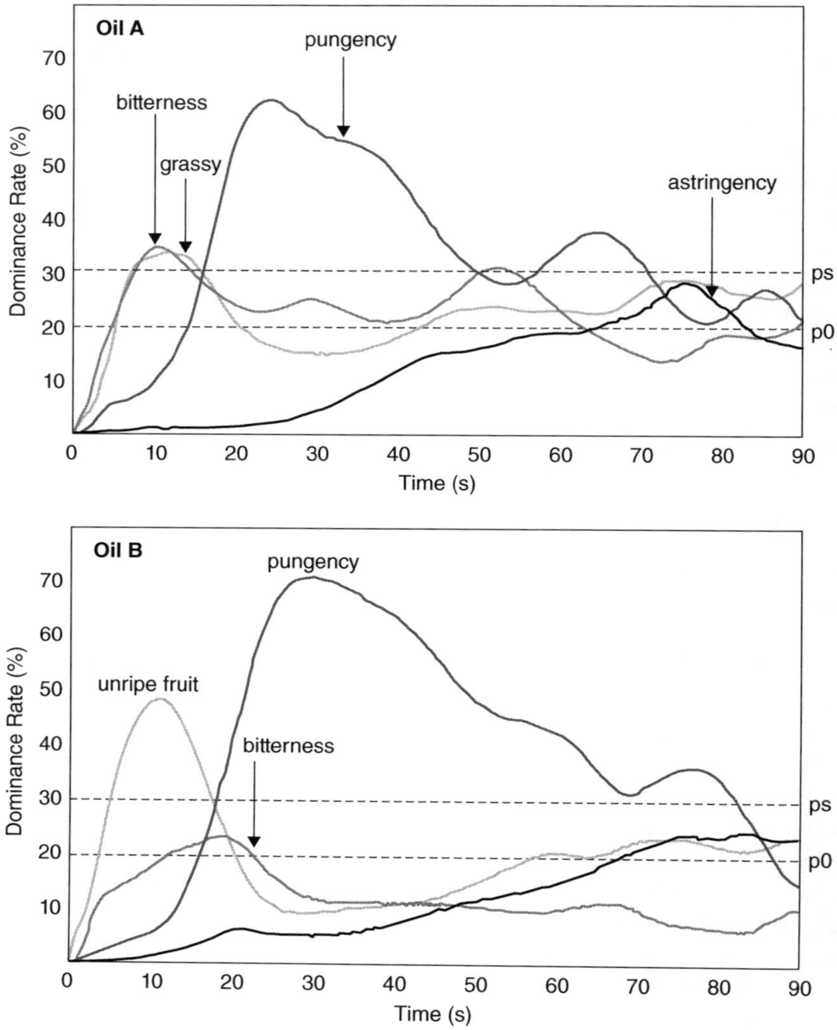

Figure 7.1 Example of TDS curves for two extra virgin olive oils from two Italian regions (A from Tuscany; B from Lake Garda region, Veneto). Evaluations were performed by a trained panel of 13 subjects. A list of five attributes was used: bitterness, pungency, astringency, grassy, and unripe fruit flavors. p0 represents the chance level and ps represents the 95% significance level. *Source:* Reprinted from Dinnella, C., Masi, C., Zoboli, G., and Monteleone, E. (2012a) with kind permission from Elsevier.

to study sensory interactions occurring in oil–food combinations (Dinnella *et al.*, 2012a).

In order to understand better the potential of this method in studying the sensory functionality of extra virgin olive oils, some evidence from the cited work is reported. Two oils differing in their sensory profile, obtained by means of descriptive analysis, were considered (Figure 7.2). The two oil were

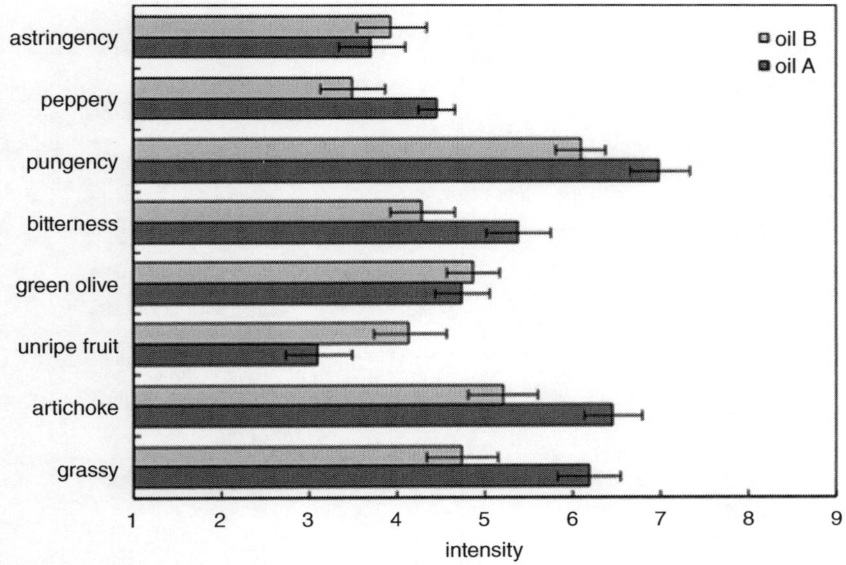

Figure 7.2 Sensory profile represented as a bar plot for two extra virgin olive oils from two Italian regions (A from Tuscany; B from Lake Garda region, Veneto) selected to be combined with a tomato sauce. Evaluations were performed by a trained panel of 13 subjects. Intensity scale: 1 = extremely weak; 3 = weak; 5 = moderate; 7 = strong; 9 = extremely strong.

added (at 10% w/w) to a tomato sauce and the resulting combinations analyzed by the TDS method (Figure 7.3). Furthermore, consumer responses on liking and perceived freshness of oils, tomato sauce, and their combinations were collected (Figure 7.4).

Sour taste, watery character and tomato flavor are the dominant sensations during the first part of the evaluation of tomato without oil. Sourness is the most dominant sensation from the beginning of the evaluation and lasts as one of the most dominant sensations for up to 40 s. The watery sensation is dominant between 10 and 20 s and the tomato sensation starts to be dominant after the first 10 s. Tomato flavor remains the only dominant sensation after 30 s and lasts until the evaluation ends. The reduction in the sour taste dominance rate value, the suppression of watery character dominance and the onset of bitter taste dominance are the main effects of combining tomato with oil A. Similar effects on sourness and watery dominances are induced by adding oil B to tomato. Moreover, oil B contributes a typical unripe fruit flavor amongst the sensations dominating the first part of the tomato–oil B evaluation. Results from consumer test (Figure 7.4a) show a significant increase in liking for tomato combined with either oil A or oil B compared with tomato without oil. Several factors relating to the general effects of adding oil rather than specific modifications of the sensory profile could account for the increased liking of tomato + oil samples compared with that of tomato without oil. In fact, no significant differences in liking are

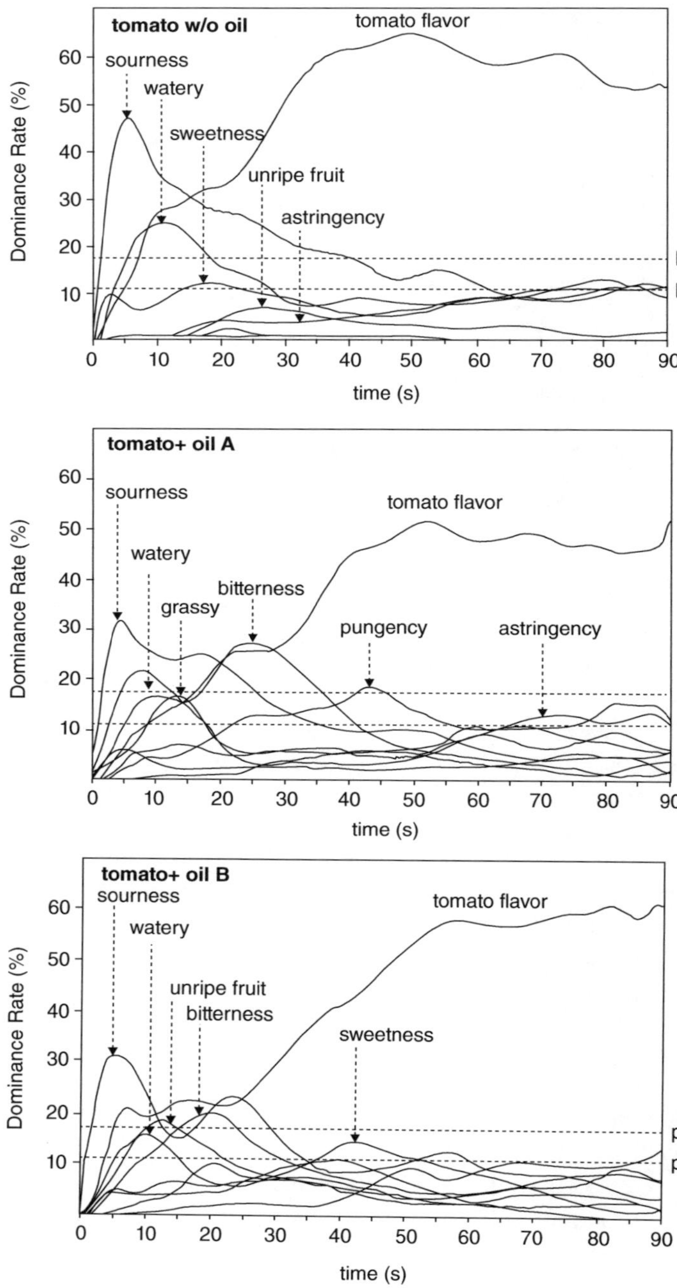

Figure 7.3 TDS curves for a tomato sauce evaluated without oil and in combination with 10% w/w of oils A and B. Evaluations were performed by a trained panel of 13 subjects. A list of eight attributes was used: sourness, bitterness, pungency, astringency, grassy, unripe fruit, canned tomato, and fresh tomato. p0 represents the chance level and ps represents the 95% significance level. *Source:* Reprinted from Dinnella, C., Masi, C., Zoboli, G., and Monteleone, E. (2012a) with kind permission from Elsevier.

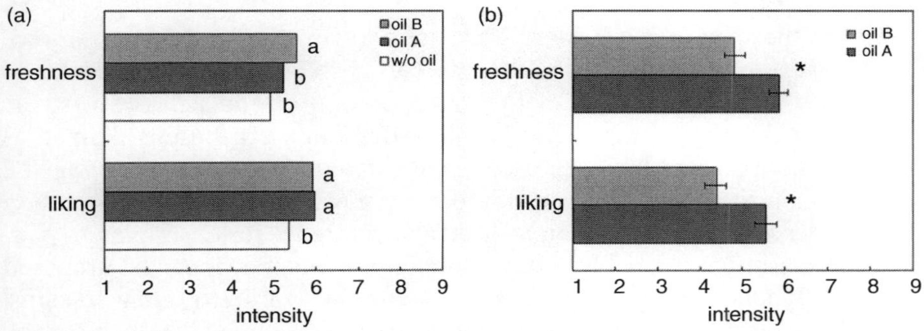

Figure 7.4 (a) Liking and freshness ratings for tomato without oil and combinations with 10% w/w of oils A and B. Values followed by different letters are significantly different ($p \geq 0.05$). (b) Liking and freshness ratings for oils A and B. *, $p \leq 0.001$. Evaluations were performed by 64 regular extra virgin olive oil consumers recruited in the Florence area (Tuscany, Italy). Intensity scale: 1 = dislike extremely; 5 = neither like nor dislike; 9 = like extremely. *Source:* Reprinted from Dinnella, C., Masi, C., Zoboli, G., and Monteleone, E. (2012a) with kind permission from Elsevier.

found when comparing tomato + oil A with tomato + oil B, despite the differences in flavors and taste. Modifications of the sensory properties of tomato combinations induced by both oils, namely the reduction in both the watery attribute and the sour taste, and also the well-known positive effect of fat addition on food palatability, might account for the increased liking of these samples. It is interesting that when oils are tested on their own, there is a significant difference in liking mean scores (Figure 7.4b). Tomato combined with oil B is perceived as fresher than both tomato without oil and tomato with oil A. The unripe fruit flavor, together with the other sensory properties specifically contributed by oil B to the tomato profile, accounts for the increased perceived freshness of tomato + oil B compared with tomato + oil A and tomato without oil. Hence the two oils have a varied sensory function. Again, it is interesting that perceived freshness in tomato–oil combinations is not predicted at all by the evaluation of the oils on their own.

Considering this evidence, the aim of this chapter is to describe in detail the TDS method and to consider further factors to be taken into consideration when studying the sensory functionality of extra virgin olive oils.

7.2 The Temporal Dominance of Sensation method

7.2.1 Language development and panelist training

The TDS method is applied with a trained panel, which means that, as for descriptive panels, qualified subjects develop a vocabulary to describe the sensations they perceive considering the temporal evolution of the sensations. They also establish a consensual definition of each attribute and are trained for the method in itself. Thus, before assessment of products, assessors participate in a series of language sessions. A dominant attribute is defined as the attribute associated with the sensation catching the attention at a given time. A main list of attributes developed by means of the term generation phase of descriptive analysis can be used as a reference for describing to panelists the sensations probably induced by samples. Standards can be prepared to train the panelists to recognize the descriptors. During training sessions, panelists are asked to taste samples and select from the main list the most relevant attributes for describing the temporal evolution of sensations. In general, three training sessions are needed to develop a consensus attribute list for TDS evaluations.

It is recommended that the attributes do not exceed 10 descriptors to enable panelists to use the whole list during evaluation sessions (Pineau *et al.*, 2012). On average, four different attributes are selected during TDS evaluation independently of the total number of attributes included in the list.

Terms describing opposites of the same sensation (e.g., soft/hard, watery/dense) should be included in the attribute list if they represent the dominant sensation at the same evaluation time in different products. Missing an attribute describing the dominant sensation at a given time might induce the overestimation of sensations still perceived at that time but actually having a weak impact on the overall perception.

An example of descriptors of dominant sensations in extra virgin olive oil and in oil–food combinations is reported in Table 7.1.

The application of the method requires dedicated software for data acquisition, now commercially available, and a couple of individual sessions, preferably under the supervision of expert personnel, are needed to familiarize panelists with software procedures for data recording. For each sample, each panelist is presented with the complete list of attributes on a computer screen. Panelists are trained to click on the "Start" button as soon as the sample is in the mouth and immediately starting the evaluation. Figure 7.5 shows an example of an evaluation sheet for oil–tomato sauce combinations. It is important that panelists are told that not all attributes necessarily have to be selected as dominant and that a given attribute can be selected as dominant several times during the evaluation. The possibility of including the option "nothing dominates" in the list can be considered.

A general instruction for panelists could be the following: "Put the sample in your mouth and immediately start the evaluation by clicking on the "Start" button. During the evaluation time you are free to select an attribute several times. When you consider that the dominant attribute has changed, you have to select the new dominant attribute, and so on, until the evaluation ends."

Individual performances of panelists are evaluated at the end of the training period by having the panel evaluate in triplicate a subset of samples to

Table 7.1 Example of attributes describing dominant sensations in oils and relevant combinations with beans and tomatoes.

Attributes		Definition	Standard	Oils	Combinations	
					Puréed tomatoes	Puréed beans
Flavor	Bean	Odor of canned white beans				×
	Bitterness	Bitter taste	Quinine dichloride (0.025 g/l)	×	×	×
	Grassy	Odor of freshly mown grass	*cis*-3-Hexen-1-ol in seed oil (140 μl/l)	×	×	×
	Metallic	Metallic taste typical of canned food	$FeSO_4$ (1.5 g/l)			×
	Pungency	Leaving a burning sensation in the back of the throat		×	×	×
	Sourness	Sour taste	Citric acid (0.25 g/l)		×	
	Sweetness	Sweet taste	Sucrose (5.0 g/l)		×	×
	Tomato	Odor of canned tomatoes			×	
	Unripe fruit	Odor of unripe, "green" fruit		×	×	×
Mouthfeel	Astringent	Dryness of the oral surface and tightening and puckering sensation of the mucosa and muscles around the mouth	Aluminum sulfate (0.3 g/l)	×	×	×
	Creamy	Soft and smooth texture				×
	Watery	Thin, weak texture, opposite of viscous			×	

Source: Adapted from Dinnella, C., Masi, C., Zoboli, G., and Monteleone, E. (2012a) with kind permission from Elsevier.

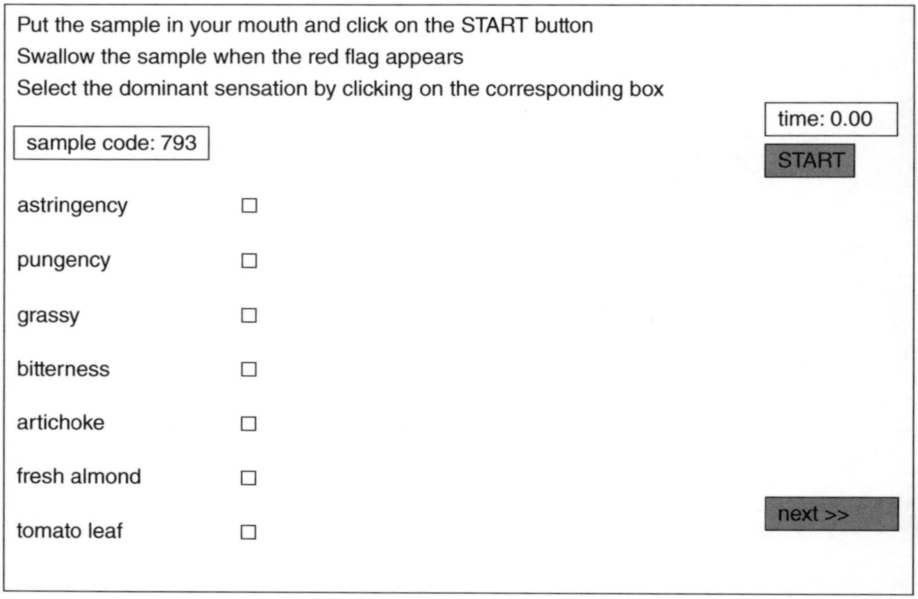

Figure 7.5 Example of TDS evaluation sheet for oil–tomato combinations.

be used for the real study. The panel leader visually inspects, for each panelist, the frequency with which each sensation is considered as dominant for each product across replications. This is a time-consuming process but, as reported in Section 7.2.4, new approaches have recently been developed in order to test panel performance by means statistical tests commonly applied in descriptive analysis.

7.2.2 Sensory procedure

The number of samples per session should be defined according to general rules of descriptive sensory tests. In particular, for olive oil samples and relevant combinations with foods, the number of samples for each session should not be greater than five (served monadically). Each sample should be presented in an appropriate container and identified by a three-digit code. For oil and liquid/semi-solid oil–food combinations, panelists should be instructed to pour the sample into a spoon for evaluation. Panelists put the sample in their mouth and start the evaluation by selecting the attribute perceived as dominant. The total evaluation time should be defined by expert personnel, accordingly to panelists' suggestions, taking into account the persistence of sensations after swallowing. The whole evolution of sensations perceived during tasting of olive oils is covered by a 60–90 s evaluation time. The time of keeping the sample in the mouth before swallowing can also be defined; 8 s before swallowing can be considered appropriate for olive oil and relevant combinations with puréed foods.

A procedure for restoring the perceptive ability between evaluations of subsequent samples should be defined. An effective procedure for this consists of rinsing the mouth with water for 30 s, eating some plain crackers for 30 s, and rinsing with water for a further 30 s. A 10 min break after evaluating 2–3 samples is suggested.

In TDS tests, each assessor is asked to replicate the evaluation of samples. The number of replicates depends on the size of the differences that the experimenter is required to detect and on the number of attributes included in the evaluation sheet. Small expected differences require a higher number of replicates. Similarly, the number of required replicates increases when the number of descriptors increases. The required number of observations (replicates × panelists) to estimate the significance of dominance rate for each attribute and for each product is usually above 30. In general, the number of replicates in olive oil and oil–food combination studies is 3–4 with 10–12 panelists involved in the test. A slightly higher number of panel members (14–16) can be used to reduce the number of evaluations per panelist (Pineau *et al.*, 2012). The order of presentation of samples should be balanced for first-order and carryover effects across replications.

7.2.3 Data acquisition

The attribute list is displayed on the computer screen. Attribute positioning can significantly affect the number of selections and the time at which attributes are selected. On average, attributes positioned at the top of the list are selected more frequently and earlier than attributes positioned at the bottom (Pineau *et al.*, 2012). Therefore, randomization of attribute positioning across panelists is advised to reduce bias due to attribute order. The same attribute order can instead be used for the same panelist in each evaluation to facilitate attribute identification in the list during the test.

Each time the dominant sensation changes, panelists select the corresponding new dominant attribute. During evaluation, software registers the time elapsed since a given attribute was selected until a new attribute becomes dominant. The frequency of selection with which each attribute has been selected as dominant over the total number of observations (panelists × replicates) is computed by the software at regular time intervals (every 0.5 s) and expressed as dominance rate (DR). The resulting frequency distribution for each attribute is converted into a continuous function, giving the dominance curve (Figure 7.6). Dominance curves of different attributes can be superimposed, giving a graphical representation of the product profile evolution during the evaluation time. Dominance curves are accompanied by estimates of the chance level (theoretical proportion of panelists selecting an attribute at random) and the significance level (the smallest proportion that

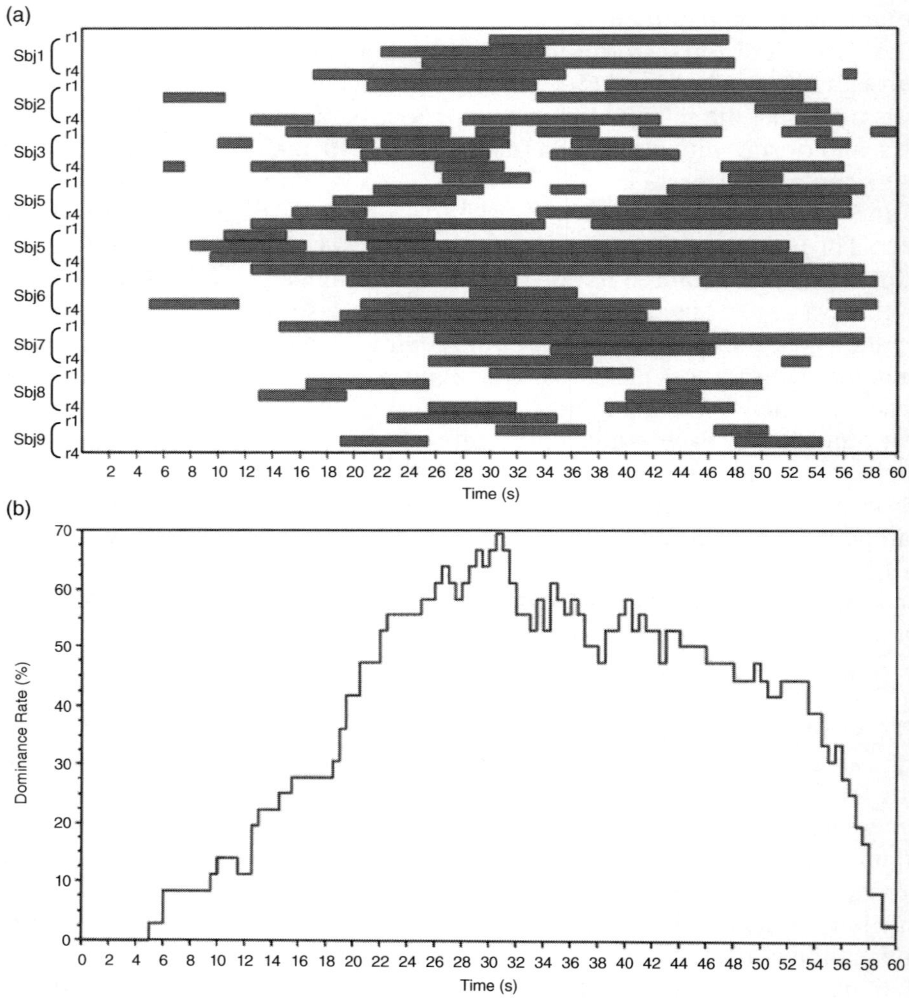

Figure 7.6 Frequency distribution of the attribute "pungency" expressed by a group of nine subjects in four replicates of one extra virgin olive oil sample: (a) raw data and (b) non-smoothed dominance curve.

is significantly greater than the chance level) based on binomial tests (Pineau *et al.*, 2009).

7.2.4 Data analysis

Differences between pairs of products can be assessed by subtracting the relevant dominance rates for each attribute at each point using a Fisher's exact test to determine the significance level (Lenfant *et al.*, 2009). Statistical tests based on randomization have been proposed as a valid statistical means to investigate overall differences between products (Meyners and Pineau, 2010) and to test panel reliability (Meyners, 2011), but the main limitations of this approach are the computation time and the dependence on the randomization used in the experiment.

In order to simplify analyses and to be able to use standard statistical methods frequently used for sensory data (Lawless and Heymann, 2003; Næs, Brockhoff, and Tomic, 2010), Pineau, Neville, and Lepage (2011) proposed the use of aggregated data in time intervals. For each time period, the binomial data (0 = attribute not selected as dominant; 1 = attribute selected as dominant) are summarized and the frequency values of attribute dominance by each panelist are computed. Frequency values are then submitted to an analysis if variance (ANOVA) model to estimate panel and assessor performance. Summarizing panelist responses as frequency values in a given number of time periods instead of considering all the acquisition time points represents one of the key features of this approach. It should be emphasized, however, that in order not to lose information regarding the dynamic nature of the test, a careful selection of the number and duration of time periods is needed. Recently, Pineau *et al.*'s method of was further validated by residual analysis of ANOVA models and criteria for selecting the most appropriate periods for TDS curve splitting were proposed (Dinnella *et al.*, 2013).

7.2.5 Sensory profile of oil and oil–food combinations by TDS and descriptive analysis

In comparing TDS curves and descriptive analysis profiles of extra virgin olive oils, Dinnella *et al.* (2012a) showed that, in general, attributes perceived with mean intensity ratings greater than moderate (pungency, bitter taste, and grassy flavor) in descriptive analysis were also the sensation with the highest dominance rate in TDS (Figure 7.1 and Figure 7.2). On the other hand, unripe fruit flavor was evaluated at moderate intensity in the descriptive analysis test but was the most dominant flavor in the dominance test. Moreover, the TDS curves clearly showed the different dynamics of the perception of stimuli, with taste and flavor perceived as dominant when the sample is in the mouth and the irritant sensations dominating after swallowing (Sinesio, Moneta, and Esti, 2005).

The general consistency between TDS curves and descriptive analysis data was also found for more complex stimuli such as oil–food combinations. The sensations dominating the dynamic sample profile were also the attributes rated with intensity equal to or higher than moderate. However, "unexpected" flavors and tastes contributed by olive oils to the descriptive profile of combinations were perceived as dominant sensations despite their weak or weak/moderate intensity.

Usually, the presence of olive oil induces an increase in the number and type of taste and flavor perceived while the sample is still in mouth. The spreading of panelists' attention among a large number of transient sensations in the early evaluation stage, and the consequent decrease in dominance values of the relevant attributes, may induce the suppression of attributes describing food texture (see the tomato without oil and oil–tomato combination TDS curves in Figure 7.3). On the other hand, the dynamic profile after swallowing tends to be less complex and the panelists focus their attention on a small number of long-lasting sensations.

The different dynamics of sensations induced by complex stimuli can account for differences between information from static and dynamic profiling (Labbe *et al.*, 2009). Unexpected and less familiar sensations tend to catch a subject's attention even when perceived at low intensity levels and they consequently become sensations dominating the perceived dynamic profile of the foods.

7.3 Comparing the sensory functionality of extra virgin olive oils with a varied sensory style

This section presents a possible practical approach to studying the sensory functionality of extra virgin olive oils with a varied sensory profile when combined with the same food. Only a limited number of papers have been published on this topic. This section refers to studies carried out in the Sensory Unit of our department (Dinnella *et al.*, 2012b; Piochi *et al.*, 2012). The question behind the interest in investigating the sensory functionality of extra virgin olive oils is simple: do oils with varied sensory characteristics differently modify the sensory properties of a food to which are combined? The answer is not immediately obvious. In fact, to prove that different oils have different sensory functionalities, the following activities are needed: (1) the description of the sensory profile of oils; (2) the evaluation of the sensory properties of the oil–food combinations; and (3) the collection of consumer liking responses for the combinations.

7.3.1 Description of the sensory profile of oils

Consider, as an example, the interest in studying the sensory functionality of oils with different origins (olive variety and region). The first step should be the evaluation of similarities and differences among the oils and the identification of different sensory styles. Recall that, in agreement with Monteleone (see Chapter 5), the term "style" indicates the sensory profile that describes an oil (or a group of oils) as different from others. The aim of this step should be the selection of a limited number of samples that represent well the sensory differences among the products of interest. This can be achieved by running a descriptive analysis and computing a perceptual map. An example is

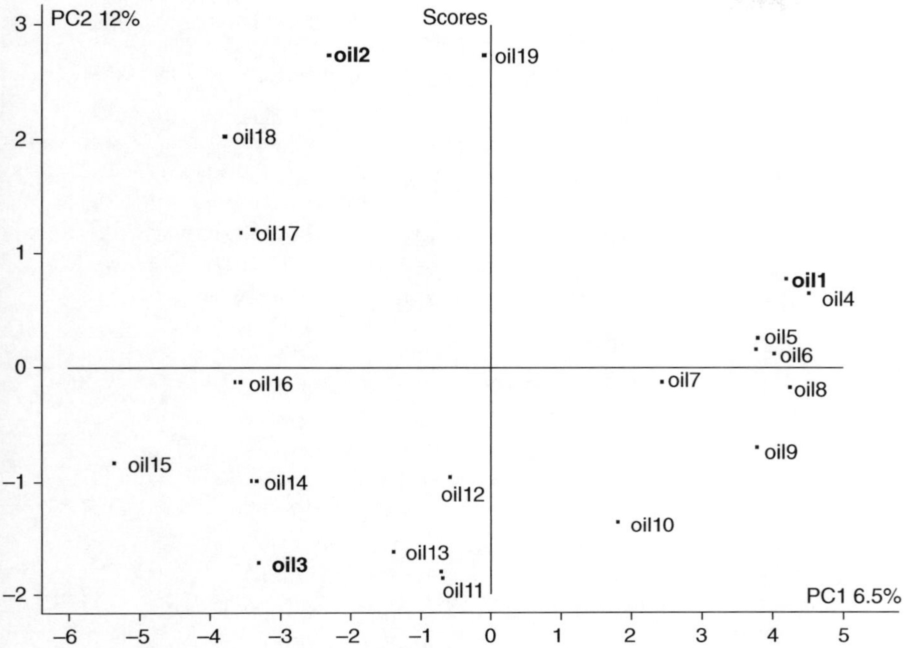

Figure 7.7 Score plot from principal component analysis of significant descriptive sensory attributes for 19 extra virgin olive oil samples. Selected oils are indicated in bold.

shown in Figure 7.7. Visual inspection of the map indicates that samples on the right of the first component (oil 1, oil 4–oil 10) are similar to each other and different from samples grouped on the left. On this side of the map, samples are widely spread along the second component (from the bottom to the top) in which samples oil 3 and oil 11–oil 16 are separated from samples oil 2, oil 17, and oil 18. In other words, selecting, for example, oil 1, oil 2, and oil 3 and representing their sensory profile by means a spider plot (Figure 7.8), it is possible to identify the three sensory styles accounting for the main sensory differences among 19 oils.

7.3.2 Evaluation of the sensory properties of the oil–food combinations

Once the oils have been selected and their profiles described, their effect on the sensory properties of a given oil–food combination can be investigated. The complex network of physicochemical and perceptual interactions between oil and food components is responsible for the perceived profile of oil–food combinations. Thus, the question arises of whether the intensity of attributes perceived in the oil can be considered a predictor of its contribution to the sensory profile of an oil–food combination. Relationships between

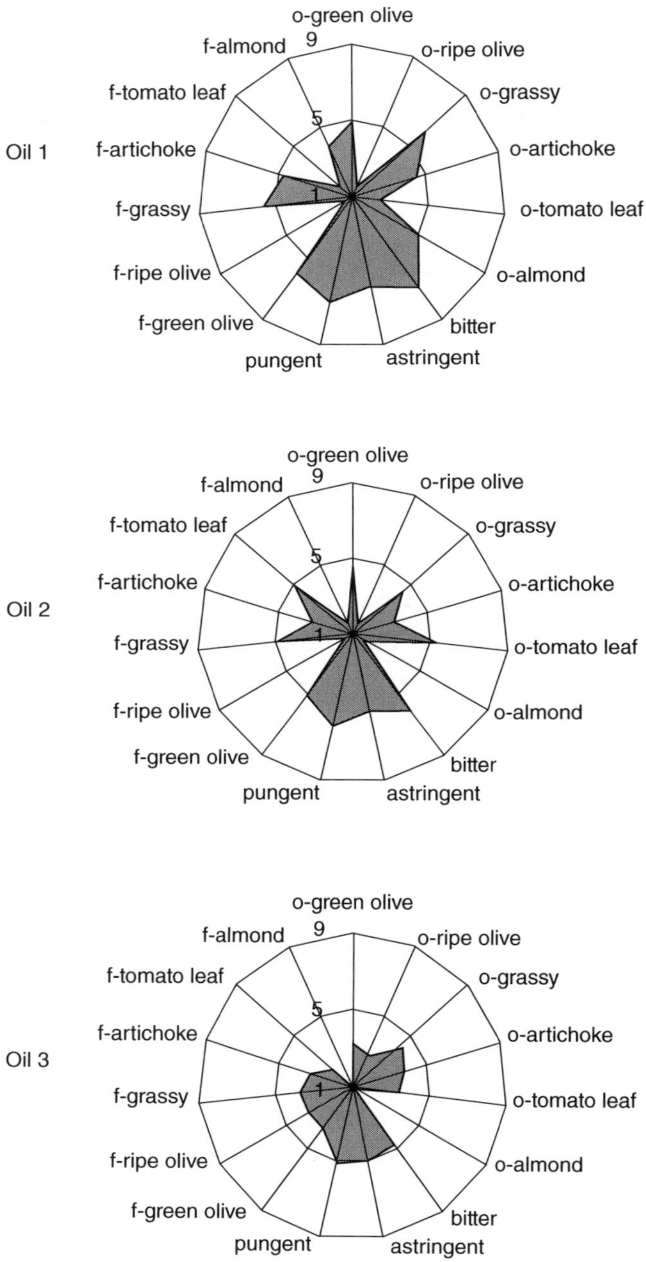

Figure 7.8 Sensory profile represented as a spider plot of three oils selected as representative of different styles (oil 1, northern Italy, cv. Drizar 100%; oil 2, Spain, cv. Piqual 100%; oil 3, southern Italy, cv. Nocellara del Belice 70%, Cerasuola 10%, Coratina/Galatina 20%). Evaluations were performed by a trained panel of 13 subjects. Intensity scale: 1 = extremely weak; 3 = weak; 5 = moderate; 7 = strong; 9 = extremely strong.

chemical concentration, detection threshold, and suprathreshold intensity are extremely complex even in the case of simple model taste solutions (Keast and Breaslin, 2002; Keast and Roper, 2007). The same set of complex rules determines the connections occurring between the sensory properties of an oil–food combination and both the amount of oil used and the intensities of its sensory attributes. Identifying the amount of oil needed to induce perceivable modifications to the sensory profile of the food to which it is added represents a key step in sensory functionality assessment. The simplest way to investigate this aspect is to run discrimination tests on oil–food combinations obtained with extra virgin olive oils representative of different styles. As an example, oil 1, oil 2, and oil 3 (Figure 7.8) added to a tomato sauce can be considered (Piochi *et al.*, 2012). Oil 1 can be described as an oil with a very complex style: several attributes perceived at strong intensities, its sensory profile is characterized by intense green odors and flavors mainly described as fresh almond, grassy, and green olive, intense bitter taste, and irritant sensations (peppery and pungency). Oil 2 can be described as an oil with a moderately complex style: few attributes perceived at more than moderate intensity; the descriptor tomato leaf characterizes the profile along with a moderate intensity of bitterness, pungency, and peppery sensations. Oil 3 can be described as an oil with a very simple style: green notes, bitter taste, and irritant sensations are perceived at weak–moderate intensity and its sensory profile is mainly characterized by a ripe olive sensation.

The three oils were added to a tomato sauce at three levels of concentration: 5, 10, and 20%. A triangle test was performed to identify the amount of oil needed to induce perceivable differences among combinations prepared at same concentration with different oils.

The triangle test results (Table 7.2) indicated that combinations of tomatoes with oil 2 are perceived to be significantly different from those with both oil 1 and oil 3 in the whole range of tested concentrations. On the other hand, significant differences between combinations with oil 1 and oil 3 are perceived only at the highest oil content (20%).

Table 7.2 Triangle test: amount of oil needed to induce perceivable differences among oil–tomato combinations prepared with two different oils at the same concentration level, percentage of correct answers. Thirty-three subjects participated in one session; nine counterbalanced triads (oil 1 vs oil 2; oil 2 vs oil 3; oil 1 vs oil 3 at concentrations of 5, 10 and 20%) were evaluated. Subjects were asked to mark the odd sample.

Oil concentration (%)	Oil 1 vs oil 2[a]	Oil 2 vs oil 3[a]	Oil 1 vs oil 3[a]
5	51*	51*	33
10	57*	57*	41
20	51*	55*	54*

[a]For values marked with *, $\alpha \leq 0.05$.

The results obtained from the triangle test clearly show that the possibility for an oil to modify significantly the perceived profile of the food with which is combined is marginally related to the intensity of sensations induced by the oil when tasted in isolation. In fact, the oil concentration needed to induce perceivable differences in combination appears to be much more related to the quality of oil sensory descriptors than to their intensity.

Once the most appropriate oil concentrations have been identified, the sensory properties of the combinations can be described. The evaluation of the dynamic profile of the oil–food combinations by means of TDS allows the observer to describe the differences induced by oils with a varied sensory style in a given combination. The information that it is possible to obtain from the analysis of TDS curves in the same example is considered (Figure 7.9). The comparison between TDS curves for tomato sauce and relevant combinations with oil 1, oil 2 and oil 3 shows that the main effect of oils, irrespective of their sensory profiles, is the suppression of the dominant sensation "canned tomato." A change of lipid level can alter the proportioning of aroma compounds between aqueous and nonaqueous phases, thus modifying the aroma profile in the headspace of combinations compared with that of the same food without oil. This physicochemical interaction induces a flavor modification. On the other hand, the dynamic profiles are very different in the three combinations, indicating that oils with varied sensory profiles modify differently the sensory properties of the food with which are combined. Taste, flavors, and irritant sensations perceived in oils are also perceived in combinations but their intensity in oils does not necessarily relate to their dominance in combinations.

In general, results from the TDS test indicate weak relationships between quality categories based on the intensity of oil sensory attributes and the oil's capacity to contribute to the sensory profile of an oil–food combination. Communicating to consumers oil sensory properties as intensity-based quality categories can induce expectations about the oil's potential capacity to complement the aroma and taste of foods that are likely to be disappointing.

7.3.3 Collection of consumer liking responses for the combinations

The last step in comparing the sensory functionality of oils with a varied sensory style when combined with a given food is the collection of consumer responses (liking) for the combinations. In general, the capacity of oils with a varied sensory style to affect significantly consumer liking for the relevant combinations should be analyzed considering the following possible cases:

1. Likings expressed for food without oil added and for combinations are not different, meaning that oil-induced sensory changes are not relevant for the acceptability of that food.

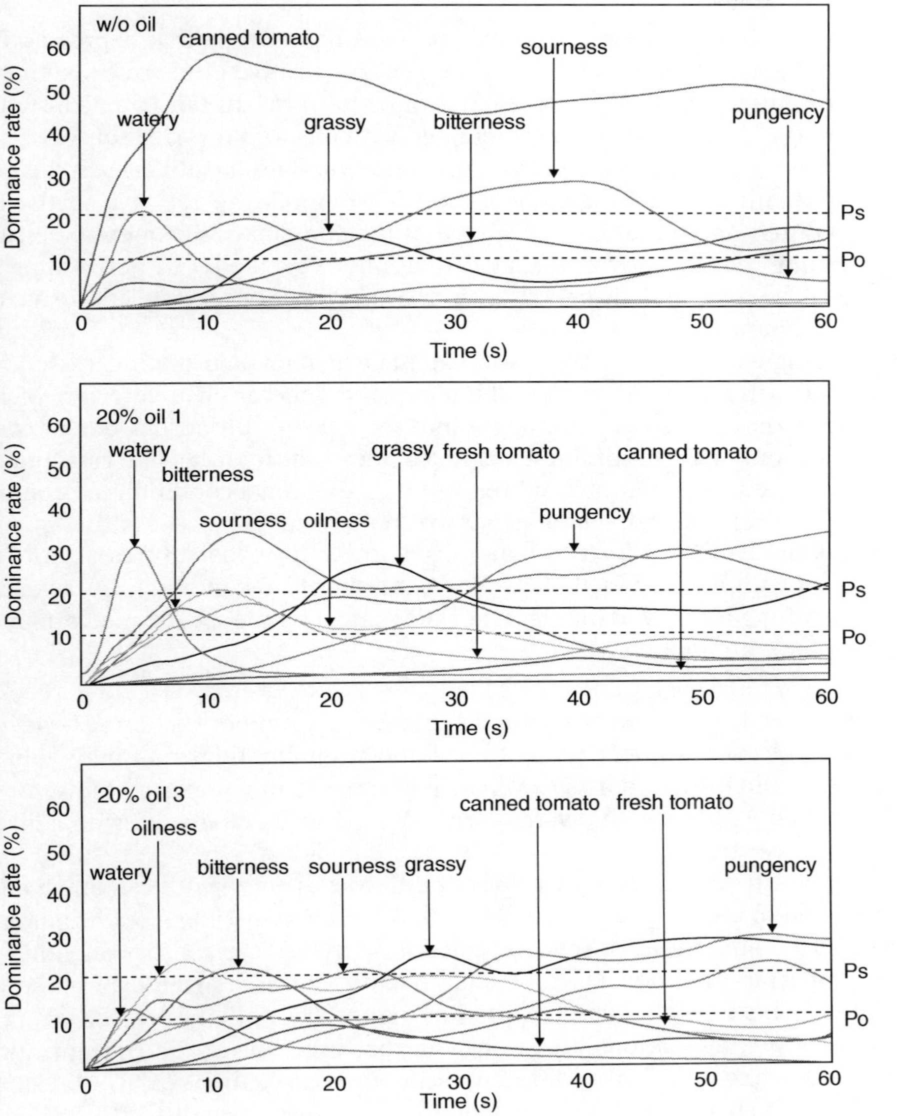

Figure 7.9 TDS curves for a tomato sauce evaluated without oil and in combination with 20% of oil 1 and oil 3. Evaluations were performed by a trained panel of 13 subjects. A list of eight attributes was used: sourness, bitterness, pungency, watery, oiliness, grassy, canned tomato, and fresh tomato. p0 represents the chance level and ps represents the 95% significance level. For a color version of this figure, see the color plates section.

2. Liking expressed for the combinations is higher than that expressed for food without oil added due to the general positive effect of fat addition or more generally to a so-called "context effect." In this case, consumer likings for oil–food pairings obtained with oils with a varied sensory style are not significantly different. This condition is not positive when the aim of creating a oil–food combination is to emphasize the role of the oil in determining the sensory appeal of the combination. An extreme case would be a defective oil (or even another vegetable oil) performing as well as an extra virgin oil with a distinctive sensory style in a given combination.
3. Liking expressed for the combinations is higher than that expressed for food without oil added due either to the general positive effect of fat addition or to some specific oil-induced sensory properties of the combinations. This condition leads to significant differences in liking among proposed combinations and represents the optimal condition to promote the sensory diversity among extra virgin olive oils.
4. Liking expressed for combinations is lower than that expressed for the food without oil added. This represents a critical condition. In fact, sensory functionality of oils has a negative effect on the sensory performance of the combination.

How to conduct consumer testing in practice is not reported here. However, an example of consumer liking for oil–food combinations that may help in understanding the importance of such a response in comparing the sensory functionality of oils with a varied sensory style when combined with a given food is presented.

It is worth emphasizing that sensory differences among oils can affect consumer liking for the relevant oil–food combinations depending on the amount of oil used and on the specific interactions between the sensory and chemical properties of the oil and food components. Consider again the case study of oils 1, 2, and 3 already presented in Figure 7.8. A liking test was run on oil–tomato combinations prepared with the three oils at the concentration levels inducing perceivable differences among combinations (5, 10, and 20%) (Table 7.3). It can be noted that there is a general positive effect of oils,

Table 7.3 Mean liking ratings for oil–tomato combinations prepared with three different oils (oil 1, oil 2, and oil 3) at three concentration levels. Sixty-one regular consumers of extra virgin olive oil, recruited in the Florence area, participated in one evaluation session. Intensity scale: 1 = dislike extremely; 5 = neither like or dislike; 9 = like extremely.

Oil concentration (%)	Oil 1[a]	Oil 2[a]	Oil 3[a]
5	5.59	5.80	5.93
10	6.08[a]	5.43[b]	5.98[a]
20	5.64[b]	5.02[c]	6.30[a]

[a]Different superscript letters indicate significantly different values ($p \leq 0.05$).

independently of their specific sensory properties, on the consumer liking scores for the tomato sauce. It appears that the effect of oil on liking for a given combination is not necessarily related to the intensity of attributes describing its sensory profile. In fact, the same strong effect on liking for tomato sauce is induced by both oil 1, characterized by intense flavor and mouthfeel, and oil 3, characterized instead by weak–moderate intensity attributes.

Factors relating to the general effects of adding oil rather than specific modifications of the sensory profile are likely to account for the increased liking for combinations with respect to that of tomato without oil observed at low oil concentration. In fact, liking for combinations containing 5% oil is significantly higher than that for tomato sauce and there are no significant differences in liking among the three combinations. Hence the perceivable differences induced by 5% of oil 2 with respect to both oils 1 and 3, as indicated by the triangle test, are irrelevant in terms of consumer linking for the oil–tomato combinations. Significant liking differences among combinations prepared with the three different oils were found only at the highest concentration tested. The combination prepared with 20% of oil 3 is the most liked, meaning that the perceivable differences induced by this oil with respect to the other two specifically modify sensory properties positively related to consumer preference. On the other hand, at this concentration the combination prepared with oil 2 is the least liked. In this use condition, oil 2 does not exert any functionality in terms of consumer responses since no significant differences are observed on comparing liking for the combination and liking for tomato without oil.

Consumer responses can be relevant to identifying appropriate oil–food pairings. This is particularly important for new olive oil markets where the objective is to find the most appropriate way to expose consumers to the sensory diversity of extra virgin olive oil. For example, oil 3 shows the best expression of its sensory potentiality in combination with tomatoes whereas the use of oil 1 appears risky since high concentrations might induce a significant lowering of liking for combinations. On the other hand, pairing of oil 2 with tomato products does not seem appropriate.

7.4 Conclusion

An awareness of the complexity of perception is crucial for the sensory characterization of flavorsome ingredients such as extra virgin olive oils. Assessing oil sensory functionality could be an appropriate approach for evaluating its sensory performance and for a better understanding of oil appreciation by consumers in actual use conditions. Comparing the sensory functionality of extra virgin olive oils with a varied sensory style combined with a given food requires several sensory evaluations: (1) the description of the sensory profile of oils; (2) the evaluation of the sensory properties of the oil–food combinations; (3) the collection of consumer liking responses for the combinations.

TDS appears to be an appropriate method for describing the sensory properties of complex stimuli. The complex interactions occurring between oil and food components represent one of the most important factors responsible for the perceived profiles of combinations and can be more clearly described by TDS than by other analytical descriptive methods.

Acknowledgments

The author is grateful to Annamaria Recchia and Camilla Masi for their technical assistance in the sensory tests. The author is also grateful to Maria Piochi who carried out the sensory tests on oil–tomato combinations during the preparation of her master's degree thesis.

References

Aaslyng, M.D., and Frost, M.B. (2008) The effect of basic taste solutions and selected vegetable accompaniments on the sensory properties of pork. *Journal of Sensory Studies*, **23**, 720–742.

Aaslyng, M.D., and Frost, M.B. (2010) The effect of the combination of salty, bitter and sour accompaniment on the flavor and juiciness of pork patties. *Journal of Sensory Studies*, **25**, 536–548.

Blake, A.A. (2004) Flavour perception and the learning of food preference, in *Flavour Perception* (eds A.J. Taylor and D.D. Roberts), Blackwell, Oxford, pp. 172–202.

Bongartz, A., and Oberg, D.G. (2011) Sensory evaluation of extra virgin olive oil (EVOO) extended to include the quality factor "harmony." *Journal of Agricultural Science and Technology A*, **1**, 422–435.

Caporale, G., Policastro, S., and Monteleone, E. (2004) Bitterness enhancement induced by cut grass odorant (*cis*-3-hexen-1-ol) in a model olive oil. *Food Quality and Preference*, **15**, 219–227.

Caporale, G., Policastro, S., Carlucci, A., and Monteleone, E. (2006) Consumers expectations for sensory properties in virgin olive oils. *Food Quality and Preference*, **17**, 116–125.

Cerretani, L., Biasini, G., Bonoli-Carbognin, M., and Bendini, A. (2007) Harmony of virgin olive oil and food pairing: a methodological proposal. *Journal of Sensory Studies*, **22**, 403–416.

Déléris, I., Saint-Eve, A., Dakowski, F., Sémon, E., Le Quéré, J.L., Guillemin, H., and Souchon, I. (2011) The dynamics of aroma release during consumption of candies of different structures, and relationship with temporal perception. *Food Chemistry*, **127**, 1615–1624.

Delgado, C., and Guinard, J. (2011) How do consumer hedonic ratings for extra virgin olive oil relate to quality ratings by experts and descriptive analysis ratings? *Food Quality and Preference*, **22**, 213–225.

Dinnella, C., Masi, C., Zoboli, G., and Monteleone, E. (2012a) Sensory functionality of extra virgin olive oil in vegetable foods assessed by Temporal Dominance of Sensations and Descriptive Analysis. *Food Quality and Preference*, **26**, 141–150.

Dinnella, C., Migliorini, M., Piochi, M., Recchia, A., Gianni, G., and Monteleone, E. (2012b) Sensory functionality of extra virgin olive oil, presented at the 10th European Federation Lipid Congress, Kracow.

Dinnella, C., Masi, C., Næs, T., and Monteleone, E. (2013) A new approach in TDS data analysis: a case study on sweetened coffee. *Food Quality and Preference*, **30**, 33–46.

Gawel, R., and Rogers, D.A.G. (2009) The relationship between total phenol concentration and the perceived style of extra virgin olive oil. *Grasas y Aceites*, **60**, 134–138.

Keast, R.S.J., and Breslin, P.A.S. (2002) An overview of binary taste–taste interactions. *Food Quality and Preference*, **14**, 111–124.

Keast, R.S.J., and Roper, J. (2007) A complex relationship among chemical concentration, detection threshold, and suprathreshold intensity of bitter compounds. *Chemical Senses*, **32**, 245–253.

Krystallis, A., and Ness, M. (2003) Motivational and cognitive structures of Greek consumers in the purchase of quality food products. *Journal of International Consumer Marketing*, **16** (2), 7–36.

Labbe, D., Schlich, P., Pineau, N., Gilbert, F., and Martin, N. (2009) Temporal dominance of sensations and sensory profiling: a comparative study. *Food Quality and Preference*, **20**, 216–221.

Lawless, H.T. (2000) Sensory combinations in meals, in *Dimensions of the Meal – The Science, Culture, Business and Art of Eating* (ed. H.L. Meiselman), Aspen Publishers, Gaithersburg, MD, pp. 92–106.

Lawless, H.T., and Heymann, H. (2003) *Sensory Evaluation of Foods: Principles and Practices*, Springer, New York.

Lenfant, F., Loret, C., Pineau, N., Hartmann, C., and Martin, N. (2009) Perception of oral food breakdown. The concept of sensory trajectory. *Appetite*, **52**, 659–667.

Meillon, S., Viala, D., Urbano, C., Guillot, G., and Schlich, P. (2010) Impact of partial alcohol reduction in Syrah wine on perceived complexity and temporality of sensations and link with preference. *Food Quality and Preference*, **21**, 732–740.

Meyners, M. (2011) Panel and panelist agreement for product comparisons in studies of Temporal Dominance of Sensations. *Food Quality and Preference*, **22**, 365–370.

Meyners, M., and Pineau, N. (2010) Statistical inference for temporal dominance of sensations data using randomization tests. *Food Quality and Preference*, **21**, 805–814.

Moskowitz, H.R., and Krieger, B. (1995) The contribution of sensory liking to overall liking: an analysis of six food categories. *Food Quality and Preference*, **6**, 83–90.

Næs, T., Brockhoff, B., and Tomic, O. (2010) Quality control of sensory profile data, in *Statistics for Sensory and Consumer Science*, John Wiley and Sons, Ltd, Chichester, Chapter 3.

Nestrud, M.A., Ennis, J.M., and Lawless, H.T. (2012) A group level validation of the supercombinatorality properties: finding high quality ingredient combinations using pairwise information. *Food Quality and Preference*, **25**, 23–28.

Pfeiffer, J.C., Hort, J., Hollowood, T.A., and Taylor, A.J. (2006) Taste-aroma interactions in a ternary system: a model of fruitness perception in sucrose/acid solutions. *Perception and Psychophysics*, **68**, 216–227.

Piggott, R.J. (2000) Dynamism in flavour science and sensory methodology. *Food Research International*, **33**, 191–197.

Pineau, N., Schlich, P., Issanchou, S., Imbert, A., Cordelle, S., Mathonnière, C., Rogeaux, M., Etiévant, P, and Köster, E. (2009) Temporal Dominance of Sensations: construction of the TDS curves and comparison with time-intensity. *Food Quality and Preference*, **20**, 450–455.

Pineau, N., Neville, T., and Lepage, M. (2011) Panel performance tool for Temporal Dominance of Sensations studies, presented at the 9th Pangborn Sensory Science Symposium, Toronto.

Pineau, N., Goupil de Bouillé, A., Lepage, M., Lenfent, F., Schlich, P., Martin, N., and Rytz, A. (2012) Temporal Dominance of Sensations: what is a good attribute list? *Food Quality and Preference*, **26**, 159–165.

Piochi, M., Dinnella, C., Recchia, A., and Monteleone, E. (2012) La funzionalità sensoriale come strumento di valorizzazione delle proprietà sensoriali di oli d'oliva extravergini di pregio, presented at the 4th National Congress of the Italian Society of Sensory Science, Trieste, 22–23 November 2012.

Robertson, L., and Sagiv, N. (eds) (2005) *Synaesthesia: Perspectives from Cognitive Neuroscience*, Oxford University Press, New York.

Saint-Eve, A., Déléris, I., Panouillé, M., Dakowski, F., Cordelle, S., Schlisch, P., and Souchon, I. (2011) How texture influence aroma and taste perception over time in candies. *Chemosensory Perception*, **4**, 32–41.

Servili, M., Selvaggini, R., Esposto, S., Taticchi, A., Montedoro, G., and Morozzi, G. (2004) Health and sensory properties of virgin olive oil hydrophilic phenols: agronomic and technological aspects of production that affect their occurrence in the oil. *Journal of Chromatography A*, **1054**, 113–127.

Sinesio, F., Moneta, E., and Esti, M. (2005) The dynamic sensory evaluation of bitterness and pungency in virgin olive oil. *Food Quality and Preference*, **16**, 557–564.

Stevenson, R.J., Prescott, J., and Boakes, R.A. (1999) Confusing tastes and smell: how odours can influence the perception of sweet and sour tastes. *Chemical Senses*, **24**, 627–635.

Sudre, J., Pineau, N., Loret, C., and Martin, N. (2012) Comparison of methods to monitor liking of food during consumption. *Food Quality and Preference*, **24**, 179–189.

8 Investigating the Culinary Use of Olive Oils

Sara Spinelli
SemioSensory Research & Consulting, Carmignano – Prato, Italy

8.1 Introduction

8.1.1 Extra virgin olive oils marketing and communication: current situation and new perspectives

The promotion and sale of extra virgin olive oils in the world market is not easy. One of the main reasons for this can be identified in the extreme variety and variability among extra virgin olive oils in terms of quality: communicating to consumers the added value of an extra virgin oil of "excellence" compared with other extra virgin olive oils that can be sold at a lower price is very difficult.

Communication can be a real "key" to consumers in the market of excellence, in order to differentiate among extra virgin olive oils and to make consumers aware of the value of a specific product. However, the current ways to communicate are unable to differentiate among products and, as a result, consumers often do not perceive why they should prefer a particular extra virgin olive oil instead of another one.

At present, promotion of oils is mainly focused on indications about origins, especially in Europe: the message communicated is that oils coming from a specific country/region are better just because they come from that specific country/region. For these reasons, typicality and tradition have become two main values in the communication of oil's quality, in addition to the proliferation of Protected Designation of Origin (PDO) and similar certifications. However correct this traditional view can appear, there are strong

Olive Oil Sensory Science, First Edition. Edited by Erminio Monteleone and Susan Langstaff.

reasons to believe that such a communication has the final consequence of creating in the mind of consumers differentiated groups of products distinguished by country (Italian, Spanish, Californian, etc.) or regions (Tuscan, Sicilian, Andalusian, Cretan, etc.), but as a result every oil in the same group is perceived as equal, which makes it impossible to differentiate effectively among products.

Another mainstream tendency when communicating about olive oil is the focus on the sensory identity area, with the aim of promoting oils for their sensory specificities: "robust," "fragrant," "delicate," and "gentle" oils are proposed to consumers as general typologies that should help them to perceive the differences between oils, choosing that one best corresponding to their expectations.

If we examine the question more carefully, it will become clear that sensory information is frequently given in a confusing way because commercial typologies are often ambiguous and contradictory. On the other hand, even when sensory information is correctly given, it can be very difficult for consumers to understand what olive oil producers would hope to communicate through typologies, labels, and various publishing contents that can be downloaded from brand's websites. This happens because brands usually resort to the experts' descriptive language for their communication, but however correct, this language can be difficult or meaningless for consumers without any previous knowledge of olive oil. In this kind of communication, consumers are invited to play the role of experts, namely of skilled tasters who have the pleasure of experimenting with different sensory properties. But for consumers, getting to know which are the main sensory characteristics of a specific product (in terms of intensity of bitterness, pungency and fruitiness) may not be so important.

What generally is not considered is that oil is hardly ever consumed on its own (only in a tasting context!), in contrast to wine and cheese. Historically, olive oil has always been considered a lubricating matter, used to cook and fry as an alternative to butter or lard. Moreover, it is used as a condiment, able to modify substantially a dish's flavor, and so it could be considered as a real ingredient in culinary preparations. For this reason, it should be communicated to consumers what they can do with these different oils, suggesting and explaining the reasons why it is important to choose one oil rather than another one for its specific sensory functionality in culinary preparations.

Paradoxically, when cuisine is considered in oil communication, generally it is not to differentiate and point out diversity and variety, but to create homogenization among oils. In fact, in most cases when culinary uses are suggested in oil communication, it is to promote culinary uses of oils in general; websites of oil brands are full of recipes suggesting the use of oil in creative cooking thanks to the fantasy of chefs, but the reasons for using the specific sponsored oil rather than another one are usually not mentioned.

The case of suggesting uses in regional cooking could be even more counterproductive, especially in campaigns directed at markets not familiar with olive oils (and with a cuisine not focused on that product). Communication is focused on typicality and tradition and the message could be summarized as "each oil goes well (and best) with the regional dishes of the territory to which it belongs": so Tuscan oils are perfect in Tuscan cuisine, but also Italian oils in Italian cuisine, Greek in Greek cuisine, and so on. This communication cannot be successful for the olive oil world market: it should be clear that selling traditional cuisine to sell an oil means exporting a cultural model of cooking, and this is very hard – and perhaps futile – to do.

So, what all this suggests, and this is the thesis of this chapter, is that olive oil communication and marketing should focus on differentiating products, not only in terms of sensory characteristics but especially in terms of sensory "performance" and "sensory functionality," that is to say, in relation to the sensory effects that can be produced in culinary uses; in fact, it can be assumed that oils with very different sensory characteristics can have very different effects on the flavor of culinary preparations.

Descriptive analysis allows the study of the sensory identity of oils, developing a "sensory profile," which corresponds to a "list" of sensory characteristics evaluated by a trained panel according to their intensities (Lawless and Heymann, 1998; SISS, 2012). But what does the diversity identified by effective use of sensory analysis become when oils are used in their most common context, in cooking? As oil is not consumed on its own, new studies are required to investigate the role that oils with different sensory profiles could have in culinary preparations, showing which oil "sensory style" best highlights which food flavor [The expression "sensory style" indicates the sensory profile that describes an oil (or a group of oils) as different from others (see Chapter 5, and also Section 8.3.1)]. This information about culinary uses indicated for specific oil sensory styles could be successfully used for marketing purposes, thus allowing for the differentiation of products in terms of effective use that consumers can experience directly by preparing food or tasting it at a restaurant. Such information is very important in promoting olive oils in markets which are familiar with this product, proposing a new view on sensory diversity among olive oils; however, it is also relevant in markets not used to olive oil, to promote the contribution that different olive oils could give, for example, in Asian and other world cuisines.

This chapter aims at presenting methodological approaches in investigating the culinary use of oils by chefs, in order to gather information that could be used effectively for marketing purposes. It is divided into four parts. In the first part, sensory interactions effects are discussed (Section 8.2.1) with a focus on harmony, an important concept in food experience. In the second part, methodological approaches to food pairing studies are presented (Section 8.2.2), while the third part is focused specifically on presenting a progress report of oil–food pairing studies (Section 8.2.3). Finally, in the fourth part, a new interdisciplinary approach is presented, focused on investigating the role that oils characterized by different sensory styles can have in culinary preparations (Section 8.3).

8.2 Methodological approaches in the study of oil–food pairing

8.2.1 Interplay of ingredient combinations and the "harmony" effect

The interplay of ingredient combinations and food pairing in culinary art has always fascinated mankind, which has searched for new combinations in order to keep people interested and well nourished. The way in which ingredients can be combined and manipulated varies greatly and is potentially infinite. The effects produced on food experience can also vary. Harmony and contrast seem to be the most important features (Lawless, 2000).

Defining harmony in food experience is very difficult, because there is no overall definition in scientific literature. This word is often used with different meanings also in culinary texts. Harmony is commonly defined as "the pleasant effect made by parts being combined into a whole" (Longman, 1987), but if pleasantness is a central aspect of harmony, little is known on how it can be systematically produced in food experience.

Lawless (2000) developed the theme of "harmony with contrast" to describe what humans search for in food. According to Lawless, humans are torn between opposite tendencies, *neophilia* (interest in what is new) and *neophobia* (fear or aversion to what is new and different). For this reason, harmony in food experience is not limited to sensory balance, but has to confront the desire for sensory contrast: "Sensory contrast versus sensory balance is a real challenge for the culinary artist, to simultaneously maintain a degree of complexity while harmonizing the components parts. … The goal of achieving simultaneous harmony with contrast arises from our need to have some relief from sameness" (Lawless, 2000, pp. 92–93).

It can be stated that harmony in food experience is a pleasant effect made by ingredients and products being combined into a whole, giving rise to controlled sensory contrasts. In fact, harmony seems to be interested in the balancing of sensory contrasts, more than in avoiding them.

Sensory contrasts play an important role in the sense of pleasure, as Witherly (1987) and Hyde and Witherly (1993) pointed out when discussing texture contrasts. In fact, studies on food pleasure demonstrate that texture contrasts (e.g., firm/crisp versus soft/creamy) in a meal or on a plate "enhance the eating experience and reflect upon the excellence of food preparation" (Witherly, 1987, p. 407; see also Szczesniak and Kahn, 1984). Contrasts in sensory experience help to maintain our attention and interest in food, avoiding boredom and reducing the chance of adaptation to a specific sensory property (Hyde and Witherly, 1993; Lawless, 2000). The success of molecular gastronomy can be cited as example: part of its success is due to the interplay of contrasts among different textures and, above all, between the expected texture and the real one.

According to Lawless (1986, 2000), three principles regulate sensory perception of food mixtures: first, the *mixture suppression* principle states that individual taste and smell characteristics are less intense in a blend than if the same ingredient is tasted alone; second, the *adaptation* principle affirms that the senses become less responsive when subjected to constant stimulation; and third, the *release from suppression* principle states that once adapted to one component of a mixture, the other components are less suppressed, and so they increase in intensity or become more apparent. Even if scientific evidence points to more independence of taste and smell than interaction in a simple mixture of flavor chemicals, different results emerge with real products. This may depend on our learned pairings of certain smells and tastes: when we frequently experience two flavors together, they become associated and tend to support each other's perception (Stevenson, Prescott, and Boakes, 1995). A culinary preparation is an exemplary case of mixture interaction, devoted to the production of pleasing effects. Studying sensory interactions in a plate of food means studying the perceived sensory properties, but also the overall effects, in terms of harmony, agreement and disagreement, balance, and contrast, which contribute to pleasure and acceptability. As Lawless (2000, p. 96) pointed out, "the entire enterprise of food preparation can be seen as an attempt to make harmonious mixtures."

Extra virgin olive oils characterized by different sensory styles may play a major role in this match, effectively contributing to the production of harmony, for example, re-equilibrating the flavor of the dish, or accentuating an interesting struggle between different sensory properties. An attempt to map a different way in which two sensory properties (A and B, one of which characterizes an olive oil and the other an ingredient) can interact may be as set out in Table 8.1. An oil can act as enhancer in at least four ways:

- An oil characterized by A (e.g., sweetness) can be added to a preparation characterized by A (sweetness), with the result of the enhancement of A (the overall sweetness of the food); in this case, oil is used to intensify a sensory property already present in the culinary preparation not added by the oil, but just "enhanced" by it.
- An oil characterized by A can be added to a culinary preparation characterized by B, contributing to partially masking B (A is perceived in the culinary preparation experience). This is an effect of suppression, similar to what happens when the sweetness of a spoonful of sugar helps a bitter medicine to go down (the sweetness partially masks the bitterness of the medicine). In this way, oil can add a new sensory property to a culinary preparation.

Table 8.1 Potential effects of oils characterized by different sensory styles on culinary preparations.

Combinations in oil–food interactions	Effects	Sensory property most perceived	Possible effect on culinary preparation's experience
A ↔ A	Reinforcing of A	A	Harmonic/disharmonic
A ↔ B	Suppression of B	A (B partially or totally masked)	Harmonic/disharmonic
A ↔ B	Enhancing of B within the contrast	A (–) and B (+)	Harmonic (based on a contrast)/disharmonic
A ↔ B	Counterbalancing of B	A and B	Harmonic (based on a contrast)/disharmonic

- An oil can contribute to the production of harmony for contrast, by adding a new sensory property to a culinary preparation. In this case, an oil characterized by A can be added to a culinary preparation characterized by B, contributing to enhancing B in the contrast (A and B are both perceived in the culinary preparation but B is "enhanced" in the contrast). A and B do not just add up, but are able to produce harmonic (or disharmonic) effects.
- An oil can contribute to the production of harmony for contrast, counterbalancing an already present sensory property. In this case, a new sensory property is added to a culinary preparation, not to enhance but to equilibrate an already present one that can appear too strong.

Obviously, the above-mentioned contrasts can have varying intensities producing various effects; they can be harmonious, balanced in a way that enhances the value of both sensory properties, or not balanced and disproportionate, contributing negatively to the food experience.

From this discussion, it is clear that there is no single kind of harmony in food preparation: harmony can be characterized by different degrees of contrast, be it soft or strong and resolute, static or dynamic (that is, in continuous change). In fact, it can depend on "dynamic contrasts," namely the "moment-to-moment sensory contrast from the everchanging properties of foods manipulated in the mouth" (Hyde and Witherly, 1993, p. 1); according to the authors, dynamic contrasts contribute significantly to the palatability of food and beverages, and this is the main reason which explains why people like certain kind of foods (Witherly, 2007).

An application of this model can be easily understood if we consider basic taste interactions (sweetness and bitterness): bitterness is in fact one of the properties that commonly characterize oils. Explaining how an oil characterized by a strong or lower degree of bitterness contributes in a different way to produce harmony is extremely important. Moreover, it is important not to limit the view to taste properties, but to extend it to flavor: the degree of presence (or absence) of bitterness has to be considered in specific sensory styles, that is, combined with a specific aroma by mouth (e.g., artichoke, tomato, grass).

It is apparent that further research is needed to explore systematically the world of sensory interactions and the roles that oils can have in producing harmony.

8.2.2 Methodological approaches in the study of food pairing

Over the past 15 years, the method of combining ingredients to obtain interesting results in terms of sensory food experience has attracted much interest. Chefs and the culinary press are paying more and more attention to proposing heuristic approaches, sometimes with scientific pretensions. "Foodpairing," for example, is a technique that currently is all the rage amongst top chefs aimed at identifying which foods go well together. In the so-called "foodpairing hypothesis," ingredients sharing flavor components (determined through gas chromatography coupled with mass spectrometry, GC–MS) are more likely to taste good together than ingredients that do not. The homonym website (www.foodpairing.com) presents a database containing more than 1000 ingredients, each with a specific "tree" that is a result of an algorithm elaborated on the basis of the degree of shared components. Some monocultivar olive oils and products characterized by origin (Italian and Spanish) are also on the list, presented in combination on the basis of their GC–MS analysis. This technique, elaborated by a chef and a flavorist at Firmenich (Blumenthal, 2008), has been strongly criticized (de Klepper, 2011). First, the goodness of a combination can depend on a multitude of ingredient characteristics in addition to their flavor profile (e.g., for reasons not concerning flavor but visual aspects, texture, etc.); moreover, the reasons for combining ingredients in terms of sensory perception are completely cut off from such an approach. What does it mean that ingredients go well together? Which effects are produced in sensory food experience? Recently, Ahn *et al.* (2011) proposed a network-based approach to explore the impact of flavor compounds on ingredient combinations, identifying a series of statistically significant patterns that characterize the way in which humans choose ingredients that they combine in their food. According to the authors, only the Western cuisines show a tendency to use ingredient pairs that share many flavor compounds, supporting in a way the so-called "foodpairing" hypothesis. By contrast, East Asian cuisines tend to avoid compound sharing ingredients. The study suggests the need for further investigations that would take into account motives to combine food not limited just to shared flavor compounds.

Until now, no definite rules have been set for food pairing and the positive effect of pairing is still largely due to chefs' creativeness. In fact, scientific studies on sensory interaction effects in culinary preparations are not numerous. Several studies concern food and wine pairing, focusing on studying the reciprocal influence between the two products and on finding the best matches (King and Cliff, 2005; Bastian *et al.*, 2009; Bastian, Collins, and Johnson, 2010). Studies on wine–food pairing use descriptive analysis, suggesting that prior consumption of food before tasting a wine or vice versa modifies the intensity of several sensory descriptors within the beverage or foodstuff (wine–cheese parings, Madrigal-Galan and Heymann, 2006; Nygren *et al.*, 2002, 2003a, 2003b, Bastian, Collins, and Johnson, 2010; wine–Hollandaise sauces, Nygren *et al.*, 2001).

Only a few studies have been published on the sensory effects of the so-called "food–food pairings." Aaslyng and Frøst (2008, 2010) investigated the effects of two different vegetables (characterized by the actual basic taste) and of four of the five basic tastes on the sensory properties of pork, by means of descriptive analysis. Paulsen *et al.* (2012) studied sensory interactions in 17 different combinations of salmon and culinary sauces by using descriptive analysis within an interdisciplinary approach, combining experimental design and sensory analysis with culinary knowledge. In fact, the aim of the study was to investigate the effects of culinary sauces (characterized by different levels and combinations of the basic tastes) on the sensory attributes of salmon, in order to provide useful information for the process of developing salmon meals with tailored sensory properties. Whereas salmon was always prepared in the same manner, different sauces characterized by different levels and combinations of the basic tastes were created by using the creative design technique (Næs and Nyvold, 2004) (see Step 3 in Section 8.3.1 for a discussion of this approach). Different salmon–sauce combinations were evaluated using descriptive analysis, showing that sauces could enhance or reduce specific sensory attributes. According to the authors, this interdisciplinary approach can be used for marketing purposes to propose specific salmon-based culinary preparations to consumers characterized by different preferences.

8.2.3 Olive oil in food pairing studies

A large amount of information about appropriate oil and food pairing may be found in the popular press, on websites, and in culinary magazines, provided by food writers, chefs, and culinary experts. On the other hand, in the peer-reviewed literature only a very limited number of studies examining oil–food pairing under scientifically controlled conditions have been reported (Cerretani *et al.*, 2007, Dinnella *et al.*, 2012). In these studies, sensory methods were used to investigate the sensory functionality of olive oils paired with specific food represented by basic alimentary matrices (shrimps in the former case and tomato sauce and bean purée in the latter). Cerretani *et al.* (2007) employed sensory analysis, whereas Dinnella *et al.* (2012) used sensory analysis combined with a new methodology, Temporal Dominance of Sensations

(TDS), that is able to discriminate sensory characteristics perceived at different times in oil–food combinations (see Chapter 7 for a detailed presentation of this approach).

The aim of the work of Cerretani *et al.* (2007) was to develop a method, based on sensory descriptive analysis, to define the ideal profile of the perfectly harmonic olive oil specific for each food, thanks to the calculation of a so-called "disharmonic index."

The study was divided into three steps:

1. describing food and olive oil separately with pairing in mind;
2. evaluating the harmony of the pairing;
3. investigating the correspondence between sensory attributes of olive oil and food.

Pairing harmony was evaluated through seven attributes (fruity taste, pungent, bitter, sweet, green, ripe, fruity smell). According to the authors, a perfect harmonic pairing is a value of 5 on a 10-point scale (from too light to too intense); in contrast, values lower than 5 indicate that an olive oil is increasingly too "light" compared with the food, whereas values higher than 5 indicate that an olive oil is too intense. Examples of nonharmonic pairing could be discrepancies between fruity and pungency intensities of olive oil and saltiness, spiciness, and aroma intensities of food. In other words, in nonharmonic pairings these oil properties are too weak or too strong compared with the food. The disharmonic index (I_d) was calculated as the sum of the absolute distances of the median panel values for the seven harmony attributes from 5. A weight of 70% was attributed to the gustatory harmony and of 30% to the olfactory and green/ripe sensations.

According to Cerretani *et al.* (2007), olive oil can be used to reinforce the dominant taste of food (spicy, sweet, or bitter), and only in one case to introduce something new in the preparation (sweet, in the case of acid preparation). In fact, the authors explained that for a harmonic pairing of olive oil and food, the intensities of the fruitiness and pungency of olive oil have to balance the intensity of saltiness, spiciness, and aroma quality of the food; they also pointed out that green and bitter intensities in the oil must balance the bitter intensity of the food, and the same for the sweetness (the sweet and ripe intensities of the olive oil should balance the sweetness and fatness intensities of the food). These pairing are "in concordance", that is, that in order to obtain harmonic pairing, the sweeter the food, the sweeter the oil should be (corresponding to the first case presented in Table 8.1). Only one case of pairing "in discordance" is presented: when a food is sour, the oil should be sweet to balance the sourness. On the other hand, as Lawless (2000) pointed out, balance in contrast is a very common kind of harmony in culinary preparations. For example, an oil characterized by a bitter taste could be used in a sweet preparation to create a harmonic contrast.

In the case study presented by the authors (mantis shrimps cooked in three different ways), two oils were used, selected for their sensory overall diversity: a strong olive oil (characterized by bitterness) and a mild one (characterized by sweetness). However, neither of the oils was characterized by a specific flavor note (only fruitiness was evaluated in the pairing). But what happens if olive oils characterized by very different flavors are chosen? Could oil be used to modify the flavor of the food? In fact, oils can be very different in terms of sensory characteristics, not only according to bitterness, pungency, and fruity notes. What is lacking in such a representation is, for example, the specificity of flavor notes, which can vary widely in different oils.

It seems clear that further research is needed on sensory oil–food interactions, in addition to a clear and shared definition of "harmony" in food science, including every potential declination of this concept, from contrast to balance (see Section 8.2.1).

8.3 An original approach to studying the sensory functionality of oils in culinary preparations

Olive oils can play a relevant role, contributing to creating harmony and to achieving this "struggle" between sensory contrast and sensory balance, which is extremely important for the success of a culinary preparation. Different olive oils can be used to balance the taste and the flavor of the food or to create contrast thanks to their strong characters. Analytical descriptive studies can be very useful to test and investigate in detail oil sensory functionality in specific oil–food combinations where food is represented by basic alimentary matrices. Nevertheless, any descriptive study can in itself solve the problem regarding the selection of combinations to be further investigated, to point out which oil sensory styles best highlight which food flavor, or, vice versa, which oil sensory styles can be best highlighted in a food combination. Oil–food pairings have to be investigated in cooking, also considering culinary preparations and not only basic alimentary matrices. Studies on culinary uses of oils have to consider different kinds of preparation, treatments of ingredients, and cooking styles. Moreover, further studies are needed on sensory aspects and global effects that can be produced in terms of different kinds of harmony by experiencing a culinary preparation. Descriptive sensory approaches cannot solve this problem on their own, but they need to be included in an interdisciplinary study aimed at researching culinary uses of oils in a socio-cultural perspective.

In food pairing, there are no definite rules to refer to, but in general chefs are competent in the creation of food combinations. Their knowhow is often intuitive, corresponding to what is called "creativity," jointly with field experience and systematic processed (Vetter, 2009). For this reason, it is important to develop new approaches which include chefs' creativity in the experimental design. Only in this way is it possible to consider the world of oil–food combinations with all its richness and potentialities.

Such studies, interdisciplinary and based on qualitative methods, can be developed with two interests:

1. They are needed to explore the potential culinary uses of different oils, involving chefs' creativity in this process. These studies can help identify the best use (in terms of sensory effects) of each product in different food cultures (and culinary styles): this information can be exploited in oil marketing and communication to promote an oil, suggesting its best use, and not forgetting that this product will never be consumed on its own.
2. Results from qualitative research could also be very useful in planning the experimental design of quantitative research, focused on specific features of the sensory functionality of oils and on specific oil–food combinations. An interdisciplinary approach aimed at investigating culinary uses of oils is preliminary to studies on the sensory functionality of oils, which need previously to select food and oils to study in depth. By studying chefs' culinary uses of oils, it is possible to obtain suggestions for original combinations of interest from a sensory point of view. This allows for the planning of an experimental design with the contribution of culinary knowledge, avoiding the selection of pairings in an intuitive or not best-defined way.

Including chefs in this approach is important, because it allows the study of food pairing and combinations taking into account an overall perspective including sensory effects (in terms of perceived sensory properties, contrasts, balances, and harmony), but also pleasure, outcome, and creativity (concerned about the desire for novelty versus reassurance of sameness; cf. Lawless, 2000).

For these reasons, an approach aimed at describing the sensory role that can be played by different oils in culinary preparation needs to be multidisciplinary. First, a sensory analysis competence is needed to select sensory differences among oils and to describe the sensory characteristics of culinary preparations; second, a psycho/socio/semiotic competence is needed to study chefs' creative processes.

An original approach aimed at studying the relationship between sensory styles and culinary uses of oils is proposed. Such an approach aims at pointing out:

- how to select oils in terms of different sensory styles;
- how to study the way in which oils characterized by different sensory styles can be used in cooking to modify the flavor of a dish.

8.3.1 Design overview

In Table 8.2, an overview of the field of studies on the culinary uses of oils is presented. The need for further studies focused on exploring and

Table 8.2 Investigating culinary uses of oils – an overview.

I.	Developing and exploring new combinations between food and olive oils characterized by specific sensory styles
II.	Studying the effects of different olive oils on the sensory experience of selected culinary preparations, in terms of sensory and hedonic perceptions (acceptability, pleasure, and emotions)
III.	Developing marketing strategies based on the sensory performance specificity of different oils' sensory styles

developing new combinations between food and oils characterized by specific sensory styles is apparent. As stated above, an important feature that needs further in-depth study is the role played by oils in creating harmony in culinary preparations, and also acceptability, pleasure, and emotions. This information can be used to develop specific marketing strategies based on the sensory performance specificity of the different oils' sensory styles. A scientific approach aimed at studying the role that can be played by oils characterized by different sensory styles in culinary preparations is presented here (points I and II in Table 8.2); point III is only briefly mentioned.

In Table 8.3, the proposed approach (points I and II) is presented in detail. The first step is represented by the selection of oils with different sensory styles (1), the second consists in the selection of chefs characterized by different culinary styles (2), the third consists in interviewing chefs on oils and their culinary uses (3), the fourth is represented by the preparation of dishes (4), the fifth consists in the definition of the main sensory descriptors of each prepared dish (5), and the sixth consists in an integrative semiotic analysis on the complex bulk of data obtained (6).

In Table 8.4, suggested methods for data collection and data analysis are presented and summarized for each step, as outlined below.

Step 1. Selection of oils with different sensory styles The first and determinant step is the selection of oils with varied sensory styles to be presented to chefs. It is extremely important to select products carefully. When the interest is to compare similar products on the market, products that span the entire variation in the study must be included. This can be done by

Table 8.3 The approach step by step.

Step 1	Selection of oils with different sensory styles
Step 2	Selection of chefs characterized by different culinary styles
Step 3	Chef interviews on oils and their culinary uses
Step 4	Preparation of dishes
Step 5	Definition of the main sensory descriptors of each prepared dish Investigations of emotional experience of the prepared dishes
Step 6	Integrative semiotic analysis

Table 8.4 The approach step by step, with suggested methods for data collection and analysis.

Step	Description	Methods that can be used for data collection	Methods that can be used for data analysis	Aim
1	Selection of oils with different sensory styles	Descriptive analysis	Multivariate statistical techniques (e.g., principal components analysis, PCA)	To select extra virgin olive oils characterized by very different sensory styles
2	Selection of chefs characterized by different culinary styles	–	–	To select chefs according to the interest of the study (e.g., cooking styles, food culture)
3	Chef interviews on oils and their culinary uses	Focus group Free elicitation Repertory grid method (RGM)	Semiotic analysis Semiotic analysis Semiotic analysis	To select the oil–food combinations where oil plays a leading role in producing harmony, when the number of different oils is 3–4
		Creative design (1a, b, c)	Creative design (1a, b, c)	To select the oil–food combinations where oil plays a leading role in producing harmony, starting from the aim to be attained (in terms of sensory effects, harmony, etc.)
		Combinatorial approach	Graph theoretic approach	To screen the oil–food combinations where oil plays a leading role in producing harmony, when the number of different oils is >4 and it is important to involve a large number of chefs (e.g., in cross-cultural studies or studies in depth on a specific food culture)

(*continued*)

Table 8.4 (*Continued*)

Step	Description	Methods that can be used for data collection	Methods that can be used for data analysis	Aim
4	Preparation of dishes	Ethnographic (direct and mediate observation)	Semiotic analysis	To analyze dishes in terms of ingredients, recipe, preparation, temperature of serving, position in the meal sequence
5	Definition of the main sensory descriptors of each prepared dish	RGM	Semiotic analysis	To gather information about the main sensory properties of the dishes
		Descriptive analysis	Multivariate statistical techniques	To obtain descriptions of dishes (in terms of sensory profiles)
	Study of the emotions generated by the prepared dishes	EsSense questionnaire RGM questionnaire	Multivariate statistical techniques	To gather information about emotions generated by the prepared dishes
6	Integrative semiotic analysis	Data obtained in the previous steps	Semiotic analysis	To develop a comprehensive result and come to conclusions

selecting an appropriate number of oil samples which vary in origin and olive varieties and analyzing them by means of conventional descriptive analysis.

In fact, descriptive analysis can be used to identify similarities and differences among a group of extra virgin olive oils: this allows the selection of a restricted number of oils with a varied "sensory style." Data obtained with descriptive analysis methods can be used to achieve a perceptual map, defined as a "pictorial representation that capture the relationships among a set of products" (Lawless and Heymann, 1998). This map can be obtained by submitting sensory data from a descriptive analysis to a variety of multivariate statistical techniques, for example, principal components analysis (PCA), which can extract the most important information from complex data.

Visual inspection of perceptual maps permits the identification and grouping of samples with similar characteristics and, at the same time, different from other samples or sample groups (Næs, Brockhoff, and Tomic, 2010); in fact, results can be presented in a simplified picture or map in two or three dimensions that can be easily understood. Because of its characteristics, perceptual mapping is very effective in identifying oils with varied sensory styles (see Chapter 5).

Step 2. Selection of chefs characterized by different culinary styles Chefs characterized by different culinary styles have to be selected. Including chefs in the experimental design permits the inclusion of the creative dimension: in fact, chefs have a knowhow and a sensory sensitiveness that permit them to have good and original ideas for "mixtures" and "pairing" in order to obtain specific sensory effects.

Chefs have to be selected depending on the interests and on the size of the study: chefs belonging to the same food culture but with a varied culinary style can be selected in a study focused on a specific kind of cooking, or chefs belonging to different food cultures can be selected in a study aiming at investigating the potentiality of the use of oil independently of the culinary tradition, or vice versa in a cross-cultural study.

According to Rozin (1973), different cuisines are characterized by different "flavor principles" that are a combination of flavoring ingredients and seasoning compounds most used in a specific culinary tradition. These "flavor principles" provide powerful and characteristic flavor profiles that are familiar and pleasing (and often unconscious) to those within the system but also recognizable and replicable by those from outside. Oils characterized by different sensory styles can be used in different culinary traditions and together with different flavor principles with diverse and opposed purposes because harmony is created in a different and original way in each food culture. Studying the role played by oils characterized by different sensory styles in different food cultures and cooking traditions can be very useful in the infiltration of olive oil in markets not familiar with this product. A gradual application of the approach in the time can be suggested, proposing cultural studies taking into account cultural differences, in order to study the culinary use of oil in different cooking traditions.

Step 3. Chef interviews on oils and their culinary uses Chefs have to be interviewed in order to collect information about similarities and differences among selected oil sensory styles, in terms of perceived properties of oils *per se* and in terms of the best culinary use of different oils, to produce specific sensory effects.

Interviews must follow the tasting of oils, presented in blind conditions (without any indication of name, origin, or brand) in order not to influence chefs' evaluations with this information.

Different interview techniques can be used in order to obtain information about how each selected oil can be used with specific sensory effects. Approaches elaborated starting from the most common qualitative research methods can be used for this purpose, from group to individual interviews. Generally, in product development group discussions are usually preferred, on the basis that a group can generate more ideas than any individuals; on the other hand, in the particular case of chefs it can be argued that creativity could not be helped by the competitive background that could emerge in such a group discussion. Depth interviews, typically conducted one-to-one, allow great benefits to be obtained when insight is required, going into details. Some techniques that can be used for depth interviews are free elicitation and the repertory grid method (RGM). Table 8.5 presents a short description of each of these methods that can be used in chefs' interviews.

In a pilot study presented first in 2009 (Monteleone and Spinelli, 2009) (see Box 8.1), a revised version of the RGM was used. This approach has proven very useful in gathering data because it can be used to identify which dimensions discriminate products. Originally elaborated in psychology studies (Kelly, 1955), RGM is a methodology used since the 1980s in sensory sciences to collect information on consumer perception of food products (Thomson and McEwan, 1988; McEwan and Thomson, 1989; Scriven and

Table 8.5 Consumer research methods that can be used in chefs' interviews. See Van Kleef, Van Trijp, and Luning (2005) and Van Kleef and Van Trijp (2007) for an overview of consumer research methods and further details.

Method	Description
Focus group	Interactive group discussion technique led by a moderator, in which 8–12 respondents discuss a product or a topic
Free elicitation	Personal interviewing technique in which a respondent is asked to express the attributes they consider relevant in the perception of a particular product
RGM	Personal interviewing technique in which a respondent is asked to split the three stimuli into a pair and a single one and to explain the similarities between the paired stimuli and the dissimilarities from the single one

Mak, 1991; Green, 1992). The methodology is very simple and can be briefly summarized as follows: once they have been presented with three products, subjects are asked to indicate the two that are more similar; the standard RGM question is, "In what way are these two stimuli similar and in the same way different from the third?"

Characterized by a comparative nature, this methodology is very efficient in getting the consumers to discriminate among products and to describe their sensory properties. Different versions of RGM have been applied in sensory studies (Monteleone, Raats, and Mela, 1997; Fillion and Kilcast, 2002; Russell and Cox, 2003, 2004; Carbonell, Izquierdo, and Carbonell, 2007; Mireaux *et al.*, 2007) using, alternatively, oral or written interviews and questionnaires and asking slightly different questions.

Box 8.1 The pilot study.

The initial application of this approach is a pilot study, conducted in Florence at the Oleoteca Villa Campestri and presented first at the University of California, Davis in 2009 (Monteleone and Spinelli, 2009; Spinelli, 2011).

The specificity of this approach is to study how chefs use oils to obtain information on sensory effects of different oils in culinary preparations. For these reasons, three oils, characterized by very different sensory profiles, were selected within a group of 14 oils by means of sensory analysis. Three well-known and qualified chefs from the Florence area were selected. They worked in prestigious restaurants and represented three different approaches to cooking that could be described as "innovative," "traditional," and "focused on oil valorization." The oils were presented to the three chefs during individual interviews. Each chef was presented with the oils in three pairs and asked to describe the differences within each pair. After describing the perceived differences, the chefs were asked to indicate the best culinary use of each oil in terms of dishes that would fit with the specific sensory properties of oils (Table 8.6).

In a third session, the chefs were asked to prepare, for each oil, a proposed dish. The dishes were then studied by means of semiotic analysis. The analysis indicated that oils characterized by different sensory styles were used in different ways in culinary preparation, to obtain specific results in terms of sensory properties. In fact, there was a clear consensus among the chefs in selecting a specific oil and choosing how to use it.

Obviously, the results of this study are not relevant in themselves since they were obtained from a pilot study conducted with a limited number of both chefs and oils, but further studies are needed to explore the potentiality of olive oils representative of different sensory styles in culinary preparations.

Table 8.6 Chefs' interviews sections (pilot study; Monteleone and Spinelli, 2009).

	Section I	Section II
Aim of the session	Describing sensory differences between oils presented in pairs	Describing the culinary use of each oil
Oil presentation	A vs B B vs C C vs A	Monadic presentation (A; B; C)
Question	In what way is A different from B? In what way is B different from C? In what way is A different from C?	Which dish or recipe would fit with the sensory properties of oil A; B; C?

The method applied in this study is a modified version of RGM based on oral interviews analyzed by using a semiotic approach. This approach allows a deeper analysis than traditional studies that do not use a specific tool to analyze data in a systematic way.

In the aforementioned pilot study, each chef's interview was organized into two sections in order to obtain the most detailed description and information on the perceived identities of the oils (see Table 8.6):

I. a confrontation section, in which differences between oils were pointed out;
II. a section in which chefs were asked to think about the culinary use of each oil, according to their sensory characteristics.

The first section was based properly on RGM: each chef was first presented with the oils in three pairs (A versus B; B versus C; C versus A) and asked to describe the differences within each pair through a single question ("In which way does oil A differ from B?," etc.). Second, chefs were asked to indicate the best culinary use of each selected oil in terms of dishes that would fit with the specific sensory properties previously identified.

The method of conducting interviews is very important to obtaining consistent results: in this RGM revised version, the number of questions was limited (to three) and they were the same for each chef. Subjects were requested to "make explicit the implicit," letting the influence of the interviewer be minimal: the interviewer asked "why" in each case where the interviewee was considered to be omitting, taking for granted, or giving an answer full of gaps and implicit.

The RGM revised version demonstrated very good results in chefs' interviewing. The advantages of this method are that it encourages subjects to talk, thanks to its comparative nature, and allows, in this way, the collection of interesting preparatory information for the subsequent steps.

Because a RGM interview requires 45 min to 1 h for each subject for three products and chefs can find difficulties in tasting and concentrating on more than three or four different oils, it is suggested that this method be used with a limited number of products. On the other hand, this is not a problem because the previous step (1) allows for the selection of a limited number of products representative of different sensory styles.

Moreover, two different approaches have been presented to develop ingredient combinations in food product development and culinary preparations: creative design and a combinatorial approach based on graph theory. Although these approaches were conceived for product development, their application in a revised version to olive oils in culinary preparation studies can be useful in some cases.

Creative design (Næs and Nyvold, 2004) is a tool that combines chefs' creativity and experience with the systematic principles of experimental design, which "is used in order to ensure that the 'space' of all possibilities represented by the concept is properly spanned" (Næs and Nyvold, 2004). Creative design allows including and optimizing creativity and experience in the process of generating interesting prototypes in food product development. Once the prioritized attributes (which correspond to "concepts," namely specific descriptions of important characteristics of the product to be developed) have been selected, this method can be used to enhance communication, trying to ensure that all the main features of the concept are taken into account and also that all available creativity, experience, and sound scientific methodology are efficiently used. In fact, the key feature of this method is the use of systematic experimentation based on the selected attributes (concepts described as priorities) instead of ingredients and product process variables; only at the end are the design attributes "translated" into actual recipes. This feature could be useful when it is important to start from the aim to be achieved (e.g., contrast between flavors, delicate balancing). In this way, recipes will be developed in a systematic way, according to the kind of sensory harmony that a specific oil can produce.

The creative design method provides three main steps:

1. a. concepts discussion and selections within the group;
   b. setting up of an experimental design for the selected concepts (typically a fractional factorial design);
   c. development of the actual recipes satisfying the descriptions elaborated in (b);
2. production of prototypes according to the recipes developed by the product developer and their sensory evaluation;
3. consumer testing of prototypes.

This methodology has been applied in food pairing studies to develop culinary sauces with variations in levels (high and low) of the basic taste to pair with salmon (Paulsen *et al.*, 2012). This method allows a focus on the main aim of producing prototypes with the relevant attributes, taking into account different abilities and knowledge and setting up useful guidelines to obtain fruitful combinations.

In sensory functionality studies of oils, this method could be arranged and used to plan an experimental design of some specific features. In particular, operations presented at point 1 could help in developing recipes in which different oils sensory styles could be used to produce specific sensory effects (Step 3 in Table 8.4).

It can be argued that this method in itself does not allow for the selection of which attributes are to be chosen and of interest to investigate; for this reason, the authors suggested group discussion as an aid, but it can be claimed that other elicitation techniques could also be used for this purpose (the RGM, for example).

Recently, Nestrud's group proposed a new method, based on the graph theoretic approach, aimed at screening large numbers of possible food combinations in order to predict potentially successful combinations in a highly cost-efficient manner (Nestrud *et al.*, 2011; Nestrud, Ennis, and Lawless, 2012). This method proposed a combinatorial approach to determine compatibility amongst ingredients based on the principle of supercombinatorality (SC); this principle states that combinations that are fully pairwise compatible will be considered more compatible overall than combinations that are not fully pairwise compatible. Combinations that are fully pairwise compatible are called "cliques" and are composed of items considered compatible when evaluated in pairs. For example, if the pair A–B, the pair B–C, and the pair A–C were considered compatible, then the combination A–B–C is considered "fully pairwise compatible," and hence a clique. According to the SC principle, by eliminating combinations that are not cliques, it is possible to screen large numbers of potential food combinations, allowing a focus on a short list of fully compatible combinations.

In the first part of a study on 25 unique pizza toppings (Nestrud, Ennis, and Lawless, 2012), 124 subjects were asked to consider if 300 possible pairs of the ingredients were compatible or not (with YES or NO questions). In the second part, each subject was presented with combinations of toppings elaborated on the results of the first part of the study: 21 maximal cliques, 19 non-maximal cliques, and 15 non-cliques.

As observed by the authors, this method has some advantages and some limitations. On the one hand, it is a fast and consumer-driven method that permits the gathering of information in a cost-efficient manner directly from consumer; on the other hand, it cannot be used with "real" food but only at a conceptual level, because of the great number of combinations tested (300 in the study cited above). For these reasons, this method is not meant to supplant existing product methods, but only to augment and help them.

The application of such methodology to olive oil studies needs some discussion. This approach might be useful for finding a reduced number of combinations on the basis of preference and acceptability, but it is not concerned directly with sensory effects. Such an approach can be applied to oils in a very limited way, because it presupposes that subjects understand the differences between different oils without tasting them. Since only experts could have this knowledge of sensory differences among products, an application of this method to oils might be considered only with culinary experts, such as chefs, in studies where it is necessary to screen large numbers of potential food combinations (e.g., cross-cultural studies). On the other hand, the application of this method has to be carefully considered, weighing the pros and cons in the specific case study: more practical approaches to developing new original solutions in oil–food combination are often needed.

Table 8.4 provides an overview of the alternative methods that can be used for collecting and analyzing data in this step.

Step 4. Preparation of dishes Chefs have actually to prepare the recipes they have proposed for each specific oil. It is important in this step to study the effective culinary preparation process in terms of recipes and different treatments of the ingredients. Some ethnographic tools such as direct and mediate observation of each chef preparing a specific dish with an oil characterized by a specific sensory style can be helpful, together with reports. Observation is a strong technique that can provide unique insight into product and related consumer behavior and also into the culinary practices [see Buck (2007) for the application of ethnography in food product development].

Culinary preparations are studied in terms of the type of operation on ingredients (making, cooking, etc.), temperature of making and of serving (with raw, cooked, frozen items), recipe and positioning of the dish in the meal sequence (e.g., starter, first or second course). Semiotics can be used advantageously in the analysis of dishes (see Step 6).

Step 5. Definition of the main sensory and emotional descriptors of each prepared dish After having collected information about the preparation of dishes, it is important to identify the main sensory descriptors of each prepared dish. It is important that dishes are analyzed also in terms of harmony and overall effects produced. With this aim, various sensory evaluation techniques can be used. Descriptive analysis with a trained panel can be used to identify the principal attributes characterizing culinary preparations (Cerretani *et al.*, 2007; Aaslyng and Frøst, 2008; Dinnella *et al.*, 2012; Paulsen *et al.*, 2012). The RGM could also be used, allowing avoidance of the panel training stage. This can be done to reduce costs, when the aim of the study is not to obtain sensory profiles of specific culinary preparations but to gather information about the perceived sensory properties (with a qualitative and not quantitative interest). For the complexity of food evaluation, a RGM approach conducted with food experts, rather than with consumers, can be preferable.

Moreover, if the study is interested in investigating also the emotions generated by culinary preparations and the role of oils in this process, further methods can be used [e.g., EsSense Profile (King and Meiselman, 2010) or EmoSemio, a recently proposed approach based on the RGM (Spinelli *et al.*, 2012)].

Step 6. Integrative semiotic analysis Such a multidimensional approach based on a "multiplication of points of views" (Augé and Colleyn, 2004) by means of different methods needs an integrative approach in data analysis. A semiotic approach reveals a very useful tool in the analysis of different data such as oils' sensory profiles, RGM interviews, direct and mediates observation of dish preparation and recipes.

Semiotics is a field of studies with a linguistic, philosophical, and anthropological background, of interest in studying texts and in interpreting the meaning within social practices. Recently, this methodology has been used also in food studies for its effectiveness in verbal and visual texts (see Box 8.2). A semiotic method aims at consistent results in studying what texts mean and how to interpret them, showing their semantic structure (Greimas and Courtès, 1979; Eco, 1979, 1990). Thus, semiotics can be useful in interpreting the text's meaning with a standardized method as objectively as possible. The aim is to explain how a text can be correctly interpreted and how and why this analytical interpretation can differ from real people's interpretations.

Because of these characteristics, a semiotic approach has proven to be very effective in the study of a complex situation such as the culinary uses of different oil sensory styles, where not only sensory but also a strictly cultural dimension is implied.

Box 8.2 Semiotics in food studies.

Born to study literary texts, semiotics has developed methodological tools to study every type of "text," from visual arts (such as painting, photography, cinema) to advertisements, lifestyles, social practices and habits. Every so-called "text-object" can be studied by analyzing its deep semantic structure, that is, the relationship between semantic categories (Greimas and Courtès, 1979; Eco, 1979). Semiotics has a long tradition in marketing, communication, and consumer studies (Floch, 1990, 1995; Landowski, 1989). This approach could be efficiently used in food studies, to analyze visual and verbal-visual texts as labels, advertisements, and other communications. Moreover, it is possible to apply semiotics tools to study the packaging's communicative impact on consumers and to study purchase practices and fruition of the sale places by consumers. In two recent studies, semiotics has been applied to analyses of food labels, to determine if consumers' expectations raised by these labels are in agreement with what the labels communicate (Ares *et al.*, 2011; Piqueras-Fiszman, Ares, and Varela, 2011) and to lay down criteria for determining whether food labeling gives consumers a true and fair view of the properties of the food product in question (Smith, Møgelvang-Hansen, and Hyldig, 2010).

Semiotics tools are fit to study verbal language, so they can be used to optimize and obtain consistent results in interviews analysis, but can also help in the choice of the questions and in the formulation of interviews or questionnaires. For these reasons, combining a semiotic approach with an RGM can be very useful (Spinelli and Monteleone, 2009).

Moreover, a semiotics approach can be used to study gastronomy; initial studies have been conducted on what a dish communicates visually in terms of color and texture contrasts (Fontanille, 2004) and on what is communicated from a wider sensory point of view, considering tastes and flavors also (Spinelli, 2012).

The semiotic approach is based on the individuation of differences and relations of opposition at a semantic level, namely the level of meaning. This approach is based on the assumption that an item (e.g., a word) does not signify anything on its own, but is always inter-defined in relation to another item; in fact, a semantic category is an opposition of two concepts defined in a relational way. Therefore, this approach is based on the individuation of the semantic categories that correspond to inter-defined opposition of meaning dimensions.

For these reasons, a semiotic approach has to be applied at different levels in olive oil culinary uses studies (Table 8.4). In the pilot study mentioned previously, all the heterogeneous data obtained were studied by using a semiotic approach, which allows the study of different kinds of "documents" (such as interviews, recipes, videos, and observation) in relation to one another, leading to an aggregate and integrative result (Landowski, 2004).

A semiotic analysis on text interviews allows the identification of different kinds of semantic oppositions corresponding to different ways to discriminate and characterize products.

Moreover, using a semiotic approach, dishes can be analyzed as "semantic configurations," namely as significant constructions where different properties are opposed. In particular, dishes have to be studied in terms of recipe, but also preparation (operations, type of alimentary matrix, time, temperature), temperature of consumption, and main sensory characteristics (Spinelli, 2012). A semiotic approach can be used to single out semantic categories (oppositions of items) pertinent to characterizing the role of different oils' sensory styles used as ingredients in culinary preparations.

Table 8.7 Semantic categories describing dishes (pilot study; Monteleone and Spinelli, 2009).

Semantic categories	/fresh/ versus /elaborate/
Definition	Complexity level of preparation (in terms of time and kinds of operations needed)
(Sub) semantic categories	/partitive totality/ versus /integrative totality/
Definition	Fresh dishes were characterized by a mixing of parts that remain recognizable for flavor and consistency such as in salads (partitive totality), whereas in elaborated dishes ingredients are mixed until they become a single matter: consistencies are altered and flavor integrated, as in creams and soups (integrative totality)
(Sub) semantic categories	/cold/ versus /warm/
Definition	Dish service temperature
(Sub) semantic categories	/crunchy/ versus /creamy/
Definition	Dish tactile characteristic

In the pilot study mentioned previously, six oppositions (semantic categories) were identified by means of semiotic analysis of dish preparations and experts' evaluations (Table 8.7 and Table 8.8):

- /fresh/ versus /elaborate/;
- /partitive totality/ versus /integrative totality/,
- /cold/ versus /warm/;
- /crunchy/ versus /creamy/;
- /sweet/ versus /bitter/;
- /enhancing/ versus /contrasting/.

In Table 8.9, a semantic category concerning dish analysis is presented as an example; a so-called "semiotic square" is used to represent visually the relationship between the items. In fact, the semiotic square is a schema to represent a semantic category, an opposition of two concepts defined in a relational way by their differences. Briefly, a semiotic square works in this way: starting from a given opposition of concepts, in this case /fresh/ and /elaborate/, the square entails first the existence of two other concepts by denying them,

Table 8.8 Semantic categories describing the role of different oils sensory styles in culinary preparations (pilot study; Monteleone and Spinelli, 2009).

Semantic categories	/sweet/ versus /bitter/;
Definition	Dish tasting characteristic (basic taste)
(Sub) semantic categories	/enhancing/ versus /contrasting/
Definition	Role of oil in determining the overall flavor of the dish

Table 8.9 An example: a semiotic square used in analysis of dishes (pilot study; Monteleone and Spinelli, 2009)[a]

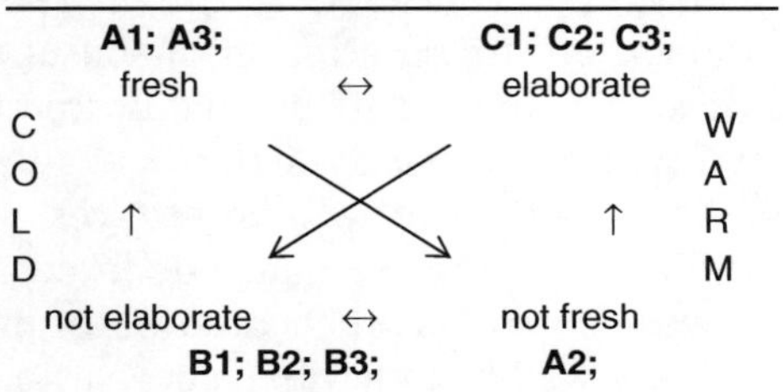

[a]A, B, C indicate oils, and 1, 2, 3 indicate chefs.

namely /not fresh/ and /not elaborate/, which are in the following relationships:

- /fresh/ and /elaborate/: basic opposition;
- /fresh/ and /not fresh/; /elaborate/ and /not elaborate/: contradiction.

It is important to note that each opposition is not valid in absolute terms but only in this context, as defined in the study. Therefore, /fresh/ and /elaborate/ are not opposed for their general meaning but for the specific one that in this study is assigned to them. Here these words are used as labels to indicate different characteristics of the dishes in terms of preliminary preparations of ingredients (partitive totality versus integrative totality), kind of process on the ingredients (in terms of cooking), texture properties of the prepared dish, and temperature. A further explanation of the semantic categories related to dish preparation is summarized in Table 8.7, where the meaning of each category is detailed in terms of the characteristics of dish preparation.

The analysis showed that some dishes could be described as /fresh/ (A1 and A3), whereas others had to be considered /elaborate/ (each dish prepared with oil C). On the other hand, some dishes could not be characterized by these two concepts and were in a less-defined condition: for example, A2 is described by the negation of the fresh, tending to the /elaborate/, and then sharing with the /elaborate/ some features even if it cannot be perfectly described by this term.

The opposition /cold/ versus /warm/ is integrated on the semiotic square of /fresh/ versus /elaborate/; it appear, in this case, at a second level:

- /cold/ is individuated on the axis of /not elaborate/ → /fresh/ and characterizes all the dishes described by these two concepts;
- /warm/ is individuated on the axis of /not fresh/ → /elaborate/ and characterizes all the dishes described by these two concepts.

The importance of introducing composite descriptors in food sensory evaluations, and also sensory descriptors used in descriptive analysis (Civille and Lawless, 1986) has been noted (Paulsen *et al.* 2012). This allows additional information to be obtained on the description of sensory changes in food–food combinations. Moreover, multi-dimensional descriptors may be closer to consumer perception (Giboreau *et al.*, 2007).

In research on the sensory effects of culinary sauces on salmon (Paulsen *et al.*, 2012) two composite descriptors were included: "complexity" and "harmony." "Complexity" was defined as "the sensation of many tastes and flavors" and "harmony" as "a well-rounded balanced taste and flavor." Relationships between composite and other one-dimensional sensory attributes have been interpreted by means of a PLSR (partial least-squares regression) model, showing that complexity and harmony were described by specific intensities of the taste attributes; for instance, in that case complexity was described by a relatively high intensity of the attributes umami and bitter and relatively low intensity of the attribute sour. According to the authors, "the challenge is to create a general definition of such composite attributes and to achieve a general consensus in the panel" (Paulsen *et al.*, 2012, p. 108). In fact, it can be very difficult to calibrate and train the panel to a defined standard and a subjective dimension can be involved. Moreover, it is important to achieve a general consensus not only within the panel but also in the scientific community: the previously mentioned case of harmony is exemplary of the ambiguity of this concept in food studies. In addition, harmony and complexity seem to depend on food culture and on the specific food: it is clear that harmony in a salmon-based preparation is different if compared with harmony in a dessert, for example.

For these reasons, further studies are needed on the meaning and on the description of culinary harmony and contrast/balance effects (harmony can be produced in different ways). Complexity, too, seems to be a key attribute to be further developed and detailed for its importance in describing food experience. In the case of olive oils, it seems to be extremely important: oils characterized by different sensory styles seem to be used with the best results in different dishes in terms of complexity, intended as freshness and degree of elaboration as defined above.

8.4 Conclusion

Further studies are necessary to investigate how different oil sensory styles can be used in combination with other ingredients in a cross-cultural perspective, in a specific culinary tradition, or in creative cookery, to obtain determinate effects, so as to create harmony. Studying the culinary uses of oils is important not just *per se*, but also to describe and communicate the sensory properties of a product normally used as an ingredient. For these reasons, sensory evaluation has to be integrated with tools from other disciplines such as semiotics to approach the study of the relationships between sensory properties of oils and their culinary uses from a multidisciplinary point of view. This sort of information would also facilitate the development of a language to communicate the sensory value of oils, because communication about oils can be focused on oil sensory styles and on oil uses. In fact, marketing the excellence of oils implies informing consumers of the best use of oils, related to what we have called their "sensory styles."

This information can be very useful to plan experimental design based on descriptive analysis: once ingredients and sensory effects for specific oils and food culture/culinary styles have been selected, quantitative methods can be used to investigate deeply some specific features and to test hypotheses emerging from qualitative research.

All this information about the sensory functionality of different oil sensory styles can also be used in marketing, to communicate a specific "sensory identity" of the product, for example, by using visual communication, in addition to verbal communication. This should be done not only in terms of what a product is (namely in terms of sensory characteristics often presented in a difficult or ambiguous way and consequently misunderstood by consumers), but also in terms of how a product can be used as an ingredient to build and modify flavor and textures of culinary preparations.

Sensory diversity has to be experienced by consumers and can be taught by means of "learning and playing routes"; for example, oil producer's websites could suggest activities to experiment with sensory diversity of oils. Instead of proposing a list of recipes, websites could suggest activities to experiment with the use of different oils in cooking. Something similar can easily be done in restaurants, for example, proposing a ludic way to experiment with sensory diversity.

A world is opening up for olive oil and its communication.

References

Aaslyng, M.D., and Frøst, M.B. (2008) The effect of basic taste solutions and selected vegetable accompaniments on the sensory properties of pork. *Journal of Sensory Studies*, **23**, 720–742.

Aaslyng, M.D., and Frøst, M.B. (2010) The effect of the combination of salty, bitter and sour accompaniment on the flavor and juiciness of pork patties. *Journal of Sensory Studies*, **25**, 536–548.

Ahn, Y.-Y., Ahnert, S.E., Bagrow, J.P., and Barabási, A.-L. (2011) Flavor network and the principles of food pairing. *Scientific Reports*, **1**, Article **196**, 1–7 (doi:10.1038/srep00196).

Ares, G., Piqueras-Fiszman, B., Varela, P., Marco, R.M., López, A.M., and Fiszman, S. (2011) Food labels: do consumers perceive what semiotics want to convey? *Food Quality and Preference*, **22**, 689–698.

Augé, M., and Colleyn, J.P. (2004) *L'Anthropologie*, PUF, Paris.

Bastian, S.E.P., Payne, C.M., Perrenoud, B., Joscelyne, V.L., and Johnson, T.E. (2009) Comparisons between Australian consumers' and industry experts' perceptions of ideal wine and cheese combinations. *Australian Journal of Grape and Wine Research*, **15**, 175–184.

Bastian, S.E.P., Collins, C., and Johnson, T.E. (2010) Understanding consumer preferences for Shiraz wine and Cheddar cheese pairings. *Food Quality and Preference*, **21**, 668–678.

Blumenthal, H. (2008) *The Big Fat Duck Cookbook*, Bloomsbury Publishing, London.

Buck, D. (2007) Methods to understand consumer attitudes and motivations in food product development, in *Consumer-Led Food Product Development* (ed. H. MacFie), Woodhead Publishing, Cambridge, pp. 141–157.

Carbonell, L., Izquierdo, L., and Carbonell, I. (2007) Sensory analysis of Spanish mandarin juices. Selection of attributes and panel performance. *Food Quality and Preference*, **18**, 329–341.

Cerretani, L., Biasini, G., Bonoli-Carbognin, M., and Bendini, A. (2007) Harmony of virgin olive oil and food pairing: a methodological proposal. *Journal of Sensory Studies*, **22**, 403–416.

Civille, G.V., and Lawless, H.T. (1986) The importance of language in describing perceptions. *Journal of Sensory Studies*, **1**, 203–215.

de Klepper, M. (2011) Food pairing theory: a European fad. *Gastronomica: The Journal of Food and Culture*, **11**, 55–58.

Dinnella, C., Masi, C., Zoboli, G.P., and Monteleone, E. (2012) Sensory functionality of extra-virgin olive oil in vegetable foods assessed by Temporal Dominance of Sensations and Descriptive Analysis. *Food Quality and Preference*, **26**, 141–150.

Eco, U. (1979) *Lector in Fabula,* Bompiani, Milan.

Eco, U. (1990) *The Limits of Interpretation*, Indiana University Press, Bloomington, IN.

Fillion, L., and Kilcast, D. (2002) Consumer perception of crispness and crunchiness in fruits and vegetables. *Food Quality and Preference*, **13**, 23–29.

Floch, J.M. (1990) *Sémiotique, Marketing et Communication*, PUF, Paris.

Floch, J.M. (1995) *Identitées Visuelles*, PUF, Paris.

Fontanille, J. (2004) À deguster des yeux. Notes sémiotiques sur la "mise en assiette." À propos de la cuisine de Michel Bras. *Visible*, **1**, 195–216.

Giboreau, A., Dacremont, C., Egoroff, C., Guerrand, S., Urdapilleta, I., and Candel, D. (2007) Defining sensory descriptors: towards writing guidelines based on terminology. *Food Quality and Preference*, **18**, 265–274.

Green, S. (1992) Validation and extension of the Repertory Grid Technique and Generalised Procrustes Analysis in food acceptability research, PhD thesis. University of Reading.

Greimas, A.J., and Courtès, J. (eds) (1979) *Sémiotique – Dictionnaire Raisonné de la Théorie du Langage*, Hachette, Paris.

Hyde, R.J., and Whiterly, S.A. (1993) Dynamic contrast: a sensory contribution to palatability. *Appetite*, **21**, 1–16.

Kelly, G. (1955) *The Psychology of Personal Constructs*, Norton, New York.

King, M., and Cliff, M. (2005) Evaluation of ideal wine and cheese pairs using a deviation-from-ideal scale with food and wine experts. *Journal of Food Quality*, **28**, 245–256.

King, S.C., and Meiselman, H.L. (2010) Development of a method to measure consumer emotions associated with foods. *Food Quality and Preference*, **21**, 168–177.

Landowski, E. (1989) *La Société Réfléchie. Essais de Socio-Sémiotique*, Seuils, Paris.

Landowski, E. (2004) *Passions Sans Nom*,PUF, Paris.

Lawless, H.T. (1986) Sensory interactions in mixtures. *Journal of Sensory Studies*, **1**, 259–274.

Lawless, H.T. (2000) Sensory combinations in the meal, in *Dimensions of the Meal: the Science, Culture, Business, and Art of Eating* (ed. H.L. Meiselman), Aspen Publishers, Gaithersburg, MD, pp. 92–106.

Lawless, H.T., and Heymann, H. (1998) *Sensory Evaluation of Food: Principles and Practices*, 2nd edn, Springer, New York.

Longman (1987) *Longman Dictionary of Contemporary English*, Longman, Harlow.

Madrigal-Galan, B., and Heymann, H. (2006) Sensory effects of consuming cheese prior to evaluating red wine flavor. *American Journal of Enology and Viticulture*, **57**, 12–22.

McEwan, J.A. and Thomson, D.M.H. (1989) The repertory grid method and preference mapping in market research. A case study on chocolate confectionery. *Food Quality and Preference*, **1**, 59–68.

Mireaux, M., Cox, D., Cotton, A., and Evans, G. (2007) An adaptation of repertory grid methodology to evaluate Australian consumers' perceptions of food products produced by novel technologies. *Food Quality and Preference*, **18**, 834–848.

Monteleone, E., and Spinelli, S. (2009), Role of sensory evaluation in describing and communicating the excellence in olive oil, presented at Beyond Extra Virgin. An International Conference on Excellence in Olive Oil, 21–23 June 2009, University of California, Davis, and Napa Valley, California.

Monteleone, E., Raats, M.M., and Mela, D.J. (1997) Perceptions of starchy food dishes: application of the Repertory Grid Method. *Appetite*, **28**, 255–265.

Næs, T., and Nyvold, T.E. (2004) Creative design – an efficient tool for product development. *Food Quality and Preference*, **15**, 97–104.

Næs, T., Brockhoff, P.B., and Tomic, O. (2010) *Statistics for Sensory and Consumer Science*. John Wiley & Sons, Ltd, Chichester.

Nestrud, M.A., Ennis, J.M., Fayle, C.M., Ennis, D.M., and Lawless, H.T. (2011) Validating a graph theoeretic screening approach to food item combinations. *Journal of Sensory Studies*, **26**, 331–338.

Nestrud, M.A., Ennis, J.M., and Lawless, H.T. (2012) A group level validation of the supercombinatorality property: Finding high-quality ingredient combinations using pairwise information. *Food Quality and Preference*, **25**, 23–28.

Nygren, I.T., Gustafsson, I.B., Haglund, A., Johansson, L., and Noble, A.C. (2001) Flavor changes produced by wine and food interactions: Chardonnay wine and Hollandaise sauce. *Journal of Sensory Studies*, **16**, 461–470.

Nygren, I.T., Gustafsson, I.B., and Johansson, L. (2002) Perceived flavor changes in white wine after tasting blue mould cheese. *Food Service Technology*, **2**, 163–171.

Nygren, I.T., Gustafsson, I.B., and Johansson, L. (2003a) Effects of tasting techniques – sequential tasting *vs* mixed tasting – on perception of dry white wine and blue mould cheese. *Food Service Technology*, **3**, 61–69.

Nygren, I.T., Gustafsson, I.B., and Johansson, L. (2003b) Perceived flavor changes in blue mould cheese after tasting white wine. *Food Service Technology*, **3**, 143–150.

Paulsen, M.T., Ueland, Ø., Nilsen, A.N., Öström, Å., and Hersleth, M. (2012) Sensory perception of salmon and culinary sauces – an interdisciplinary approach. *Food Quality and Preference*, **23**, 99–109.

Piqueras-Fiszman, B., Ares, G., and Varela, P. (2011) Semiotics and perception: do labels convey the same messages to older and younger consumers? *Journal of Sensory Studies*, **26**, 197–208.

Rozin, E. (1973) *The Flavor-Principle Cookbook*, Hawthorn Books, New York.

Russell, C., and Cox, D. (2003) A computerised adaptation of the Repertory Grid Methodology as a useful tool to elicit older consumers' perceptions of foods. *Food Quality and Preference*, **14**, 681–691.

Russell, C., and Cox, D. (2004) Understanding middle-aged consumers' perceptions of meat using repertory grid methodology. *Food Quality and Preference*, **15**, 317–329.

Scriven, F.M., and Mak, Y.L. (1991) Usage behaviour of meat products by Australians and Hong Kong Chinese: a comparison of free choice and consensus profiling. *Journal of Sensory Studies*, **6**, 25–36.

SISS (Società Italiana di Scienze Sensoriali) (2012) *Atlante Sensoriale dei Prodotti Alimentari*, Tecniche Nuove, Milan.

Smith, V., Møgelvang-Hansen, P., and Hyldig, G. (2010) Spin versus fair speak in food labelling: a matter of taste? *Food Quality and Preference*. **21**, 1016–1025.

Spinelli, S. (2011) Percorsi semiotici sul gusto. Tra percezione e pratiche (Semiotic perspectives on taste. Between perception and practices), PhD thesis, Istituto Italiano di Scienze Umane (SUM, Florence)/University of Bologna.

Spinelli, S. (2012) Il gusto del gusto. Percorsi semiotici per lo studio della gastronomia, in *En Torno a la Semiotica de la Cultura. Actas del I Congreso Internacional del GESC*, Fragua-Fundación Ortega-Marañon, Madrid, pp. 287–298.

Spinelli, S., and Monteleone, E. (2009) A semiotic approach in analyzing consumer perception of food by using the repertory grid method, poster presentation at the 7th Pangborn Sensory Science Symposium, Florence.

Spinelli, S., Masi, C., Dolce, S., Di Marzo, S., Zoboli, G.P., and Monteleone, E. (2012) How does it make you feel? An approach to measure emotions in food products experience, presented at "A sense of Inspiration," 5th European Conference on Sensory and Consumer Research, 9–12 September 2012, Bern.

Stevenson, R.J., Prescott, J., and Boakes, R.A. (1995) The acquisition of taste properties by odors. *Learning and Motivation*, **26**, 433–455.

Szczesniak, A.S., and Kahn, E.L. (1984) Texture contrasts and combinations: a valued consumer attribute, *Journal of Texture Studies*, **15**, 285–301.

Thomson, D.M.H., and McEwan, J.A. (1988) An application of the Repertory Grid method to investigate consumer perception of food. *Appetite*, **8**, 1–14.

Van Kleef, E., and Van Trijp, H.C.M. (2007) Opportunity identification in new product development and innovation in food product development, in *Consumer-led*

Food Product Development (ed. H. MacFie), Woodhead Publishing, Cambridge, pp. 321–341.

Van Kleef, E., Van Trijp, H.C.M., and Luning, P. (2005) Consumer research in the early stages of product development: a critical review of methods and techniques. *Food Quality and Preference*, **16**, 181–201.

Vetter, K. (2009) Chefs designing flavor for meals, in *Meals in Science and Practice. Interdisciplinary Research and Business Applications* (ed. H.L. Meiselman), Woodhead Publishing, Cambridge, pp. 509–526.

Witherly, S.A. (1987) Physiological and nutritional influence on cuisine and product development, in *Food Acceptance and Nutrition* (eds J. Solms, D.A. Booth, R.M. Pangborn and O. Raunhardt), Academic Press, London, pp. 403–415.

Whiterly, S.A. (2007) Why Humans Like Junk Food. The Inside Story on Why You Like Your Favorite Foods, the Cuisine Secrets of Top Chefs, and How to Improve Your Own Cooking Without a Recipe!, iUniverse, Lincoln, NE.

Part II

9 Olive Oils from Spain

Agustí Romero,[1] Anna Claret,[2] and Luis Guerrero[2]
[1]*IRTA, Mas de Bover, Olive Production, Oil Technology and Nuts, Constantí, Spain*
[2]*IRTA, Food Technology Centre, Monells, Spain*

9.1 Historical perspective

The olive tree is one of the most characteristic plants in the Mediterranean area. In fact, there is historical evidence of its presence in the Iberian Peninsula since prehistoric times. Some olive stones were found in the Neolithic site at El Garcel (2700–2400 BC), proving the existence of olive trees in this area (Almeria, Andalusia) at that period of time (Siret, 2001). Although the first olive tree orchards in Spain started in Phoenician times (1050 BC), olives, olive oil, and derivatives were integrated as part of the food culture around 600 BC. The introduction of the olive oil culture in Spain seems to have occurred simultaneously in two main areas: in the north, Ampurias (Catalonia), thanks to the commercial exchanges made by the Greeks, and in the south, Cadiz and Malaga, due to the commercial activities performed by the Phoenicians.

The Iberians rapidly adopted and extended olive tree orchards and culture. When the Romans arrived in Spain (212 BC), Hispania already had an important number of olive orchards providing olives and olive oil, especially in the Guadalquivir and Ebro Valleys. However, it was the Romans who actually improved and extended the olive oil culture, not only in Spain but also all around their colonies. During the Roman Empire, Spanish olive oil production settled mainly in the south of Hispania (Bética range), with intensive commercial exchanges with other provinces of the Empire. Olive oil from this region was highly appreciated and commercialized throughout the occidental Roman Empire. At that time, the "golden liquid" was so important that the

Olive Oil Sensory Science, First Edition. Edited by Erminio Monteleone and Susan Langstaff.

Emperor Adriano adopted the olive tree branch as a symbol of the Roman Hispania, even minting coins with this symbol. Olive orchards spread rapidly all over the regions where this plant was able to grow.

After the collapse of the Roman Empire, during the Visigoth dominance (fifth to seventh centuries), olive growing continued to increase even in mountain areas and those having less favorable climates.

The growing of olive trees increased further, especially in the Guadalquivir Valley, during the eight centuries of the Hispanic–Arabic civilization (eighth to fifteenth centuries). The Arabs introduced their varieties and increased the cultivation and culture of olive oil to such an extent that Spanish words such *aceituna* (olive) or *aceite* (oil) have an Arabic origin (e.g., *aceite* comes from the Arabic word *al-zait*, which means olive juice).

In the fifteenth and sixteenth centuries, the olive-growing area reached its maximum extent, being similar or even larger than nowadays. At that time, after the discovery of America, the first olives trees were also introduced in Peru, Chile, Argentina, and Mexico. Later, in the nineteenth century and thanks to the railways, olive trees were also grown in the inner regions of the Iberian Peninsula.

Today, olive trees in Spain can be found in 34 provinces, with a growing surface of about 25 000 km^2, of which 96% cultivate varieties for olive oil production (MAGRAMA, 2012a).

9.2 Geographic and climatic characteristics

The geographic relief of Spain is especially diverse and rugged, having numerous mountain ranges, a central plateau, two main depressions (Figure 9.1) and an average altitude of 660 m. The Spanish mainland has a surface area of 493 458 km^2 and about 4600 km of coastline. The mountain ranges are mainly west–east oriented and have an important influence on climate characteristics. Both climate and topography determine different types of vegetation, water, and soils.

The Spanish climate is also varied as a result of the latitudinal location and irregular relief. The Iberian Peninsula is located in a temperate zone with a mixture of warm and cool air zones (subtropical and polar). Within the country it is relatively common to simultaneously and at the same altitude temperatures of 15 and 40 °C in summer time and to find places with annual precipitation values higher than 2500 mm (humid oceanic climate) and lower than 200 mm (Mediterranean desert climate). There are, however, some general climatic patterns: temperatures diminish progressively from the prime coastal lines towards the inland and from North to South; January is the coldest month and August the hottest; the sea water temperature is higher on the Mediterranean Sea than on the Cantabrian or the Atlantic coast; and the thermal amplitude is higher in the central plateau.

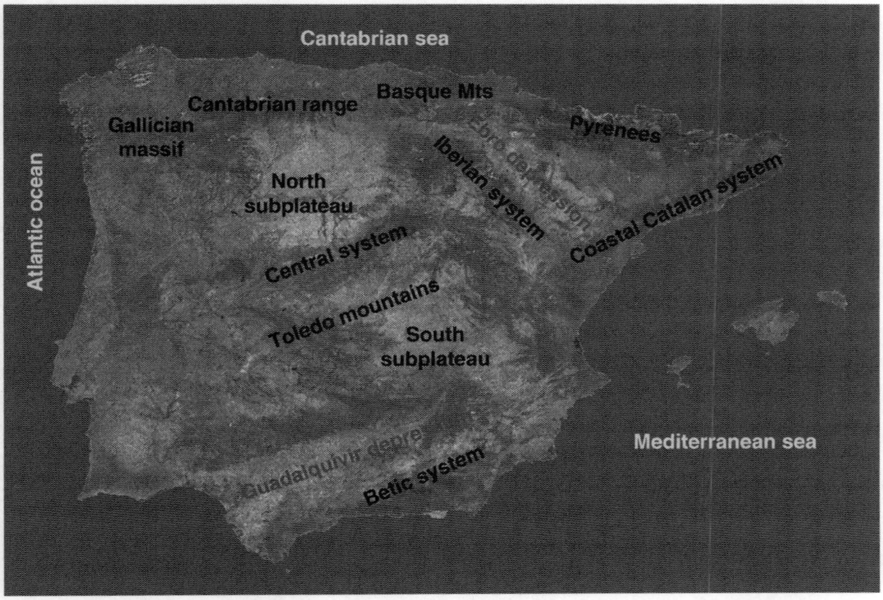

Figure 9.1 Main geographic characteristics of Spain. *Source:* own elaboration (base map from IGN, 2012). For a color version of this figure, see the color plates section.

Four main different climates can be found in Spain: Atlantic or oceanic, Mediterranean, subtropical, and mountain (IGN, 2012):

1. *Atlantic or oceanic climate:* Located in the north and northwest of the country, from the Pyrenees to Galicia. The annual rainfall can easily reach 1000 mm spread equally throughout the year and with mild temperatures varying between 12–15 °C in winter and 20–25 °C in summer because of its proximity to the sea.
2. *Mediterranean climate:* This is the predominant climate in Spain. It covers the whole Mediterranean costal zones, the interior of the peninsula and the Balearic Islands. There are, however, some differences between the different Mediterranean climate zones:
   a. Typically Mediterranean climate: erratic rains between 400 and 700 mm per year especially concentrated in spring and autumn; short and mild winters and long and hot summers.
   b. Mediterranean climate with cold winters (continental Mediterranean climate): located mainly in the central plateau and the Ebro depression, it is characterized by long and cold winters, very hot summers, and low rainfall (about 400 mm per year).
   c. Dry Mediterranean climate: this climate can be found in the southeast of Spain and its main characteristic is a very low rainfall, less than 300 mm per year.

3. *Subtropical climate:* Situated in the Canary Islands and also in a small area on the Granada coast (Almuñecar), this climate has warm temperatures all year long and low rainfall levels.
4. *Mountain climate:* Located in the large mountain systems, it has very cold winters and cool summers.

Figure 9.2 shows some climate diagrams representative of the different Mediterranean climate zones.

In Spain, large olive tree plantations can be found throughout the zones having a Mediterranean climate, which represent more than 50% of the Spanish surface area. Some plantations can be also found in the Canary Islands (subtropical climate) and in atypical zones such as Galicia with an Atlantic climate. In fact, olive trees can be grown almost everywhere in Spain except in those areas with a high altitude (higher than 700–1100 m, depending on the latitude).

According to the Spanish Ministry of Agriculture (MAGRAMA, 2012a), the total number of olive trees is Spain in 282 696 000, most of them located in Andalusia (61.8%). Overall, more than 250 different varieties of olive trees have been identified, most of them autochthonous and grown in relatively small areas.

Traditionally, the different varieties had been grown exclusively in the territory where they belonged, but nowadays it is relatively easy to find, for instance, Arbequina plantations in southern Spain and Picual and Hojiblanca plantations close to the French border in Catalonia. This geographic and consequently climatic delocalization of the different varieties, not only all over Spain but also throughout different countries (e.g., Italy, Argentina, Morocco, and the United States), can have a noticeable effect on the characteristics of the olives and on the extracted olive oil at both sensory and chemical levels (Tous *et al.*, 1997). In any case, the effects of the olive's variety and degree of ripeness on the sensory properties of the olive oil obtained could be strong enough to mask partially the effect of growing different varieties outside their traditional areas, thus reducing the effect of geographic and climate factors.

Although many cultivars are grown in Spain, only 24 are considered as main cultivars (Table 9.1 and Figure 9.3) and just five (Picual, Cornicabra, Hojiblanca, Manzanilla de Sevilla and Arbequina) represent more than 65% of Spanish olive tree growth (Rallo and Barranco, 2001). Most Spanish cultivars are for oil production and only two are for table purposes (Manzanilla de Sevilla and Gordal Sevillana) and four for double use (Hojiblanca, Empeltre, Manzanilla Cacereña, and Aloreña).

The special relationship in Spain between the main cultivars and regions supports the current distribution of Protected Designations of Origin (PDOs). In fact, in 2012 there were registered 28 PDOs in Spain, as described

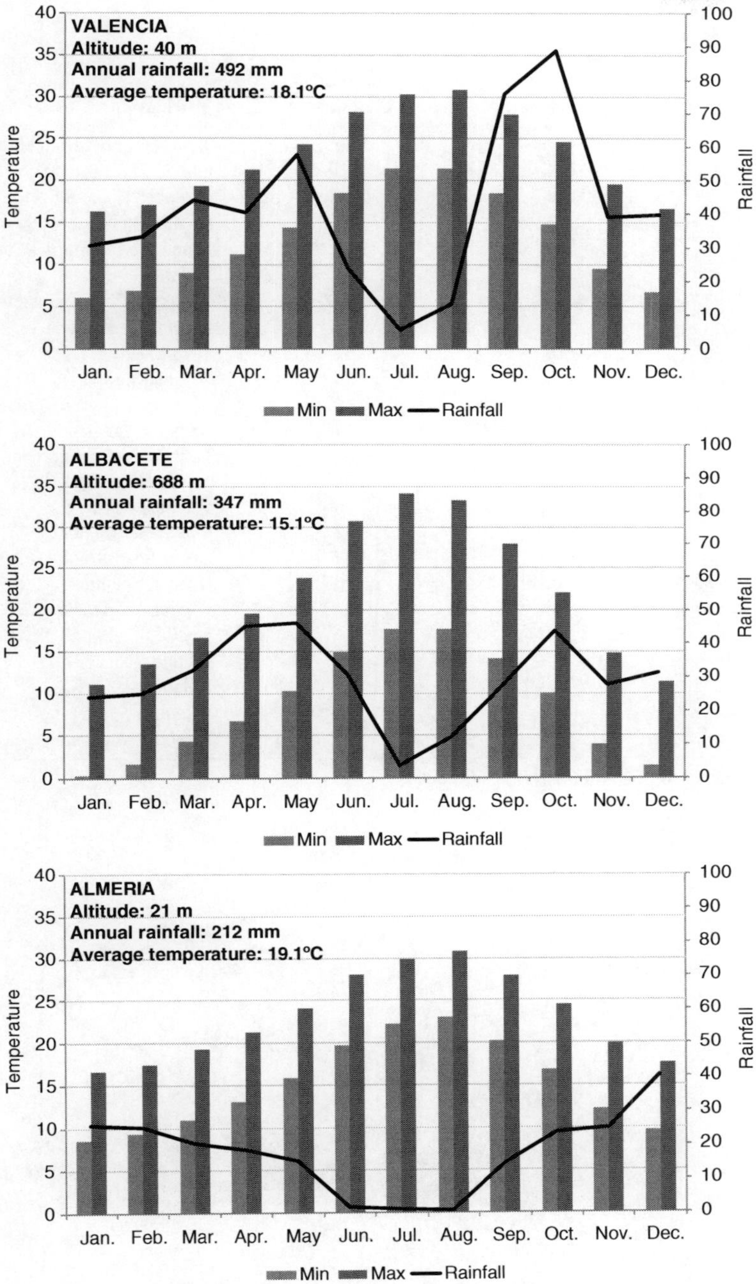

Figure 9.2 Climate diagrams of a typically Mediterranean climate (Valencia), a Mediterranean climate with cold winters (Albacete) and a dry Mediterranean climate (Almeria): average values between 2000 and 2011; temperatures in °C and rainfall in mm. *Source:* own elaboration (climate data obtained from AEMET, 2012). For a color version of this figure, see the color plates section.

Table 9.1 Main olive cultivars in Spain.

Cultivar	Main use	Surface area (× 1000 ha)	Distribution
Picual	Oil	900	Jaén, Córdoba, Granada, all Spain
Cornicabra	Oil	270	Ciudad Real, Toledo
Hojiblanca	Oil–table	267	Córdoba, Málaga, Sevilla
Manzanilla de Sevilla	Table	95	Sevilla, Badajoz, all Spain
Arbequina	Oil	91	Lleida, Tarragona, all Spain
Morisca	Oil	74	Badajoz
Empeltre	Oil–table	72	Zaragoza, Teruel, Baleares
Manzanilla Cacereña	Oil–table	64	Cáceres, Salamanca
Lechín de Sevilla	Oil	51	Sevilla, Cádiz
Picudo	Oil	35	Córdoba, Granada
Lechín de Granada	Oil	30	Granada, Almería, Murcia
Verdial de Badajoz	Oil	29	Badajoz, Cáceres
Morrut	Oil	28	Tarragona, Castellón
Sevillenca	Oil	26	Tarragona, Castellón
Villalonga	Oil	26	Valencia
Castellana	Oil	22	Guadalajara, Cuenca
Farga	Oil	21	Castellón, Tarragona
Verdial de Huévar	Oil	20	Huelva, Sevilla
Blanqueta	Oil	17	Alicante, Valencia
Gordal Sevillana	Table	12	Sevilla
Verdial de Vélez-Málaga	Oil	11	Málaga
Aloreña	Oil–table	11	Málaga
Changlot Real	Oil	5	Valencia
Alfafara	Oil	4	Valencia, Albacete

Source: Barranco, Trujillo, and Rallo, 2005.

Figure 9.3 Main Spanish cultivars. For a color version of this figure, see the color plates section.

Table 9.2 Main olive cultivars in the Spanish North-East growing regions.

Regions	PDO	Main cultivars	Most typical sensory properties
Catalonia	Les Garrigues	Arbequina (main), Verdiell	Medium to high greenly fruity. Balanced with a medium to high pungency and bitterness
	Siurana	Arbequina (main), Rojal, Morrut	Half greenly fruity. Balanced, with medium pungency and half to low bitterness
	Oli del Baix Ebre-Montsià	Morrut, Sevillenca, Farga	Half greenly fruity. Balanced, with medium pungency and bitterness
	Oli de Terra Alta	Empeltre (main), Arbequina, Morrut, Farga	Medium ripe fruity. Balanced, with medium pungency and light bitterness
	Oli d'Empordà	Argudell (>51%), Curivell, Llei de Cadaqués	Medium to high green fruity. Balanced, with half pungency and bitterness
Aragón	Aceite del Bajo Aragón	Empeltre (>80%), Arbequina, Royal	Half ripe fruity. Sweet, with very light bitterness and low pungency
Baleares	Oli de Mallorca	Empeltre, Arbequina, Picual	Medium green fruitiness. Medium to low pungency and low bitterness
La Rioja	Aceite de La Rioja	Arróniz, Verdial, Empeltre, Arbequina, Hojiblanca, Picual, Redondilla, Macho, Negral, Royal, Cornicabra, Manzanilla, Blanqueta	High greenly fruity. Medium to low pungency and bitterness
Navarra	Aceite de Navarra	Arróniz (>90%), Arbequina, Empeltre	Balanced. Half to high fruitiness. Medium to low pungency and bitterness

in Table 9.2, Table 9.3, and Table 9.4 and shown in Figure 9.4; most of them are inside larger olive growing areas that usually produce the same cultivars but with different specificities (related to particular climatic or soil conditions). Some important Spanish olive growing areas, not included in any PDO, are also reported in the tables.

9.3 Main sensory properties of Spanish olive oils

The sensory properties reported below are mean profiles from Spanish official panels for testing virgin olive oils (when that information exists, as summarized in Table 9.5). Official regulations of the International Olive Oil

Table 9.3 Main olive cultivars in the Spanish East and Central growing regions.

Region	PDO	Main cultivars	Most typical sensory properties
Comunidad Valenciana	Aceite de la Comunidad Valenciana	Villalonga, Blanqueta, Farga, Serrana de Espadán, Morrut, Cornicabra, Alfafara, Changlot Real, Rojal, Canetera, Nana, Arbequina, Empeltre, Cuquillo, Sollana, Callosina, Llumeta, Millarenca, Borriolenca	High green fruitiness with medium to low bitterness and pungency
Extremadura	Gata-Hurdes	Manzanilla Cacereña	Medium greenly fruity with low pungency and very light bitterness
	Aceite de Monterrubio	Cornezuelo and Picual (>90%), Mollar, Corniche, Pico-Limón, Morilla, Cornicabra	Medium fruity, pungent and bitter
Castilla-La Mancha	Montes de Toledo	Cornicabra	Half fruity. Balanced, with half bitterness and pungency
	Campo de Calatrava	Cornicabra, Picual	Medium greenly fruity. Medium to high pungency and bitterness
	Campo de Montiel	Cornicabra, Picual, Manzanilla, Arbequina and local types	Medium to high green fruity. Balanced with half to high bitterness and pungency
	Aceite de La Alcarria	Castellana	High greenly fruity and pungent
	Madrid (no PDO)	Cornicabra, Manzanilla	Medium to high fruitiness, both green and ripe. Balanced with half to low pungency and bitterness

Council (IOOC) [International Olive Council (IOC) since 2006] have been used, mainly IOOC T.20/Doc. No. 15 (2002) that defines standard sensory analysis for virgin oils and IOOC /T.20/Doc. No. 22 (2005) that describes particular descriptors for PDO virgin oils.

Most descriptions of PDO virgin olive oils are based on Official Specifications reported for each PDO to the Spanish Agriculture Ministry (MAGRAMA, 2012b). Most sensory profiles for specific cultivars come from experimental data from several comparative cultivar trials and collections from IRTA in Catalonia over several years (Tous and Romero, 2001; Romero, Tous and Guerrero, 2005; Romero and Tous, 2006); in addition, some published reports on cultivar virgin olive oil descriptions have been

Table 9.4 Main olive cultivars in the Spanish Andalusia growing regions.

Region	PDO	Main cultivars	Most typical sensory properties
Jaén	Sierra Mágina	Picual (>90%), Manzanillo de Jaén	High greenly fruity and medium bitter
	Sierra de Segura	Picual, Verdala, Royal, Manzanillo de Jaén	Medium fruity and bitter
	Sierra de Cazorla	Picual, Royal	High greenly fruity with medium bitterness
	Jaén (no PDO)	Picual and others	High green fruitiness with medium to high bitterness and pungency
Cádiz	Sierra de Cádiz	Lechín de Sevilla, Manzanilla, Verdial de Huévar, Verdial de Cádiz, Hojiblanca, Picual, Alameña de Montilla, Arbequina	Medium to high greenly fruity. Light bitterness and pungency
Córdoba	Priego de Córdoba	Picudo, Hojiblanca, Picual	Medium to high green fruity. Picudo produces sweet oils whereas Picual produces bitter oils
	Montoro-Adamuz	Picual, Nevadillo Negro, Lechín, Picudo, Carrasqueña	Medium to high fruitiness and bitterness, with half pungency
	Lucena	Hojiblanca (>90%), Arbequina, Picual, Lechín, Tempranilla, Ocal, Campanil, Chorrúo	Medium green fruity. Light bitterness and pungency
	Baena	Picudo, Lechín, Chorrúo, Hojiblanca, Picual	Blends richer in Picual are of high fruitiness with light bitterness whereas blends higher in Picudo are ripely fruity and sweet
Granada	Poniente de Granada	Lucio, Nevadillo de Alhama de Granada, Loaime, Hojiblanca, Picual, Picudo	Medium green or ripe fruitiness (depending on the blend) with light bitterness and pungency
	Montes de Granada	Picual, Loaime, Lucío, Negrillo de Iznalloz, Escarabajuelo, Gordal de Granada, Hojiblanca	Oils from an early harvest are intense green fruity, slightly bitter and pungent whereas late harvest implies soft ripe fruitiness
Sevilla	Aceite de Estepa	Hojiblanca, Arbequina, Manzanilla, Picual, Lechín de Sevilla	Medium fruity, bitterness and pungency. The richer the Hojiblanca ratio in the blend, the greener is the kind of fruity
Málaga	Antequera	Hojiblanca (90%), Gordal de Archidona	Greenly fruity of medium to high intensity. Bitterness and pungency range from low to half

Figure 9.4 Registered Protected Designations of Origin in Spain. For a color version of this figure, see the color plates section.

considered (Uceda and Aguilera, 2005; Pérez and Vergara, 2006; González Coloma, 2009).

9.3.1 Main Spanish olive-growing areas

9.3.1.1 North-East (Catalonia, Aragón, La Rioja, Navarra, and Balearic Islands) This area includes three main different kinds of virgin olive oils, depending on the main olive cultivar, as described below. In addition, there are some local olive cultivars, such as Arroniz in Navarra and Argudell in North Catalonia.

- Empeltre: traditional cultivar in Aragon, La Rioja, Navarra, and Terra Alta (Catalonia), all of them growing inside inland areas, whereas the Balearic Islands are exposed to a Mediterranean influence. Virgin olive oils from this cultivar are ripe fruity of half intensity, balanced with medium to light pungency and bitterness. As the fruits ripen fairly fast, a sweet sensation is common. When the fruit quality and extractive process are sufficiently accurate, a secondary aroma can be identified as retronasal perceptions, reminiscent of walnuts, green grass, and soft fruits.

Table 9.5 Mean sensory profiles for main olive cultivars in Spain.

Cultivar	No. of data	Fruitiness	Apple	Soft fruits	Green	Bitterness	Pungency	Sweetness	Astringency	Other
Arbequina	1489	5.11	1.31	1.29	3.57	2.84	3.56	4.14	1.59	3.29
Blanqueta	32	4.89	1.40	1.14	3.94	5.54	5.40	2.69	3.78	3.77
Cornicabra	8	5.78	0.38	0.56	4.31	4.88	5.52	2.39	3.29	3.13
Empeltre	194	4.62	1.29	1.85	3.24	3.45	4.20	3.95	2.04	3.41
Farga	28	4.96	1.59	1.48	3.84	3.92	4.39	3.59	2.24	3.71
Hojiblanca	15	5.74	1.38	0.61	4.63	3.57	4.18	3.25	2.89	4.11
Lechín de Sevilla	2	6.34	0.00	0.00	4.63	4.83	5.01	3.16	3.28	3.43
Manzanilla Cacereña	2	5.25	0.00	0.00	3.45	3.90	4.10	4.25	1.50	3.20
Manzanilla de Sevilla	16	5.16	1.52	1.21	4.54	6.06	4.80	2.72	3.16	4.22
Morrut	37	4.92	1.85	1.53	3.74	4.29	4.43	3.47	2.70	3.48
Picual	61	5.97	1.49	0.80	4.89	4.84	4.84	2.81	3.34	4.32
Picudo	8	5.23	0.73	0.96	4.08	4.43	5.06	2.65	2.83	3.97
Sevillenca	9	4.84	2.48	2.47	3.18	2.84	3.47	3.96	0.59	3.25
Villalonga	16	5.56	1.31	1.21	4.55	5.61	4.89	3.07	4.19	4.02

Source: International Olive Council, Sensory analysis of olive oil-method for the organoleptic assessment of virgin olive oil, COI/T.20/Doc. No. 15/Rev. 2, 2007.

- Arbequina: spreads from the Mediterranean coast of Catalonia, where oils are smoother, to the inland areas of Catalonia, with sharper oils from the same cultivar. Virgin olive oils are medium fruity with green and ripe notes, slight bitterness, and pungency. The final perception in the mouth is clearly sweet, while retronasal aromas are reminiscent of grass, herbs, and soft fruits. This profile tends to be greener and sharper when trees are grown at high altitude or far from the coast.
- Morrut: in the South of Catalonia, sometimes blended with Sevillenca or Farga, as a transition to the East area. Virgin olive oils are green fruity of medium intensity, balanced with medium pungency and bitterness. Secondary aromas are related to green grass and vegetables such as artichoke or green apple notes, with a slight final astringent mouthfeel. In off-years, when fruits ripen faster, or in blends with Farga or Sevillenca, oils are smoother with some soft fruit notes.
- PDOs include:
  - Les Garrigues: since the main cultivar is Arbequina and the climate and soils promote olive stress, virgin olive oils from an early harvest are balanced with medium or high greenly fruity and medium bitterness and pungency. Virgin olive oils from late harvest are medium greenly fruity, balanced with medium pungency and medium or low bitterness. Secondary aromas are mainly related to mown grass, green olives, and walnuts; sometimes found are retronasal perceptions of artichoke, exotic fruits (banana), fresh almonds, or tomato with a final almond-like mouthfeel (average description from more than 180 commercial samples analyzed by the Official Virgin Oil Tasting Panel from Catalonia, considering 2002–2011 harvests).
  - Siurana: the main cultivar is Arbequina and the climate is smooth. Virgin olive oils from an early harvest are greenly fruity of medium to high intensity, balanced with medium bitterness and pungency. Virgin olive oils from a late harvest are medium greenly fruity and pungent with medium or low bitterness. Secondary aroma is reminiscent of fresh mown grass and green olives, but also some retronasal aroma can be perceived, related to walnuts, artichoke, exotic fruits (mainly ripe banana), tomato, or green fruits. Usually, there is a final sweet and almond-like clear mouthfeel. The general profile tends to be sharper in higher orchards (average description from more than 520 commercial samples analyzed by the Official Virgin Oil Tasting Panel from Catalonia, considering 2002–2011 harvests).
  - Oli del Baix Ebre-Montsià: a natural blend of Morrut with Farga or Sevillenca, producing greenly fruity oils with a balanced mouthfeel. Secondary aroma is mostly reminiscent of freshly mown grass and green olives; other retronasal aromas relate to exotic fruit (ripe banana or kiwi), walnut, chamomile, tomato, almonds, or artichoke. Usually a final almond-like mouthfeel can be perceived (average description

from more than 15 commercial samples analyzed by the Official Virgin Oil Tasting Panel from Catalonia, considering 2002–2011 harvests).

- Oli de Terra Alta: balanced virgin olive oils half ripely fruity with medium pungency and light bitterness. Secondary aromas are mostly walnut (from the main cultivar Empeltre), soft fruits, and mown grass; in addition, chamomile or some green notes (e.g., green almond, artichoke, tomato) can be perceived with a final almond-like mouthfeel (average description from more than 57 commercial samples analyzed by the Official Virgin Oil Tasting Panel from Catalonia, considering 2005–2011 harvests).
- Oli d'Empordà: since the main cultivar is Argudell, virgin olive oils are medium greenly fruity, balanced with medium bitterness and pungency. Secondary aroma are mainly related to mown grass and green walnut, but exotic notes, green fruits, or artichoke can also sometimes be perceived with a final almond-like mouthfeel (average description from more than 50 commercial samples analyzed by the Official Virgin Oil Tasting Panel from Catalonia, considering 2004–2011 harvests).
- Aceite del Bajo Aragón: with Empeltre as the principal cultivar, virgin olive oils from early harvest are medium green fruity, whereas late harvest produces ripely fruity oils of medium intensity with low pungency, very light bitterness, and clearly sweet. Particular secondary notes are related to walnuts with a light almond-like mouthfeel (description adapted from the PDO website).
- Oli de Mallorca: fruity virgin olive oils mostly consist of medium green intensity, with medium to low pungency and low bitterness; secondary aromas are related to the main cultivar (tomato and almonds for Arbequina, walnuts for Empeltre, or fig leaves and herbs for Picual). On the other hand, sweet virgin olive oils are flat with very low retronasal perceptions (mainly soft fruits or apple) and clearly sweet (average description from more than 60 commercial samples analyzed by the Official Virgin Oil Tasting Panel from Catalonia, considering 2005–2011 harvests).
- Aceite de La Rioja: a natural blend of several cultivars (Table 9.2), virgin olive oils are high greenly fruity with medium to low pungency and bitterness. Secondary aromas are mainly related to green olives, artichoke, tomato, or green pepper. Sometimes, soft fruits or exotic fruits (banana) can be perceived (description adapted from the PDO website).
- Aceite de Navarra: virgin olive oils are well balanced with at least medium fruitiness. Pungency and bitterness are medium to low, while retronasal notes related to cultivar can be perceived (artichoke for Arroniz, banana for Arbequina, or sweet and almond-like for Empeltre) (description adapted from the PDO website).

9.3.1.2 East (Comunidad Valenciana, Murcia) This is a long and narrow zone along the Mediterranean coast that joins the North-East and Andalusian areas. It includes four subzones, according to the main cultivars and virgin olive oils:

- Farga: in the North, connecting Southern Catalonia. Sometimes blended with Sevillenca (also called Serrana). Virgin olive oil is medium fruity with both green and ripe notes, balanced with medium to slight pungency and bitterness, with a final sweet perception. Slight retronasal perceptions are related to grass, herbs, soft fruits, and green apples.
- Villalonga: mainly in Valencia. Virgin olive oils are green fruity of medium to high intensity, well balanced with medium to high bitterness and medium pungency. Secondary notes reminiscent of green leaves and artichoke or green banana that enhances medium astringency.
- Blanqueta: in Alicante. Virgin olive oils are medium fruity with both green and ripe notes. Although fruitiness is balanced enough with bitterness and pungency, these mouth attributes used to be of higher intensity. Secondary notes of green grass or herbs, joined with final notes of soft fruits.
- Lechín de Granada: in Murcia, common cultivar with Granada and Almeria (both in Andalusia). Virgin olive oils are medium fruity with green and ripe notes, well balanced with low bitterness and pungency. Secondary aroma reminiscent of apples, showing a final almond-like mouthfeel (adapted from González Coloma, 2009).
- PDOs include:
  - Aceites de la Comunitat Valenciana: since virgin olive oils are a natural blend of several cultivars (Table 9.3), the sensory profile shows a high intensity of green fruitiness with medium to low bitterness and pungency. Secondary aromas mostly reminiscent of green olives and almonds or herbs (description adapted from the PDO website).

9.3.1.3 Central zone (Extremadura, Castilla-La Mancha, Madrid) This is a wide area from Portugal to Valencia that includes several main cultivars:

- Cornicabra: virgin olive oil is of medium to high fruitiness, with green and ripe notes, well balanced with pungency and bitterness. Secondary aromas include both grass and soft fruits, with a final slight astringency. In off-years, fruits ripen faster and perceptions are softer, with some soft fruit notes.
- Manzanilla Cacereña: virgin olive oils are half to high greenly fruity, balanced with medium pungency and bitterness, although with a clear sweet perception. Retronasal aromas are related to grass.
- Castellana: virgin olive oils are medium fruity, with low bitterness and pungency. Secondary aromas are related to soft fruits and mown grass with a final sweet and almond-like mouthfeel (adapted from González Coloma, 2009).

- PDOs include:
  - Gata-Hurdes: medium greenly fruity with low pungency and very light bitterness (description adapted from the PDO website).
  - Montes de Toledo: medium fruitiness, well balanced with medium bitterness and pungency (description adapted from the PDO website).
  - Aceite de Monterrubio: medium fruity, pungent and bitter with final almond-like mouthfeel (description adapted from the PDO website).
  - Campo de Calatrava: medium greenly fruity with pungency and bitterness of medium to high intensity. Secondary aroma reminiscent of apples and fresh fruits (description adapted from the PDO website).
  - Aceite Campo de Montiel: medium to high greenly fruity, well balanced with medium or high bitterness and pungency. Secondary aromas are mainly related to olive leaves and tomato, with retronasal notes of soft fruits. Sometimes, aromatic sensations reminiscent of green almonds, apples, fig leaves, banana, or artichoke can be perceived. In addition, a late harvest usually produces ripely fruitiness (description adapted from the PDO website).
  - Aceite de la Alcarria: high greenly fruity and pungent. Secondary aromas are mainly related to olive leaves with slight perceptions of mown grass or green banana (description adapted from the PDO website).

9.3.1.4 Andalusia This includes different regions where particular cultivars are dominant. Also, there are two areas growing cultivars for table purposes (Sevilla growing Manzanilla de Sevilla, and Gordal Sevillana, and Malaga growing Aloreña):

- Picual: Virgin olive oils are of high-green fruitiness with medium to high bitterness and pungency. Secondary aromas reminiscent of fig leaves or tomato leaves. Usually, a final astringency and clear retronasal persistence are perceived.
- Lechín de Granada: described earlier (East area).
- Hojiblanca: Virgin olive oil is green fruity of medium to high intensity, balanced with medium pungency and medium to low bitterness. Secondary notes reminiscent of green apples or freshly mown grass.
- Lechín de Sevilla: Virgin olive oil is greenly fruity of high to medium intensity, balanced with medium pungency and bitterness. Retronasal perception reminiscent of grass or green fruits, with a final slight astringency.
- Verdial de Huévar: Virgin olive oils are of half fruitiness with both green and ripe notes, with medium to slight pungency and bitterness, usually of lower intensity that sweetness. Retronasal perceptions include soft fruits, grass, and herbs.
- PDOs include:
  - Sierra Mágina: Picual is more than 90%, producing virgin olive oils with high greenly fruity and medium bitter (description adapted from the PDO website).

- Sierra de Segura: natural blend of up to four cultivars (Table 9.4), mean profile is medium fruity and bitter with several secondary notes (description adapted from the PDO website).
- Sierra de Cazorla: since Picual is the main cultivar, virgin olive oils are characterized by high greenly fruity with medium bitterness and secondary aroma reminiscent of fig leaf, apples, and almonds (description adapted from the PDO website).
- Sierra de Cádiz: natural blend of several cultivars (Table 9.4) with medium to high greenly fruity with balanced and light bitterness and pungency. Secondary aromas reminiscent of soft fruits and herbs (description adapted from the PDO website).
- Priego de Córdoba: three kinds of virgin olive oils are produced, depending on the main cultivar in the blend (description adapted from the PDO website):
  - Picudo (at least 50%): medium greenly fruity with sweet perceptions.
  - Hojiblanca (at least 50%): medium greenly fruity with both fresh mown grass and sweet perceptions.
  - Picual (at least 50%): medium to high green fruitiness with medium bitterness.
- Montoro-Adamuz: a natural blend of several cultivars (Table 9.4), showing medium to high fruitiness and bitterness, with medium pungency. Secondary aromas are related to olive leaves and soft fruits (description adapted from the PDO website).
- Lucena: with Hojiblanca representing more than 90%, virgin olive oils show medium green fruitiness with light bitterness and pungency. Secondary aromas mostly reminiscent of mown grass with a final almond-like mouthfeel (description adapted from the PDO website).
- Baena: includes up to five cultivars (Table 9.4) and two kinds of virgin olive oils are produced (description adapted from the PDO website):
  - High fruitiness with light bitter almond-like perception.
  - Ripely fruity and sweet.
- Poniente de Granada: since virgin olive oils are natural blends of several cultivars (Table 9.4), the mean profile shows medium green or ripe fruitiness with light bitterness and pungency. Sometimes a final sweet mouthfeel can be perceived (description adapted from the PDO website).
- Montes de Granada: a natural blend of several cultivars (Table 9.4); depending on the harvest time, two kinds of virgin olive oils are produced (description adapted from the PDO website):
  - Intense fruitiness, from an early harvest, green and slight pungent and bitter.
  - Soft ripe fruitiness, from a late harvest.

- Aceite de Estepa: three kinds of virgin olive oils are produced, depending on the particular blend of cultivars (description adapted from the PDO website):
  - A blend of Hojiblanca (at least 50%), Arbequina (20–30%), Manzanilla, Picual, and Lechín de Sevilla: medium greenly fruity with some ripe notes and medium intensity of bitterness and pungency.
  - A blend of Hojiblanca (40–60%) and Arbequina (60–40%): medium green and ripe fruitiness, with medium bitterness and pungency.
  - Hojiblanca (100%): medium greenly fruity with medium pungency and bitterness. Secondary aromas are related to freshly mown grass and green fruits.
- Antequera: although the Hojiblanca content is more than 90%, three different kinds of virgin olive oils are produced (description adapted from the PDO website):
  - Greenly fruity of medium to high intensity and secondary notes related to almonds, exotic fruits (mainly banana), and even soft fruits.
  - Bitter and pungent virgin olive oils.
  - Sweet oils.

Figure 9.5 shows examples of the external aspect of some Spanish cultivars.

Figure 9.5 Examples of some Spanish cultivars. For a color version of this figure, see the color plates section.

References

AEMET (2012) *Servicios Climáticos*, Agencia Estatal de Meteorología. Ministerio de Agricultura, Alimentación y Medio Ambiente, Madrid, http://www.aemet.es/es/serviciosclimaticos (last accessed May 2012).

Barranco, D., Trujillo, I., and Rallo, L. (2005) Elaiografía hispánica, in *Variedades del Olivo en España. Libro I: Elaiografía Hispánica* (eds L. Rallo, D. Barranco, J.M. Caballero, C. Del Río, A. Martín, J. Tous, and I. Trujillo), Junta de Andalucía, MAPA y Ediciones Mundi-Prensa, Madrid, pp. 47–231.

González Coloma, A. (2009) Variedades de olivo y aceites, in *Los Aceites de Oliva en la Gastronomía del Siglo XXI* (ed. R. Ansón), Academia Española de Gastronomía y Editorial Everest, León, pp. 75–94.

IGN (2012) Instituto Geográfico Nacional, Madrid, http://www.ign.es/ign/main/index.do (last accessed May 2012).

MAGRAMA (2012a) *El Olivar Español*, Ministerio de Agricultura, Alimentación y Medio Ambiente, Madrid.

MAGRAMA (2012b) *Denominaciones de Origen e Indicaciones Geográficas Protegidas*, Ministerio de Agricultura, Alimentación y Medio Ambiente, Madrid, http://www.magrama.gob.es/es/alimentacion/temas/calidad-agroalimentaria/calidad-diferenciada/dop/ (last accessed May 2012).

Pérez, M.A., and Vergara, G. (2006) Caracterización del aceite de oliva: el aceite de oliva virgen de Madrid, in *El Aceite de Oliva Virgen de Madrid* (eds G. Vergara and M.A. Pérez), IMIDRA, Madrid, pp. 175–204.

Rallo, L., and Barranco, D. (2001) Olivo. Producción de aceite de oliva, in *La Horticultura Española* (eds F. Nuez and G. Llácer), SECH, Barcelona, pp. 352–355.

Romero, A. and Tous, J. (2006) Características sensoriales de los aceites de oliva virgen producidos en Cataluña, in *Los Aceites de Oliva de Cataluña* (eds J. Boatella and J. Contreras), Generalitat de Catalunya, DAAM, Barcelona, pp. 116–127.

Romero, A., Tous, J., and Guerrero, L. (2005) Caracterización sensorial (Banco de Germoplasma de Cataluña), in *Variedades de Olivo en España. Libro II: Variabilidad y Selección* (eds L. Rallo, D. Barranco, J.M. Caballero, C. Del Río, A. Martín, J. Tous and I. Trujillo). Junta de Andalucía, MAPA y ediciones Mundi-Prensa, Madrid, pp. 373–382.

Siret, L. (2001) *España Prehistórica*, Arráez Editores, Mojácar.

Tous, J., and Romero, A. (2001) Evaluación sensorial de variedades de olivo. *Fruticultura Profesional*, No. **120** (Especial Olivicultura III), 9–11.

Tous, J., Romero, A., Plana, J., Guerrero, L., Díaz, I., and Hermoso, J.F. (1997) Chemical and sensory characteristics of Arbequina olive oil obtained in different growing areas of Spain. *Grasas y Aceites*, **48**, 415–424.

Uceda, M. and Aguilera, M.O. (2005) Caracterización sensorial (Banco de Germoplasma Mundial de Córdoba), in *Variedades de Olivo en España. Libro II: Variabilidad y Selección).* (eds L. Rallo, D. Barranco, J.M. Caballero, C. Del Río, A. Martín, J. Tous and I. Trujillo), Junta de Andalucía, MAPA y Ediciones Mundi-Prensa, Madrid, pp. 373–382.

10
Olive Oils from Italy

Marzia Migliorini
Metropoli, Azienda Speciale della Camera di Commercio di Firenze – Divisione Laboratorio Chimico Merceologico, Florence, Italy

10.1 Introduction

Italy ranks second among the olive oil producers in Europe and worldwide; its national average production per year is more than 440 000 t, of which extra virgin olive oil accounts for two-thirds. Olive trees are planted on about one million hectares (2.4 million acres) as primary cultivation and on slightly less land as secondary crops, associated with sowable land or with other arboreal species (vines, citrus trees, almond trees, etc.). About 2% of the olive trees are widely planted at mountain altitudes, 53% are planted in the hills, and 44% are planted on the plains. There are approximately 170 million productive trees and more than one million farms involved in olive growing. On average, the area of olive orchards is around 1 ha (Unaprol, 2012). Geographically, the prime area of Italian olive production is located between 30 and 45°N, which, in general, is the geographical distribution of suitable areas for olive tree production in the northern hemisphere. In the Mediterranean area, Italy represents the central point of olive production because of its history and environmental conditions (Fontanazza, 2005).

Because of the typical Mediterranean climate, olive tree cultivation is very widespread in the southern regions (79.2%) and in the central regions (18.7%). In the north of Italy, production is more limited (2.1%), although it is increasing and is focused in areas where the climate is more temperate, such as the coasts of the Liguria region and the hilly areas near Lake Garda (2010–2011 crop season data from Unaprol, 2012).

Almost 75% of olive oil production is concentrated in the southern and insular areas: Apulia (31.4%), Calabria (30.8%), Sicily (9.4%), Sardinia (1.2%), and Basilicata (1.2%). Campania represents 7.7%. These regions are

Olive Oil Sensory Science, First Edition. Edited by Erminio Monteleone and Susan Langstaff.

considered the warm climate subarea for olive cultivation. Because of the favorable climate for olive production, it is possible to obtain high yields where plant–water needs are satisfied through irrigation (Fontanazza, 2005). However, some internal areas of Sicily, Sardinia, Calabria, Basilicata, and the coastal area of Apulia are considered medium climate areas.

The central–southern regions provide around 18% of the olive production and include lower Lazio, Abruzzo, Campania (coastal areas in Campania are included in the warm subarea) and Molise. In general, the environmental conditions of this area ensure optimum productivity. Maximum yields are gained with supplemental irrigation, based on seasonal conditions and plant requirements.

The total olive oil production from the central–northern olive-producing regions (Lazio, Marche, Umbria, Tuscany, Liguria, Emilia Romagna, Lombardy, and Veneto) is much lower than in the other two areas. This area can be included in the cold subarea, except for the coastal areas of Liguria and the south of Tuscany. The environment is not suited to maximum productivity in terms of olive quantity and annual bearing. It is possible to observe recurring damage from cold weather, which can sometimes destroy flowering buds and, partially or totally, the canopy (Fontanazza, 2005).

With respect to the olive production system, the country can be divided into two areas: marginal and suitable growing areas. The former includes the coastal and internal hilly areas of Liguria, the Gargano promontory in Puglia, the Cilento area in Campania, the Locride area in Calabria, and the areas of Nebrodi and Madonia in Sicily, and, in addition, the pre-Apennine area from the towns of Assisi to Spoleto in Umbria and the upper part of Lake Garda.

The suitable growing area is characterized by optimum climate and pedologic conditions; these are located mainly in the south of the country. Here olive production is characterized by different kinds of groves. The presence of old and sometimes very old plants of low efficiency is frequent, in addition to plantations with irregular distances between rows. However, as reported by Fontanazza (2005), in the last 30 years technological evolution has moved from traditional olive groves to new olive production systems classifiable into two fundamental models: (i) medium-density plantations, adapted for mechanical harvesting using trunk shakers combined with umbrella or with nets put under trees to catch the fruits; and (ii) high-density plantations adapted for full mechanization and especially for continuous harvesting using straddle machines.

Olive species are considered to have originated in the Near-East Mediterranean area (specifically Minor Asia, between Pamir and Turkistan). However, Italy (and in particular Sicily and Magna Grecia) can be considered the area of greatest economical importance. Accessible literature and archaeological excavations converge on dating the cultivation of oil trees in Roman territory at least to the first half of the sixth century BC (Cherubini, 2012). At the same time, olive production developed along the coastal and subcoastal areas of the eastern Mediterranean Sea, including southern European and northern African countries, advancing later with the Romans to the northern areas of Italy, Spain, France, and the Balkans (Blázquez Martínez, 1996). Italian agri-food history and economy, and also food and culinary culture and the evolution of olive oil production, consumption, and culture are strongly related to each other.

Italy is the country with the highest number of recorded olive cultivars: to date approximately 540 varieties have been recorded, equal to 42% of the total worldwide (Franca, 2010). The availability of such a large "biodiversity" is an extraordinary source of sensory differences among Italian extra virgin olive oils. In fact, it is known that the genetic origin of the olives is the prevalent factor which affects their volatile composition and both their phenolic content and profile. In order to demonstrate the potential diversity of Italian oils, Figure 10.1 maps monocultivar olive oils produced during the

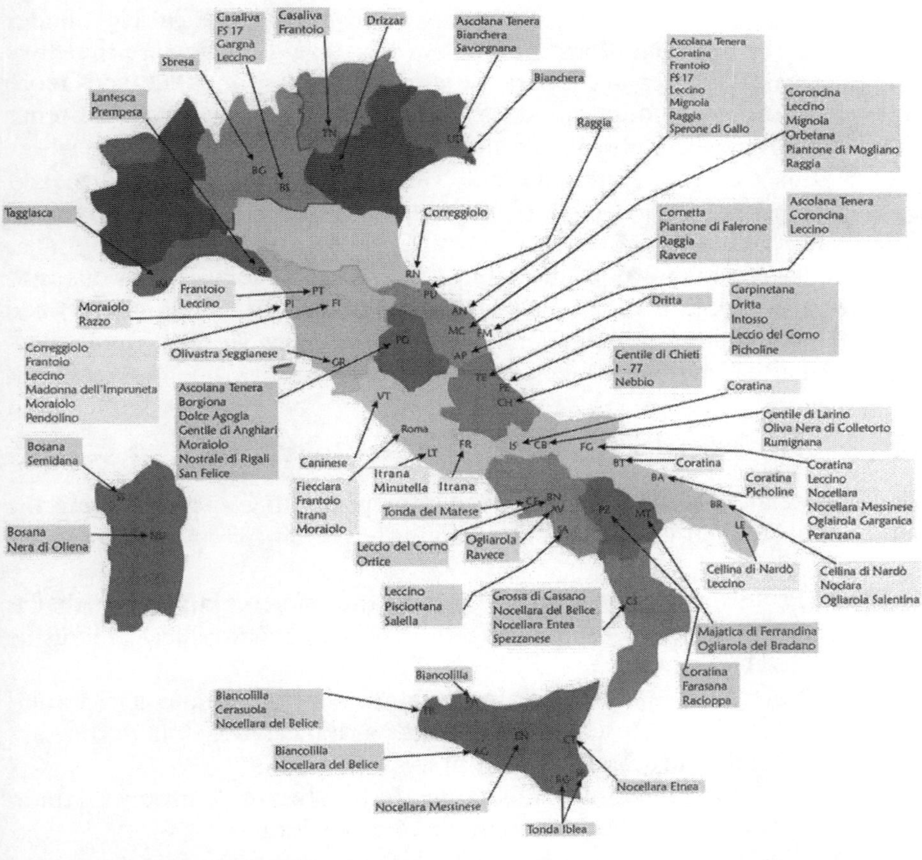

Figure 10.1 Areas of production of Italian single cultivar oils, presented during the 9th National Fair of the Single Cultivar Oils promoted and organized by ASSAM – Agenzia Servizi Settore Agroalimentare delle Marche (Office of Services for the Agro-Food Sector in the Marche region). For a color version of this figure, see the color plates section.

2011 crop season. From the figure it can also be noted that some cultivars (e.g., Leccino and Frantoio) are widely spread across the country whereas others are grown just in a specific region or even a specific area of a region. The expression of the chemical and sensory characteristics of monocultivar olive oils varies depending on the influence of many factors, such as environmental conditions, olive ripeness, and processing conditions. Thus differences among oils induced by cultivars and environmental conditions can be emphasized or suppressed. Alfei (2012a) observed that, on average, Frantoio monocultivar oils from the Marche region differ from the same oils produced in Tuscany in relation to their phenolic content. This difference also results in sensory differences. Tuscan oils tend to be characterized by a higher strength of green fruitiness, pungency and bitterness. However, the whole picture of differences among Italian extra virgin olive oils is much more complex then could be assumed from the comparison of oils obtained from different cultivars under the same producing conditions or from the same cultivars under different producing conditions. In fact, in each olive-growing area there are normally main varieties, secondary varieties, and sometimes ecotypes more or less represented (Fontanazza, 2005). Very few areas grow specific varieties alone. There are examples in the northern Bari province (Apulia), where the Coratina variety is almost the only variety grown, or in the Belice Valley (Sicily), where 90% of olive plants are represented by the Nocellara del Belice variety.

The most effective way of describing differences among Italian oils is to consider a description of the Protected Designation of Origin (PDO) and Protected Geographical Indication (PGI) certified extra virgin olive oils, paying special attention to their sensory profiles.

10.2 PDO and PGI extra virgin olive oils in Italy

The European Union has three systems for promoting and protecting the names of quality agricultural products and foods:

- PDO (Protected Designation of Origin): the name of a product that is produced, processed, and prepared in a defined geographical area using recognized techniques.
- PGI (Protected Geographical Indication): used to denote agricultural products and foods closely linked to a geographical area, where they are produced and/or processed, and/or prepared.
- TSG (Traditional Speciality Guaranteed): emphasizes a product's traditional composition or traditional mode of production.

These systems encourage the diversification of agricultural production, protect the product names from misuse and imitation and help consumers by providing information on the specific characteristics of the products.

Private and national systems for certifying the quality of agricultural products and foods ensure that the product has the characteristics and has been produced in accordance with the production method defined in the specification. These systems cover a wide range of measures applied at different stages of the food supply chain.

In Italy there are 43 PDO and one PGI extra virgin olive oils, which, together, cover 20 regions. In 2010, certified production amounted to ~10 500 t, matching the level achieved in 2009 (Unaprol, 2012).

Commonly, PDO product specifications are related to the parameters used for virgin olive oils classification (e.g., free acidity, peroxide number, K_{232}) set at a lower level that those for extra virgin olive oils, plus the total phenol and/or tocopherol content and the description of the main sensory properties. Hence each batch of certified olive oil, before being put on the market with the PGI or PDO mark, has to undergo chemical and sensory tests. The latter have to be carried out by a panel consisting of at least eight tasters registered in a specific National List. The sensory test includes the evaluation of the presence/absence of defects, and the correspondence of the intensities of fruitiness, bitterness, and pungency with the product specifications. Furthermore, a description of the flavor profile of the oils is performed.

At present, no official method has been defined to profile the flavor of PDO and PGI certified products. In order to fill this gap, several Italian olive oil tasting panels, acknowledged by the Italian Agricultural Ministry (MIPAAF), have developed a "profile sheet" according to both the regulation (EEC) No. 2568/91 and the recommendations of the International Olive Oil Council in COI/T.20/Doc. No. 22 (2005). Figure 10.2 shows the profile sheet used for the "selection of PDO and PGI extra virgin olive oil of Tuscany" (Regione Toscana, 2012). The activity of the recognized panels is an important source of information on the main sensory characteristics of Italian extra virgin olive oils.

This chapter continues with a description of PDO and PGI certified oils, paying special attention to their sensory profiles. The order of presentation of the regions is based on their annual olive oil production: for each region, the data on the production area, cultivars, and sensory profiles are taken from the production regulations.

10.2.1 Apulia

From north to south, from the countryside to the coastline, olive trees give the Apulia landscape some unique distinguishing features. Everything emphasizes the social and cultural value of extra virgin olive oil, its usefulness for the community, for the protection of the environment and the landscape, and its contribution to the quality of life. Apulia is the first region in Italy and worldwide in terms of land area covered by olive groves and of production of olives and of high-quality extra virgin olive oil: 60 million olive trees,

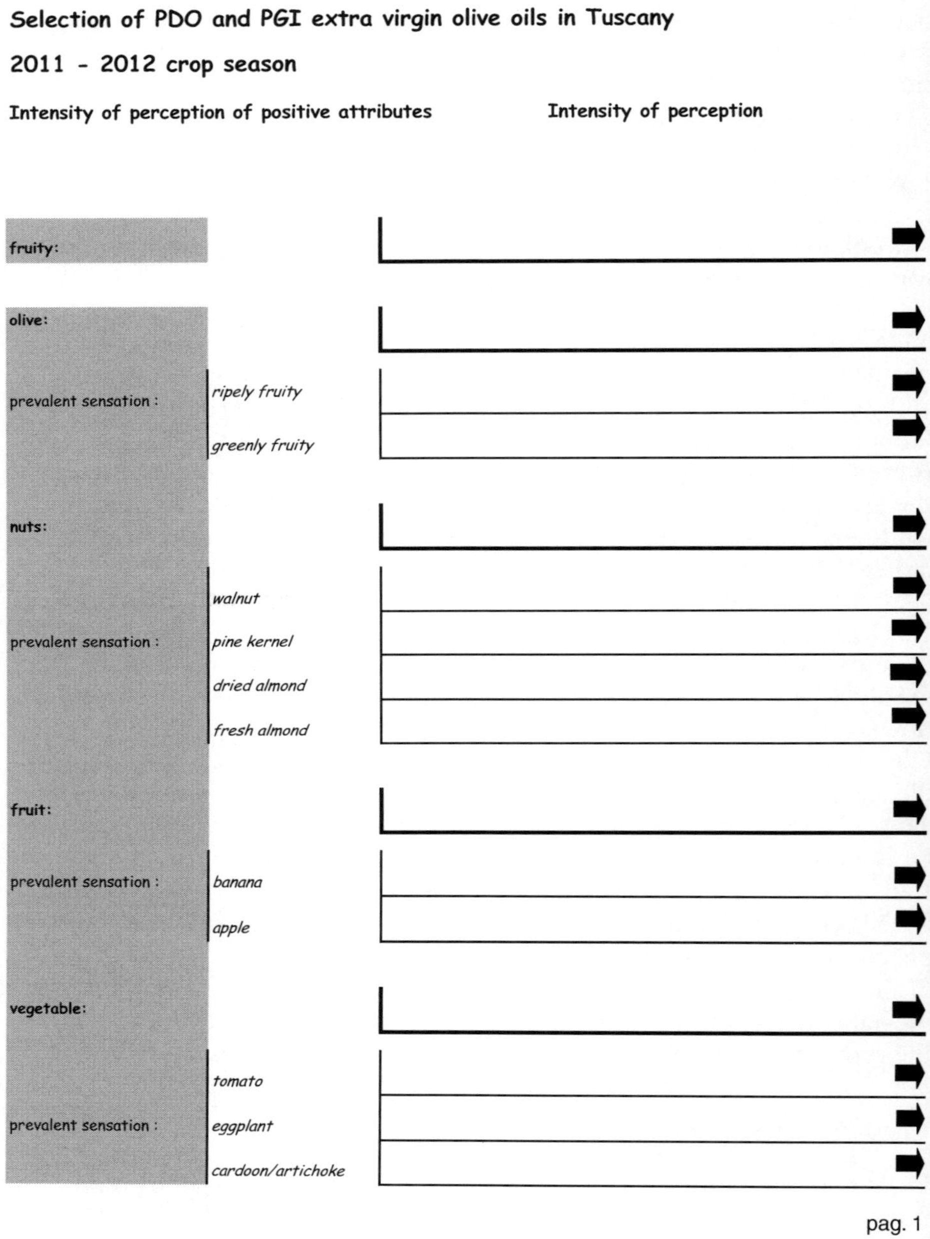

Selection of PDO and PGI extra virgin olive oils in Tuscany

2011 - 2012 crop season

Intensity of perception of positive attributes **Intensity of perception**

fruity:

olive:

prevalent sensation : *ripely fruity*, *greenly fruity*

nuts:

prevalent sensation : *walnut*, *pine kernel*, *dried almond*, *fresh almond*

fruit:

prevalent sensation : *banana*, *apple*

vegetable:

prevalent sensation : *tomato*, *eggplant*, *cardoon/artichoke*

pag. 1

Figure 10.2 Score card used for the 2012 selection of Tuscan PDO and PGI extra virgin olive oils.

grass:

prevalent sensation :

fresh grass

herbs

tomato leaf

olive leaf

bitter:

prevalent sensation :

olive bitter

woody bitter

pungent:

astringent/tannic:

fluidity:

overall hedonic value *from 0 to 10*

Discordant *Harmonic*

harmony

Date:	***Name of taster:***
Sample code:	***Panel head:***

pag. 2

Figure 10.2 (*Continued*)

5 million old trees, about 60 varieties of olive trees, and 360 000 ha of regional land devoted to this crop for an annual production of ~190 000 t of extra virgin olive oil (Centro Studi Confagricoltura Puglia, 2012; Unione Italiana delle Camere di Commercio *et al.*, 2012).

The Apulia territory is almost completely covered by PDOs. There are currently five PDO-producing regions: Collina di Brindisi, Dauno, Terra di Bari, Terra d'Otranto, and Terre Taratine (Table 10.1).

PDO Terre di Bari represents the most important designation for the Apulia region and the second in Italy after the Tuscan PGI for production, domestic sales, consumption, and value of exports. On its own, it represents 21% of the national certified product put on the market (Unaprol, 2012). The other Apulia PDOs have a very limited market.

Figure 10.3 shows a typical radar chart of an oil produced in the province of Bari, made from olives of the Coratina and Ogliarola cultivars.

Table 10.1 PDO extra virgin olive oils produced in the Apulia region.

PDO	Place of origin	Olive cultivar	Description of main sensory properties
PDO Collina di Brindisi (Province of Brindisi)		Ogliarola Salentina (min. 70%), Cellina di Nardò, Coratina, Frantoio, Leccino, Picholine (max. 30%)	Medium fruitiness with light bitterness and pungency
PDO Dauno (Province of Foggia)	Alto Tavoliere Basso Tavoliere Gargano Sub-Apennines	Peranzana, Coratina, Ogliarola Garganica and Rotondella	Medium fruitiness with fresh fruit, sweet almonds and herbaceous notes, light bitterness and pungency
PDO Terra di Bari (Province of Bari)	Castel del Monte Bitonto Murgia dei Trulli and delle Grotte	Coratina, Cima di Bitonto, Cima di Mola	From light to intense fruitiness with fresh herbs and almonds notes; medium bitterness and pungency
PDO Terra d'Otranto (Provinces of Lecce and Taranto)		Cellina di Nardò, Ogliarola Salentina (min. 60%)	Medium fruitiness with leaf notes; light bitterness and pungency
PDO Terre Taratine (Province of Taranto)		Leccino, Coratina, Ogliarola, Frantoio (min. 80%)	Medium fruitiness and bitterness; light pungency

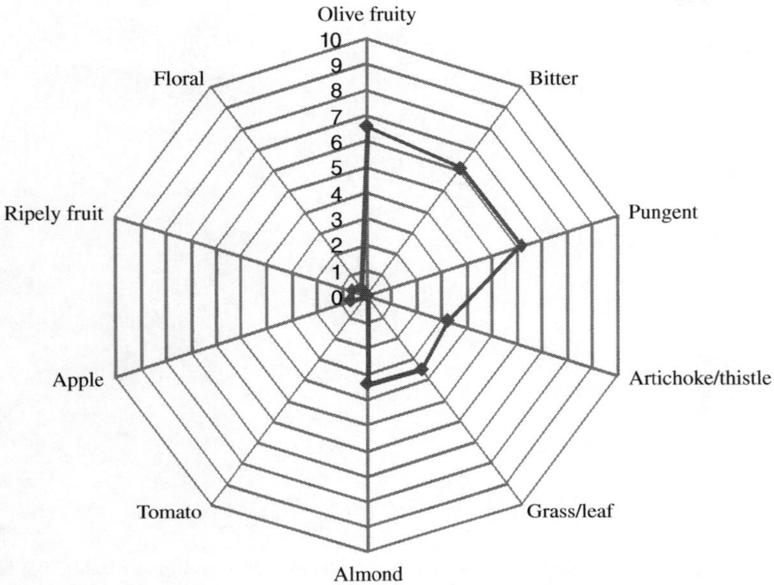

Figure 10.3 Typical radar chart of an extra virgin olive oil produced in the province of Bari, Apulia, from Coratina and Ogliarola cultivar olives (data processing by Comitato Premio Nazionale Montiferru, 2012).

10.2.2 Calabria

The Calabria Region has one of the most important olive heritages in the Mediterranean. The olive oil culture can be dated back to the fifteenth century BC.

At present, about 180 000 ha are devoted to olive groves and there are approximately 1300 oil mills for an average annual production of extra virgin olive oil over the past 3 years of ~170 000 t. As a result, at the national level, the Calabria region ranks amongst the first regions in terms of quantity of olive oil produced (Unione Italiana delle Camere di Commercio *et al.*, 2012).

Three olive oils with a PDO are currently produced in Calabria: PDO Alto Crotonese, PDO Bruzio, and PDO Lametia (Table 10.2).

Figure 10.4 shows a typical radar chart of extra virgin olive oil produced in the province of Cosenza.

10.2.3 Sicily

In Sicily, about 160 000 ha are devoted to olive trees, with an annual production of extra virgin olive oil of ~47 000 t. Sicily is one of the most important

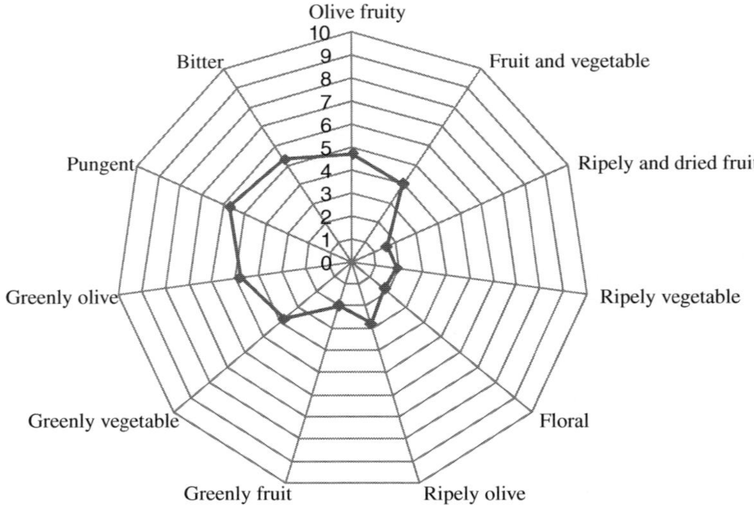

Figure 10.4 Typical radar chart of an extra virgin olive oil produced in the province of Cosenza, Calabria (data processing by Comitato Premio Nazionale Montiferru, 2010).

Table 10.2 PDO extra virgin olive oils produced in the Calabria region.

PDO	Place of origin	Olive cultivar	Description of main sensory properties
PDO Alto Crotonese (Province of Crotone)		Carolea (min. 70%)	Light fruitiness
PDO Bruzio (Province of Cosenza)	Belt of Pre-Pollino Crati Valley Ionic Pre Sila Hills Sibaritide	Tondina, Carolea, Grossa di Cassano, Dolce di Rossano	Light-medium with a sweet almond note; light bitterness
PDO Lametia (Province of Catanzaro)		Carolea (min. 90%)	Light fruitiness

areas in the world regarding the quality and the variety of the olive oil produced, for which it boasts the following strong points:

- rich autochthonous genetic pool;
- prevalence of artisanal production rather than industrial production;
- favorable pedoclimatic conditions,
- widespread production of PDO and organic olive oil;
- large percentage of olive oils with a high nutritional content determined by the olive cultivars and also by pedoclimatic conditions.

About 70% of Sicilian olive cultivation falls within the PDO area (Unione Italiana delle Camere di Commercio *et al.*, 2012). The following designations

Table 10.3 PDO extra virgin olive oils produced in the Sicily region.

PDO	Place of origin	Olive cultivar	Description of main sensory properties
PDO Monte Etna (Provinces of Catania, Enna, and Messina)		Nocellara Etnea (min. 65%)	Light fruitiness, bitterness and pungency
PDO Monti Iblei (Provinces of Siracusa, Ragusa, and Catania)	Monte Lauro Val d'Anapo Val Tellaro Frigintini Gulfi Valle dell'Irminio Calatino Trigona Pancali	Tonda Iblea, Moresca, Nocellara Etnea	Medium fruitiness with sensations of herbs; light–medium pungency
PDO Valli Trapanesi (Province of Trapani)		Cerasuola, Nocellara del Belice (min. 80%)	Medium fruitiness with herbaceous notes; light bitterness and pungency
PDO Val di Mazara (Provinces of Palermo and Agrigento)		Biancolilla, Nocellara del Belice, Cerasuola (min. 90%)	Medium fruitiness with almond notes and a sweet aftertaste
PDO Valdemone (Province of Messina)		Santagatese, Ogliarola Messinese, Minuta (min. 70%)	Medium fruitiness. Flavor descriptors: herbs, almond, fresh fruit, tomato, and thistle
PDO Valle del Belice (Province of Trapani)		Nocellara del Belice (min. 70%)	Medium–high fruitiness; medium bitterness and pungency

are present: PDO Monte Etna, PDO Monti Iblei, PDO Valli Trapanesi, PDO Val di Mazara, PDO Valdemone, and PDO Valle del Belice (Table 10.3).

Figure 10.5 shows a typical radar chart of an extra virgin oil from the Ragusa area (Tonda Iblea cultivar).

10.2.4 Campania

Olive trees dominate the inland areas of Campania and have an irreplaceable function in protecting the soil and in preserving the environment. In Campania, olive growing covers a total of ~73 000 ha, corresponding to more than 8.5 million trees. Most of the olives are grown in the province of Salerno, which by itself represents 53% of the total amount, followed by the province of Benevento with 23%. Annual olive oil production is 40 000 t (Unione

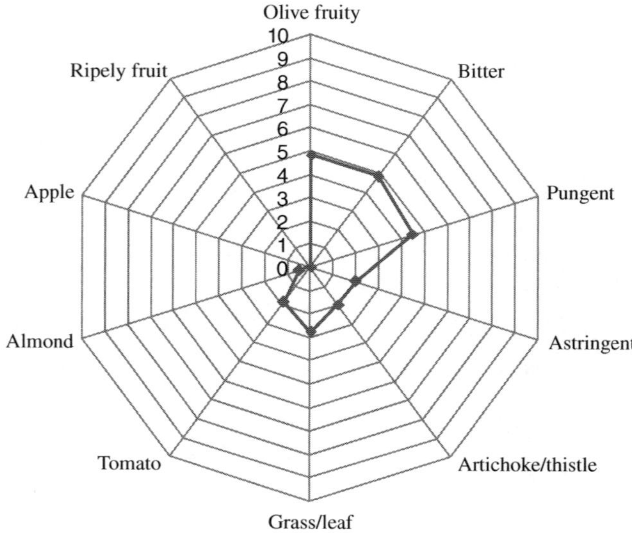

Figure 10.5 Typical radar chart of an extra virgin olive oil produced in the province of Ragusa, Sicily, produced with olives of the Tonda Iblea cultivar (data processing by Comitato Premio Nazionale Montiferru, 2011).

Italiana delle Camere di Commerci *et al.*, 2012). A very important element, which contributes to the success of olive growing in Campania, is the regional variety of the cultivars, which is very rich and diversified. The Campania region produces five different PDO oils: PDO Cilento, PDO Colline Salernitane, PDO Irpinia Colline dell'Ufita, PDO Penisola Sorrentina, and PDO Terre Aurunche (Table 10.4).

Table 10.4 PDO extra virgin olive oils produced in the Campania region.

PDO	Olive cultivar	Description of main sensory properties
PDO Cilento (Province of Salerno)	Pisciottana, Rotondella, Ogliarola, Frantoio, Salella, Leccino (min. 85%)	Light-medium bitterness and pungency
PDO Colline Salernitane (Province of Salerno)	Rotondella, Frantoio, Nostrale (min. 65%)	Medium-high fruitiness; light-medium bitterness and pungency
PDO Irpinia Colline dell'Ufita (Province of Avellino)	Ravece (min. 60%)	Medium fruitiness with tomato notes; light-medium bitterness and pungency
PDO Penisola Sorrentina (Province of Naples)	Ogliarola (min. 65%)	Medium fruitiness; light-medium bitterness and pungency
PDO Terre Aurunche (Province of Caserta)	Sessanta (min. 70%)	Medium fruitiness with artichoke and almond notes; medium bitterness and pungency

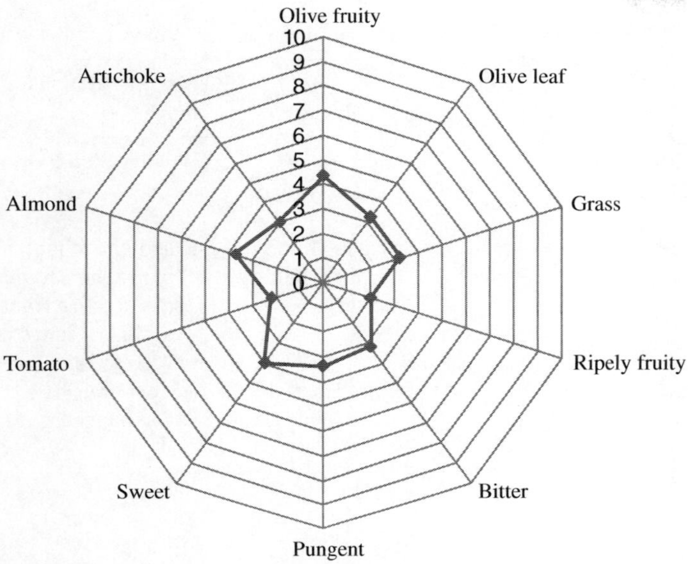

Figure 10.6 Typical radar chart of an extra virgin olive oil produced in the province of Salerno, Campania (data processing by Laboratorio Chimico Merceologico della Camera di Commercio di Napoli, 2011).

Figure 10.6 shows a typical radar chart of extra virgin olive oils produced in the province of Salerno. These products are characterized by fruitiness of medium intensity, followed by bitterness and pungency, which are of less intensity and with prevailing sensations of almond that give it overall harmony and sweetness.

10.2.5 Abruzzo

At the national level, Abruzzo ranks seventh for olive growing (~43 000 ha). Olive groves are present mainly in hilly areas, both near the coast and inland, but there are also olive trees at the foot of mountainous areas. The main cultivars are Dritta, Leccino, and Gentile di Chieti. However, today the olive sector is facing complex problems owing to both the difficulties of cultivation and the poor profitability margin. At present, the annual production of olive oil is ~19 000 t from ~450 oil mills (Unione Italiana delle Camere di Commerci *et al.*, 2012).

Three olive oils with PDOs are produced in the Abruzzo region: PDO Aprutino Pescarese, PDO Colline Teatine, and PDO Pretuziano delle Colline Teramane (Table 10.5).

Figure 10.7 shows a typical radar chart of an olive oil produced in the province of Chieti, the main area in the Abruzzo region suitable for olives.

Table 10.5 PDO extra virgin olive oils produced in the Abruzzi region.

PDO	Place of origin	Olive cultivar	Description of main sensory properties
PDO Aprutino Pescarese (Province of Pescara)		Dritta, Leccino, Toccolana (min. 80%)	Medium–high fruitiness
PDO Colline Teatine (Province of Chieti)	Frentano, Vastese	Gentile di Chieti, Leccino, Moraiolo, Nebbio	From light to high fruitiness; typical herbaceous notes characterize Frentano, while leaf notes describe Vastese
PDO Pretunziano delle Colline Teramane (Province of Teramo)		Leccino, Frantoio, Dritta (min. 75%)	Medium fruitiness, bitterness, and pungency

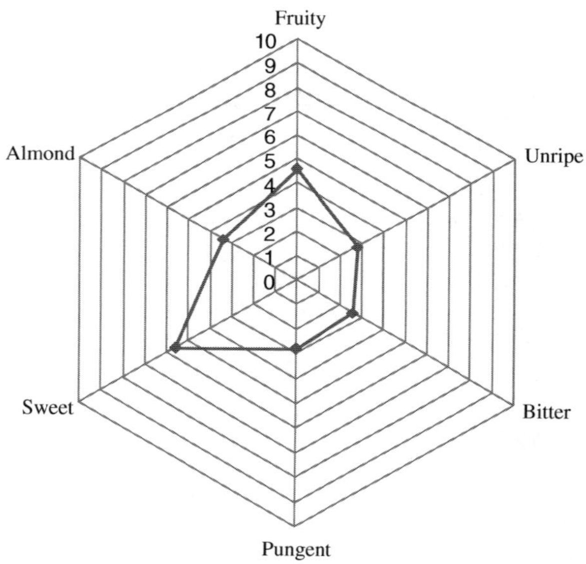

Figure 10.7 Typical radar chart of an extra virgin olive oil produced in the province of Chieti, Abruzzo.

10.2.6 Sardinia

In Sardinia, ~40 000 ha are covered by olive groves, with ~100 oil mills, which mainly work with the integral continuous system. About 8000 t of olive oil are produced each year.

The presence of olive trees in Sardinia, at least in the wild form, dates back without any doubt to the remote past: there are still many wild olive

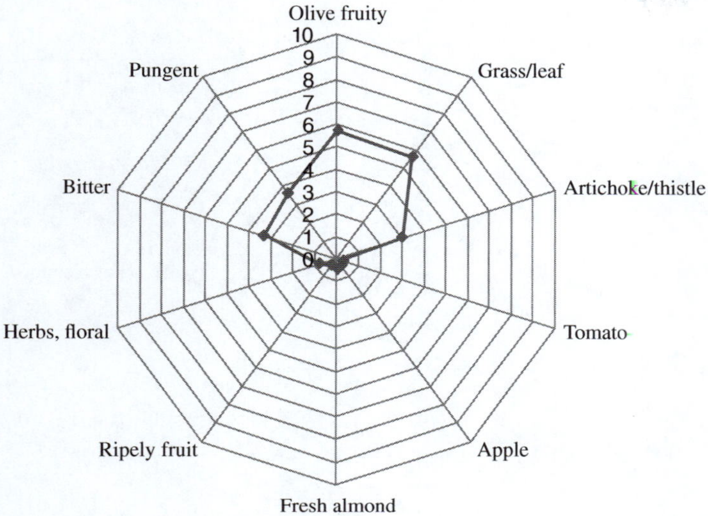

Figure 10.8 Typical radar chart of an extra virgin olive oil produced in the province of Nuoro, Sardinia (data processing by Comitato Premio Nazionale Montiferru, 2011).

trees whose age is calculated at several thousand years. The exploitation of this tree for the extraction of olive oil is also very old, taking into account that rooms and equipment to extract oil from olives have been found during archaeological excavations in some nuraghes (tower fortresses built from the Bronze Age to the second century AD).

At present, olive cultivation in Sardinia is excellent thanks to the outstanding technical skill and motivation of the sector operators, making this region one of the best areas for growing olives at the national level. The most widespread cultivars are Bosana, Semidana, Tonda di Cagliari, Olianedda, Nera di Gonnos, Nera di Villacidro, and Pizz'e Carroga. Sardinian olive oils are very complex and usually have very intense green fruity notes.

PDO Sardegna olive oil is produced throughout Sardinia, with an average annual production of 100 t (Unione Italiana delle Camere di Commerci *et al.*, 2012). The cultivars are Bosana, Tonda di Cagliari, Nera di Villacidro, and Samidana, comprising at least 80% of the total olives. Olive oils are characterized by a clear olive fruitiness with herbaceous green notes, bitterness and pungency of medium intensity, and are harmonious and pleasant overall.

Figure 10.8 shows a typical radar chart of an extra virgin olive oil produced in the province of Nuoro.

10.2.7 Marche

About 7000 ha of land are devoted to olive cultivation, mainly in the areas of Ancona, Pesaro, Jesi, and Ascoli, with a total annual production of ~3700 t.

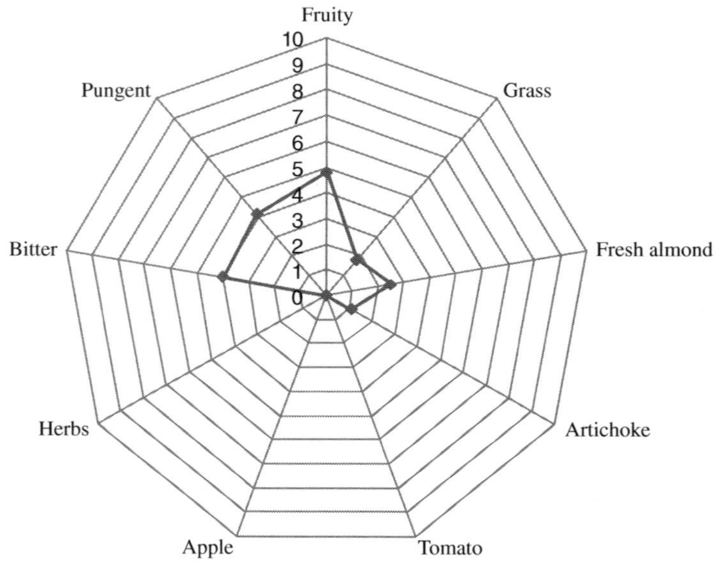

Figure 10.9 Typical radar chart of an extra virgin olive oil produced in the province of Ascoli Piceno, Marche (data processing by Alfei, 2012b).

The quality of extra virgin olive oil has recently improved, with increased production of monocultivar olive oils for an increasingly demanding market.

Monocultivar production can rely upon a wide range of olive oils in the territory, with 20 different cultivars existing in the Marche region. The main ones are Raggiola, Raggia, Coroncina, Mignola, Orbetana, Piantone di Mogliano, Ascolana Tenera, Carboncella, Piantone di Falerone, Sargano di Fermo, Frantoio, and Leccino.

Only one PDO oil is produced in the Marche region, PDO Cartoceto, which produces on average ~400 t of olive oil every year (Alfei, 2012b; Unione Italiana delle Camere di Commerci *et al.*, 2012).

The PDO Cartoceto olive oil includes the province of Pesaro-Urbino. The main cultivars are Raggiola, Frantoio, and Leccino, comprising at least 70% of the total olives. The olive oil has a green olive fruitiness, from light to medium, with flavors of herbaceous green, fresh almond, and unripe apple. In the taste there are sensations of green fruitiness, sweetness, bitterness, and pungency.

Figure 10.9 shows a typical radar chart of the oil produced in the province of Ascoli Piceno, the most suitable in the Marche region for olives.

10.2.8 Tuscany

In Tuscany, olive growing and the production of extra virgin olive oil play a significant role not only from the economic point of view, but also from

Table 10.6 PDO extra virgin olive oils produced in the Tuscany region.

PDO	Olive cultivar	Description of main sensory properties
PDO Chianti Classico (Provinces of Siena and Florence)	Frantoio, Correggiolo, Moraiolo, Leccino (min. 80%)	Light–medium fruitiness (green olive), bitterness, and pungency
PDO Lucca (Province of Lucca)	Frantoio (max. 90%), Leccino (max. 30%)	Light–medium fruitiness (green olive), bitterness, and pungency
PDO Terre di Siena (Province of Siena)	Frantoio, Correggiolo, Leccino, Moraiolo (by themselves min. 10% or combined max. 85%)	Medium fruitiness, bitterness, and pungency
PDO Seggiano (Province of Grosseto)	Olivastra Seggianese (min. 85%)	Medium green fruitiness, with artichoke notes; balanced bitterness and pungency

environmental, landscape, social, and tourism perspectives. This sector includes 97 000 ha in hilly areas and low mountains where about 15 million trees grow, with an annual production of between 14 000 and 20 000 t during the past 5 years (Bertini, Rocchi, and Tudini, 2012; Unione Italiana delle Camere di Commerci *et al.*, 2012). Tuscan olive oils can generally be recognized by their intense bitterness and pungency, followed by almond and artichoke flavors which on average are less intense.

Four PDO olive oils (Chianti Classico, Lucca, Terre di Siena, and Seggiano) and one PGI olive oil (Tuscan PGI) are produced in Tuscany, with a total amount of ~4000–4500 t produced between 2009 and 2010 (Bertini, Rocchi, and Tudini, 2012). Table 10.6 gives the characteristics of the PDO olive oils produced in Tuscany.

The Tuscan PGI olive oil includes the provinces of Arezzo, Lucca, Grosseto, Massa Carrara, Siena, Firenze, Prato, Pistoia, and Pisa. The following geographical places of origin are available: Seggiano, Colline Lucchesi, Colline della Lunigiana, Colline di Arezzo, Colline Senesi, Colline di Firenze, Montalbano, and Monti Pisani. Thirty-one local varieties are used and they have to represent at least 95% of the total olives. Each geographical location sets limits for the varieties admitted and their own sensory characteristics. For example, the cultivars admitted for the geographical regions of Montalbano are Leccino, Moraiolo, Frantoio, Pendolino, Rossellino, and Piangente up to 100%, and the sensory characteristics are fruitiness, together with sensations of almond, artichoke, and pungency.

By processing the data in "Selection of PDO and PGI Extra Virgin Olive Oil of Tuscany 2011/2012" (Regione Toscana, 2012), it has been possible to identify the profile of the extra virgin olive oils produced in Tuscany during the 2011 oil campaign, as shown in Figure 10.10.

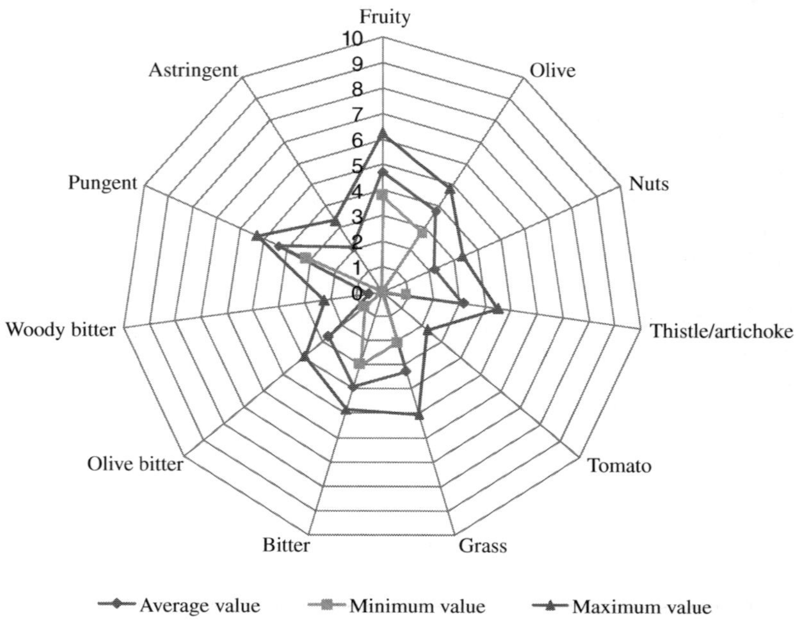

Figure 10.10 Typical radar chart of an extra virgin olive oil produced in the Tuscany region (data processing by Regione Toscana, 2012). For a color version of this figure, see the color plates section.

10.2.9 Umbria

Despite its adverse climate, Umbria is perhaps among the Italian regions that most identifies itself with the olive tree. Indeed, on the one hand, an adverse climate determines modest production, but on the other, it helps to produce high-quality olive oil.

The hilly areas at altitudes between 250 and 450 m characterize the landscape where 90% of the olive trees of the region are concentrated, mainly with the typical cultivars of Central Italy: Moraiolo, Leccino, Frantoio, Leccio del Corno, Dolce di Agogia, Correggiolo, Rosciola, and San Felice.

Approximately 27 800 ha are devoted to olive growing, with a total of 6 500 000 olive trees and 250 oil mills in the whole region (Unione Italiana delle Camere di Commercio *et al.*, 2012).

There is only one PDO olive oil in Umbria, PDO Umbria, producing a total amount of 800 t of certified olive oil every year.

PDO Umbria includes the provinces of Perugia and Terni. The following geographical locations are available: Colli Assisi-Spoleto, Colli Martani, Colli Amerini, Colli del Trasimeno, and Colli Orvietani. The main cultivars

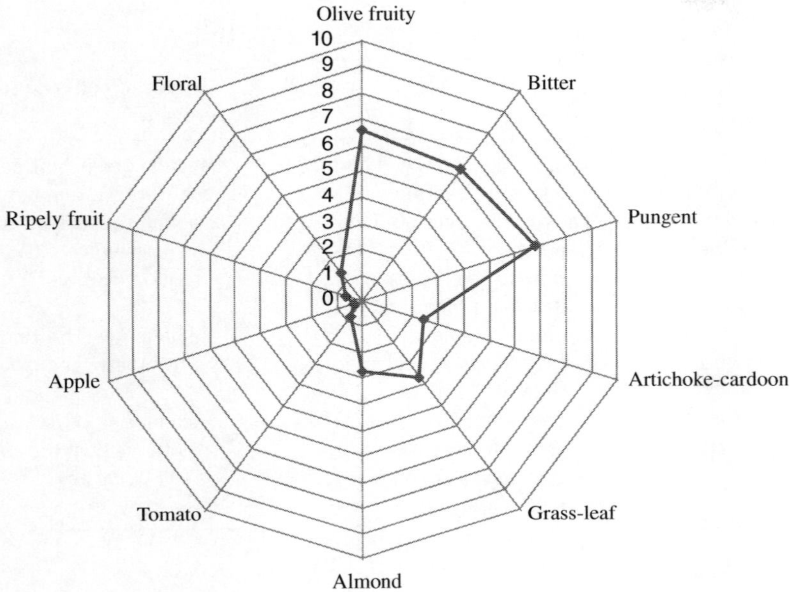

Figure 10.11 Typical radar chart of an extra virgin olive oil produced in the province of Perugia, Umbria (data processing by Comitato Premio Nazionale Montiferru, 2012).

admitted are Moraiolo, Frantoio, Leccino, San Felice, and Rajo, in different percentages depending on the place of origin. The olive oils are characterized by fruitiness, bitterness, and pungency of medium–high intensity.

Figure 10.11 shows a typical radar chart of an oil produced in the province of Perugia.

10.2.10 Lazio

Olive oil and olive trees have always characterized the Lazio landscape, with more than 86 000 ha of olive groves. The production of extra virgin olive oil has deep roots in the history and culture of this region and helps to support the local economy by developing food and wine tourism as well. Olive oils were already appreciated by the Etruscans, but it was the Romans who spread cultivation with all its different applications. The annual production of olive oil is currently ~26 000 t, with 127 000 farms and 370 oil mills (Unione Italiana delle Camere di Commercio *et al.*, 2012).

Four PDO olive oils are produced in the Lazio region: PDO Canino, PDO Sabina, PDO Tuscia, and PDO Colline Pontine (Table 10.7).

Table 10.7 PDO extra virgin olive oils produced in the Lazio region.

PDO	Olive cultivar	Description of main sensory properties
PDO Canino (Province of Viterbo)	Canino, Leccino, Pendolino, Maurino, Frantoio	Medium green fruitiness, bitterness, and pungency
PDO Sabina (Provinces of Rieti and Roma)	Carboncella, Leccino, Raja, Frantoio, Moraiolo, Olivago, Olivastrone, Salviana, Rosciola (min. 75%)	Medium ripe olive fruitiness with medium bitterness and pungency
PDO Tuscia (Province of Viterbo)	Frantoio, Caninese, Leccino (by themselves or combined) (min. 90%)	Medium green fruitiness, bitterness, and pungency
PDO Colline Pontine (Province of Latina)	Itrana from 50 to 100%; Frantoio, Leccino (by themselves or combined) (max. 50%)	Medium ripe olive fruitiness with medium bitterness and pungency

10.2.11 Liguria

The Liguria region is fortunate enough to be one of the olive-growing regions best known by consumers, although it represents just 0.8% of national production. Undoubtedly this success is the result of the efforts of the sector's operators, age-old tradition, and a temperate climate thanks to the sea breezes. The cultivar that prevails in most of the olive groves in Liguria is Taggiasca, successfully introduced by Benedictine monks. About 16 000 ha are covered by olive trees with an annual production of extra virgin olive oil of 3850 t.

Only one PDO olive oil is produced in Liguria, PDO Riviera Ligure, which has recently produced on average 580 t of olive oil per year and covers 2000 ha, employing about 480 olive growers (Unione Italiana delle Camere di Commercio *et al.*, 2012).

The PDO Riviera Ligure includes the provinces of Imperia, Savona, Genoa, and La Spezia. The following geographical locations are available: Riviera dei Fiori and Riviera del Ponente Savonese, characterized by medium intensities of ripe fruitiness with sweet sensations, and Riviera di Levante, characterized by bitter and pungent sensations.

The Taggiasca cultivar is the only one admitted for Riviera dei Fiori and Riviera del Ponente Savonese, whereas Riviera di Levante admits the Lavagnina, Razzola, and Pignola cultivars for at least 65% of the total olives.

10.2.12 Basilicata

The olive oil production of this small region is only ~6 t from two main areas of production: Vulture and Collina Materana. The main cultivars are Ogliarola del Vulture and Maiatica (Collina Materana).

The production and processing area of the PDO Vulture comprises the entire administrative territory of the municipalities of Melfi, Rapolla, Barile, Rionero in Vulture, Atella, Ripacandida, Maschito, Ginestra, and Venosa. This geographical area is characterized and known by the name Mount Vulture, an extinct volcano located in the central area of the Southern Apennines, ~60 km from the sea. The olive groves used for the production of Vulture oil are situated on the eastern and south-eastern slopes of Mount Vulture, as the mountain influences the microclimate and protects them from cold winter winds. The area is at an altitude between 400 and 700 m and has a particular, characteristically continental microclimate with winters that are generally long and cold and summers that are short and often dry. The oil is produced from the following cultivars: Ogliarola del Vulture, which must account for at least 70%, and, provided they account for no more than 30% of the trees in the olive groves, either individually or collectively, Coratina, Cima di Melfi, Palmarola, Provenzale, Leccino, Frantoio, Cannellino, and Rotondella.

The main sensory properties of the PDO Vulture oil are an amber–yellow color with green tints, medium fruitiness characterized by a grass note, and a medium-intense sensation of tomato. Bitterness and pungency are low–moderate in intensity.

10.3 Conclusion

The olive oil production system in Italy has been described as an "open laboratory" (Fontanazza, 2005) producing the latest and most advanced technologies able to support, for itself and for the world, the development of olive production techniques and practices. The differences among Italian extra virgin olive oils described in this chapter should be seen as a result of these innovations.

References

Alfei, B. (2012a) Catalogo degli oli monovarietali. *Olivo e Olio*, Suppl. 6, 16–47.

Alfei, B. (2012b) Dalle Marche l'olio giusto per ogni gusto e per ogni piatto. *Olivo e Olio*, Suppl. 6, 7–10.

Bertini, S., Rocchi, B., and Tudini, L. (eds) (2012) *Rapporto sul Sistema Rurale Toscano. Economia, Politiche, Filiere e Produzioni di Qualità*, Edizioni Edifir, Florence.

Blázquez Martínez, J.M. (1996) Origine e diffusione della coltivazione, in *Enciclopedia Mondiale dell'Olivo* (ed. Consiglio Oleicolo Internazionale), Agra, Milan.

Centro Studi Confagricoltura Puglia (2012) *l'Olivicoltura Pugliese. Criticità e Sviluppo*, Edizioni Centro studi Confagricoltura Puglia, Bari.

Cherubini, G. (2012) The long history of olive groves, in *Olivi di Toscana/Tuscan Olive Tree* (ed. P. Nanni), Edizione Polistampa, Florence, pp. 11–33.

Comitato Premio Nazionale Montiferru (2010) *Montiferru 2010. Premio Nazionale per l'Olio Extravergine d'Oliva. XVII Edizione. Catalogo Selezione degli Oli in Concorso*, Camera di Commercio di Oristano, Oristano.

Comitato Premio Nazionale Montiferru (2011) *Montiferru 2011. Premio Nazionale per l'Olio Extravergine d'Oliva. XVIII Edizione. Catalogo Selezione degli Oli in Concorso*, Camera di Commercio di Oristano, Oristano.

Comitato Premio Nazionale Montiferru (2012) *Montiferru 2012. Premio Nazionale per l'Olio Extravergine d'Oliva. XIX Edizione. Catalogo Selezione degli Oli in Concorso* Camera di Commercio di Oristano, Oristano.

Fontanazza, G. (2005) Importance of olive-oil production in Italy, in *Integrated Soil and Water Management for Orchard Development: Role and Importance,* Proceedings of the International Seminar "The Role and Importance of Integrated Soil and Water Management for Orchard Development" organized by the FAO Land and Water Development Division and the College of Agricultural Sciences, University of Teramo, Italy, 9–10 May 2004, Mosciano S. Angelo, Italy (eds J. Benites, M. Pisante, and F. Stagnari), Food and Agriculture Organization (FAO), Rome, pp. 13–49.

Franca, E. (ed.) (2010) *OLEA. Gli Olivi e l'Olio*, Edizioni OLEA, Pesaro.

Laboratorio Chimico Merceologico della Camera di Commercio di Napoli (2011) *Premio ExtraBio per gli Oli 2011 Biologici della Campania. Catalogo degli Oli Selezionati della Sesta Edizione*, Laboratorio Chimico Merceologico della CCIAA di Napoli, Naples.

Regione Toscana (2012) *Selection of PDO and PGI Extra Virgin Olive Oil of Tuscany 2011/2012*, Camera di Commercio di Firenze, Florence.

Unaprol (2012) *Lo Scenario Economico di Settore (2012)*, http://www.unaprol.it/images/stories/Lo%20Scenario%20Economico%20di%20Settore%202012.pdf (last accessed 1 August 2012).

Unione Italiana delle Camere di Commercio, Camera di Commercio di Perugia, MIPAAF, and Ministero dello Sviluppo Economico (2012) *Ercole Olivario 2012: Premio per la Valorizzazione delle Eccellenze Olearie Territoriali Italiane*, Camera di Commercio di Perugia, Perugia.

11 Olive Oils from Greece

Vassilis Zampounis,[1] Kostas Kontothanasis,[2] and Efi Christopoulou[3]
[1]*Axion Agro Ltd, Athens, Greece*
[2]*Olive and Olive Oil Magazine, Kalamata, Greece*
[3]*Ministry of Development, Athens, Greece*

11.1 Historical perspective

Olive trees and their cultivation in ancient Greece go back millennia to a time when mythology blends with history. Evangelos Velitzelos, Professor of Geology and Paleobotany at the University of Athens, unearthed fossilized olive leaves 50 000–60 000 years old in the area of the volcano of Santorini (Lampropoulos *et al.*, 2004). References to the olive tree and to olive oil were made by Homer (Homer, 8th century BC) and by many ancient Greek writers, such as Solon ("Greatest of remedies for all life's afflictions is the olive tree's fruit").

Ancient Greeks, deeply aware of the unique properties of olives and olive oil, considered these products to be gifts from the Gods.

Olive oil produced from the sacred olive trees of Athens was presented, as a prize, to winners at the Panathenaic Games. The victor of chariot races was offered 140 amphorae of olive oil, which in today's measures represents more than 5 tons (Faklaris and Stamatopoulos, 2003). Zeus, God of the Gods, was worshipped in ancient Olympia and victors at Olympic Games were crowned with an olive wreath, also known as *kotinos* (Faklaris and Stamatopoulos, 2003).

The cultivation and harvesting of the olive started in the Copper Age, around 3000 BC. with the use of means available at that time and in ways similar to those of today – in other words, either by beating the tree with a pole, handpicking the olives one-by-one, or gathering the mature fruit from the ground.

Olive Oil Sensory Science, First Edition. Edited by Erminio Monteleone and Susan Langstaff.

The grinding and extraction of olive oil also followed the same principles as today. The Latin term *trapetum* corresponds to a popular type of olive press, which consisted of two millstones rotating vertically in a large stone mortar (Faklaris and Stamatopoulos, 2003).

Specially designed amphorae were used for maritime transport, so that the sediment would settle in their pointed base. Olive oil was primarily used as a staple, either raw in salads or in cooking. Along with bread, olives, and cheese, it was considered necessary for survival.

The olive and its by-products were used as a basis for the preparation of many medicaments, such as fire wood, lamp oil, for personal hygiene, and for a multitude of other uses. Ancient Greeks used both the wood and the leaves of the olive tree.

Overall, the olive tree and olive oil were of paramount importance for ancient Greeks in terms of diet and trade, closely related to religion and civilization, touching upon all forms of everyday life (Faklaris and Stamatopoulos, 2003).

As the poet Odysseus Elytis wrote (Elytis, 1985): "If you break down Greece at the end you will be left with an olive tree, a vine, and a boat. That means with these you reconstruct."

11.2 Geographical and climatic characteristics

Greece is located in southeast Europe, in the Balkan Peninsula and the East Mediterranean Basin, connecting Europe with Asia (Turkey) and Africa and the countries of the Middle East. Its total surface is 13.2 million ha (Maps of World, 2013), of which about 3.8 million ha is cultivated land (28.8%) (ELSTAT, 2008), and 1.015 million ha of the latter (27%) is olive groves (YPAAT, 2005).

Greece has an extended coastline of 13 676 km and more than 2000 islands (Mediterrana, 2013). According to the Greek National Tourism Organization (EOT), Greek sovereign territory includes 6000 islands and islets and the country's coastline is ~16 500 km (Greek National Tourism Organization, 2013). Most of its land is hilly and semi-mountainous, thus offering a considerable environmental and climatic variety.

Greece's climate is typically Mediterranean, with a concentration of rainfall during the cold winter months. Climate change affects Greece through high temperatures during the summer, especially in the inlands, and reduced rainfall, especially in the semi-arid areas, in the southern parts of the country. The average annual rainfall is 402 mm, which is lower than in Italy (771 mm) and almost the same as in Spain (436 mm) (climatemps.com, 2013).

11.3 Overview of olive-producing regions

If the main mountain ranges are excluded, olive trees are cultivated throughout all of the rest of the country, especially in the southern regions where

Table 11.1 Distribution of olive trees for oil production in the various geographical areas of Greece (%).

Area	Proportion (%)
Peloponnese	31.57
Crete	23.40
Central Greece	14.99
Aegean Islands	9.09
Ionian Islands	4.90
Northern Greece	3.70
Total	87.65

Source: ELSTAT (2012), Census of Agriculture.

the climate is warmer. The remaining 12.35% is covered by olive trees for the production of table olives, although these percentages vary each year due to trade conditions and dual-use varieties. The distribution of olive trees for oil production and the distribution of olive producers in the various geographical areas of Greece are shown in Table 11.1 and Table 11.2, respectively. Figure 11.1 and Figure 11.2 are maps of the Protected Designation of Origin (PDO) and Protected Geographical Indication (PGI) olive-producing regions of Greece.

11.3.1 The 27 recognized PDO/PGI olive oils

The European Union has recognized 27 PDO/PGI olive oils in Greece, which form a representative example of the country's olive oil production capacity.

The data presented in Table 11.3, Table 11.4, and Table 11.5 were derived by the author from the approved folders of the Greek Ministry of Agriculture and the European Commission and additional data were provided by the

Table 11.2 Distribution of olive producers for both olive oil and table olives in the various geographical areas of Greece (%).

Area	Proportion (%)
Peloponnese	25.99
Crete	24.80
Central Greece	22.21
Aegean Islands	11.45
Northern Greece	10.71
Ionian Islands	4.83
Total[a]	100

[a]500 986 producers (4.6% of the total population according to the 2011 census).
Source: Ministry of Agriculture.

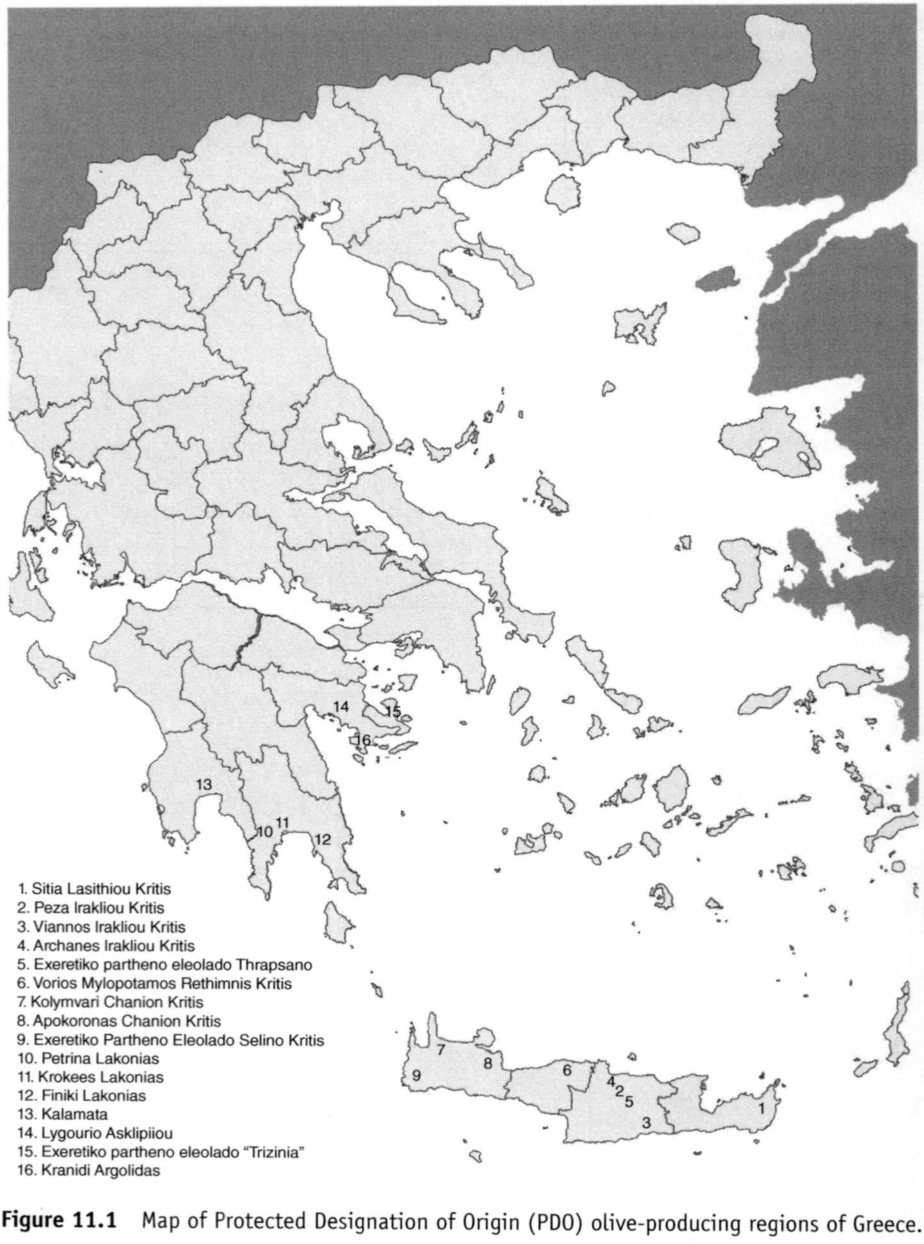

Figure 11.1 Map of Protected Designation of Origin (PDO) olive-producing regions of Greece.

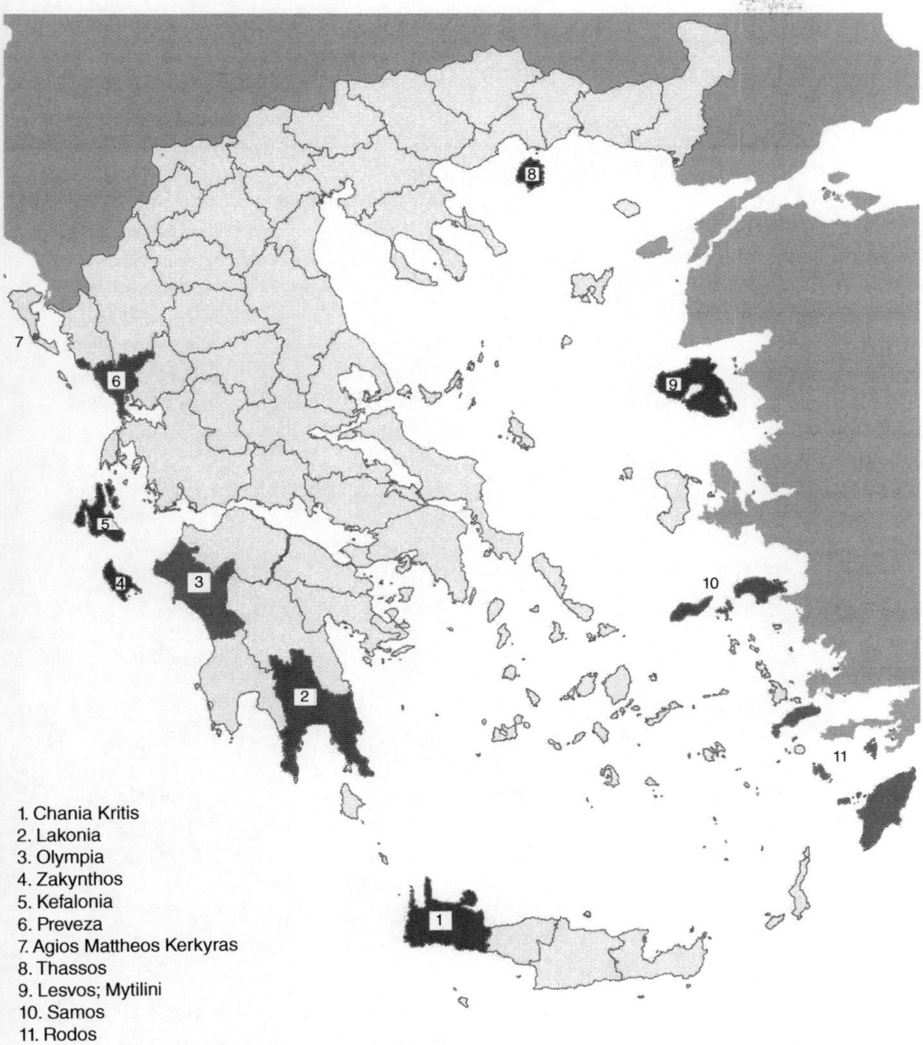

Figure 11.2 Map of Protected Geographical Indication (PGI) olive-producing regions of Greece.

International Olive Council (Technical Legal Study on Geographical Indications, Annex V) (International Olive Council, 2011).

11.3.2 Remarks

1. The harvest period for almost all varieties and regions extends from November to February. The harvest of the same variety may differ depending on the farming techniques (pruning) and the commercial

Table 11.3 Main olive cultivars in the Peloponnese, Greece.

Peloponnese region	PDO or PGI[a]	Main cultivars	Most usual sensory properties
South-western	Messinia–Kalamata	Koroneiki (90%), Tsounati (10%)	Aromas of freshly mown grass, olive leaf, and artichoke; green pepper aftertaste. Medium intensity of bitterness and pungency with a predominance of green fruitiness
South-eastern	Lakonia (PGI)	Koroneiki, Tsounati (Athinoelia), Koutsourelia, Asprolia	Not available
	Petrina Lakonias	Koroneiki	Not available
	Krokees Lakonias	Koroneiki	Not available
	Finiki Lakonias	Koroneiki (70%), Tsounati (Athinoelia) (25%), Asprolia (5%)	Not available
Central-eastern	Lygourio Asclepieion Argolis	Manaki	Not available
	Kranidi Argolis	Manaki (90%), Koroneiki (10%)	Not available
	Trizinia	Manaki (50%), Koroneiki (50%)	Not available
Central-western	Olympia (PGI)	Koroneiki (90%), Kollyreiki (10%)	Not available

[a]PDO unless specified as PGI.

purpose. For example, the trees of the Tsounati (Athinoelia) variety in Laconia are short and the olives are collected very early (September, October) for the production of very low-acidity ($<0.3^{\circ}$) and intensely fruity extra virgin olive oil. In other places (e.g., eastern Crete) farmers leave the trees to grow taller and collect the olives later, when they are fully mature, producing mild olive oil of higher acidity, but with lower fruitiness and bitterness and a lighter aroma (Mpalatsouras, 1999).

2. Olive orchards are characterized by a traditional planting system of low to medium intensity, namely 100–250 trees per hectare (Vemmos, 2007).
3. The olives are collected by hand, using a variety of small portable equipment such as combs. The olives fall on to plastic nets that cover the whole area around the tree. The olives are placed in bags or plastic containers and are transported to the olive mills (Mpalatsouras, 1999).
4. All olive mills use two- or three-phase centrifugation technology.

Table 11.4 Main olive cultivars in Crete, Greece.

Crete region	PDO or PGI[a]	Main cultivars	Most usual sensory properties
Chania (western Crete)	Kolymvari Chania	Koroneiki	Not available
	Apokoronas Chania	Koroneiki (90%), Tsounati (10%)	Not available
	Selino Chania	Koroneiki (40%), Tsounati (60%)	Not available
	Chania (PGI)	Koroneiki (75%), Tsounati (25%)	Not available
Rethymnon (central-western Crete)	Vorios Mylopotamos Rethymno	Koroneiki (90%), Chondroelia (10%)	Not available
Heraclion (central Crete)	Peza Heraclion	Koroneiki	Not available
	Thrapsano Heraclion	Koroneiki	Not available
	Archanes Heraclion	Koroneiki	Not available
	Viannos Heraclion	Koroneiki	Not available
Lasithi (eastern Crete)	Sitia Lasithion Crete	Koroneiki	Aromas of green apple, freshly mown grass; weaker aromas of artichoke, olive leaf, and ripe fruit. Moderate bitterness and pungency with a predominance of green fruitiness

[a]PDO unless specified as PGI.

11.4 Messinia–Kalamata

11.4.1 Brief historical overview

The history of olive cultivation in the Peloponnese begins somewhere in prehistory. The first written facts about olives belong in very early history regarding Messinia, south western Peloponnese, the capital of which today is Kalamata. In the mid-twentieth century, American archaeologist Carl Blegen discovered in western Messinia, near the Ionian Sea, a palace that he determined had belonged to Nestor, King of Pylos. Nestor took part in the Trojan War, as his city was the second largest center in the Mycenaean world after the city of Mycenae. The palace was destroyed in the early twelfth century BC by a fire that spread quickly due to the large quantities of olive oil

Table 11.5 Main olive cultivars in Aegean Islands, Ionian Islands, and Central Western Greece.

Region	PGI	Main cultivars	Most usual sensory properties
Aegean Islands (northern Aegean Sea)	Lesvos Island	Kolovi (70%), Adramytini (30%)	Tomato leaf aroma dominates followed by green fruit; weaker aromas of apple, freshly mown grass, and ripe fruit. Moderate bitterness and pungency with a predominance of green fruitiness
	Samos Island	Throumboelia (Chondroelia) 90%	Not available
	Thassos Island	Thassitiki	Not available
Dodecanese Islands (southern Aegean Sea)	Rhodes Island	Koroneiki, Throumboelia	Not available
Ionian Islands (western Greece)	Agios Matthaios Kerkyras (Corfu Island)	Koroneiki	Not available
	Kefalonia Island	Koroneiki (55%), Ntopia (45%)	Not available
	Zakynthos Island	Koroneiki (95%), Ntopia Lianoelia (5%)	Not available
Central-western Greece	Preveza	Lianoelia of Corfu	Not available

in the storerooms, some of which were located behind the throne room – this indicates the importance attached to olive oil. The fire had a positive outcome, however: thousands of clay tablets, covered in oil and baked at the high temperature of the fire, were saved. These tablets were written in Linear B, and in 1952, architect Michael Ventris and philologist John Chadwick, using decoding methods, were able to interpret them and determined that they were annual accounting records. Ventris was able to discern the word E LA WON, which is the modern Greek word *elaion*, meaning oil (Karpodini-Dimitriadi, 2003, Nileas, 2012).

11.4.2 Geography and climate

Messinia occupies the south-western section of the Peloponnese and consists of an area of 2991 km^2. To the north-west, it borders on Ilia, in the region of

Olympia, where the Olympic Games were held in antiquity; the Neda River serves as a natural boundary. To the north-east, the border with Arcadia winds through the slopes of mountains. The eastern boundary of Messinia with Laconia, the capital of which is Sparti, is formed by the Mount Taygetus range running north to south in the middle promontory of the Peloponnese. To the south and west, Messinia's coast lies along the Mediterranean Sea (Encyclopaedia Papyros – Larousse – Britannica, 1990).

Messinia has plains, such as that of Pamisos between Kalamata and Messini. A group of hills separate the Plain of Pamisos from the Plain of Meligala, which is located south of the mountains forming a boundary with Arcadia. There are other notable lowlands on the west coast and the coastal areas of the Gulf of Messinia (Encyclopaedia Papyros – Larousse – Britannica, 1990).

Approximately one-third of the land is flat and the other two-thirds is extensively hilly or mountainous (Matsouka, 2007). The tallest mountain is Mount Taygetus, with an altitude of 2404 m, which is mainly responsible for the microclimate of the Messinia area.

The microclimate in this zone is described as mild Mediterranean (hot and dry – temperate) to subtropical. Winters are mild and summers are long and hot. The cold period lasts from November to April and the hot period from May to October. The average annual rainfall is ~750–800 mm, with most of the rainfall coming in winter (~330 mm). Autumn receives the second highest rainfall at 250 mm, then spring with 146 mm and summer with just 23 mm. The driest month is July, with only 5.2 mm, and the wettest is November, with 138.2 mm. Concerning the average monthly temperatures through the year, the lowest temperatures are usually recorded in December and January and average 10 °C and the highest are in July and August at 28 °C. The region receives more than 3000 h of sunshine annually.

This microclimate is ideal for olive cultivation, as there are no extreme variations in temperatures and there is well-distributed and adequate rainfall to create optimum conditions for the annual olive cycle.

The Messinia soil is also the most suitable for cultivating olive trees, as it is sandy clay with a neutral to alkaline pH. The terrain is mostly hilly, with medium water permeability and satisfactory drainage to facilitate the transport of water and soil solutions (Union of Agricultural Cooperatives of Messinia, 2009).

11.4.3 Olive cultivation significance

Olive cultivation has expanded significantly in recent years; according to 1983 statistics, there were almost 11.3 million olive trees in Messinia that year, producing 33 021 t of oil and 6188 tons of table olives (Encyclopaedia Papyros – Larousse – Britannica, 1990). The Directorates of Agricultural

Economy for Messinia and Trifylia estimate that there are 15.6 million olive oil-producing trees of every variety in the Messinia region covering an area of about 86 800 ha. The average annual olive oil production is estimated at 52 000 t, but is subject to variation, mainly because olive trees are alternate bearing, and second because adverse weather conditions, the most serious of which is frost, also affect yields.

From the coast to the plains and from the small hills to the tallest mountains, wherever one looks, there are olive trees.

11.4.4 Varieties

As the name indicates, the Koroneiki variety originated from the area of Koroni, a historical city of Messinia located on the eastern side of the western-most promontory in the south Peloponnese. It is by far the dominant variety at over 95% and is cultivated throughout Messinia. The tree grows to a height of 5–7 m, with deep-green leaves about 5.47 cm long (±0.52 cm) and about 1.03 cm wide (±0.12 cm). The fruit is an elongated conical shape, with an average weight of 1.3 g and a very small nipple. The ratio of flesh to stone is 6.6:1. The oil content can be as high as 27%. It is used exclusively to produce exceptional quality olive oil. It is a very productive variety, resilient in hot, dry, and windy conditions, but vulnerable to the *Pseudomonas savastanoi* bacterium, which causes olive knots, to the olive fruit fly, and to the curculio beetle (Pontikis, 1981). In Messinia, the greatest enemy of olives is the olive fruit fly, which is combated by spraying bait from the ground. This means that a relatively small amount of the chemical is sprayed on to the tree trunk if necessary to control the insect and avoid leaving chemical residues in the olive oil. Such efforts are generally successful.

The low rainfall in Messinia during spring and summer and thus the reduced water uptake by the olive trees increase the aromatic level of the fruit and result in the production of a balanced and complex olive oil. The moderate winds and hilly relief of the area, combined with the cup-shaped pruning of the trees (three or four branches per tree and the removal of part of the foliage from its center, allowing sufficient exposure to light and air), are also factors that help the olives to mature properly. As most olive groves are located on hilly slopes and are pruned by growers into cup shapes, they are exposed to ample sunlight (Union of Agricultural Cooperatives of Messinia, 2009). As a result, photosynthesis is intensified and higher concentrations of carbohydrates are produced, which add a high percentage of aromatic compounds to the olive oil. In this way, the olive trees of the local Koroneiki variety, planted in their unique environment, are able to bear properly matured fruit and produce an olive oil with an intense green color and a pleasant flavor and aroma. Depending on the area, the olives are harvested from mid-November until the end of January and they are mostly made into extra virgin olive oil.

The Matsolia or Mastoidis tree has upright branches and grows to a height of 6–8 m. Its leaves are light green, about 6.18 cm (±0.67 cm) long and 1.18 cm (±0.13 cm) wide. The fruit is oval shaped, with an average weight of 2.6 g and a nipple. The ratio of flesh to stone is 6:1. The oil content varies around 20%. The Mastoidis is used mainly to produce exceptional quality olive oil (Pontikis, 1981). This variety is resilient to cold, and as such is mainly cultivated in the mountain villages of Mount Taygetus, where the winter frosts prevent the Koroneiki variety from thriving.

The Mavrolia trees are larger and more vigorous than the Koroneiki trees. The leaves of the Mavrolia are green, about 5.87 cm (±0.50 cm) long and 1.44 cm (±0.17 cm) wide. The olive has an elongated conical shape, with an average weight of 2.4 g and a small nipple. The ratio of flesh to stone is 6.5:1. The oil content varies around 19%. It is used mainly for the production of exceptional quality olive oil (Pontikis, 1981). This oil is mainly produced in western Messinia in early autumn, as it is an early variety and harvest begins before November. Its sensory properties are generally very good, but it has been observed that oil made from the Mavrolia varietal does not have the same shelf life as the Koroneiki under the same conditions. Good-quality Mavrolia olive oil that is properly stored has a shelf life of about 6 months, whereas Koroneiki oil can be stored for at least 18 months.

11.4.5 PDO Kalamata

The Kalamata name, which refers to the olive oil produced from the Koroneiki and Matsolia varieties in the region of the former Kalamata province, has been a registered PDO product in the European Union since 1997. The Koroneiki variety is far more prevalent, as the Mastoidis or Matsolia is mostly cultivated in a smaller area in the mountain villages of Taygetus. The PDO Kalamata zone extends from the outlying areas of Kalamata to the north for about 15 km and in the Messinian Mani region along Mount Taygetus.

Given that the olive oil from the rest of Messinia is very similar to the PDO Kalamata oil, an application has been submitted to amend the PDO Kalamata characteristics and extend the PDO zone from the former province of Kalamata to include all of Messinia. The application was submitted to the European Union in December 2009.

According to the amendment application in reference to the new zone that would include all of Messinia, the PDO Kalamata olive oil is extra virgin and is produced from the Koroneiki and Mastoidis varieties. It mainly originates from Koroneiki olives and uses a maximum of 5% Mastoidis olives. The PDO Kalamata oil resulting from the zone expansion would not include oil made from the Mavrolia variety.

The PDO Kalamata olive oil has the following chemical characteristics. The total acidity, expressed as oleic acid by weight, does not exceed 0.50 g

Table 11.6 Standard indices for the presence of various types of oxidized substances in the oil at bottling.

Index	Value
K_{232}	Max. 2.20
K_{270}	Max. 0.20
Peroxide value	$\leq$14 meq. O_2/kg
Total sterols	$>$1100 mg/kg
Oleic acid	70–80
Linoleic acid	4.0–11.0
Stearic acid	2.0–4.0
Palmitoleic acid	0.6–1.2
Palmitic acid	10.0–15.0

Source: Union of Agricultural Cooperatives of Messinia (2009).

per 100 g of oil. Standard indices for the presence of various types of oxidized substances in the oil at bottling are shown in Table 11.6.

11.4.6 Sensory properties

According to the amendment application, the sensory properties of PDO Kalamata olive oil are largely attributable to the aromatic substances formed as a result of the soil characteristics (slightly calcareous soils) and decreased water intake by the trees because of the low rainfall and the lack of irrigation in most areas of Messinia. At the same time, analysis of the fatty acids in olive oil found that this profile is very specific and constitutes a unique characteristic of PDO Kalamata olive oil. Specifically, the percentage of oleic acid is very high, whereas the levels of linoleic, stearic, palmitoleic, and palmitic acid are very specific and present in a particular ratio, setting them apart from other olive oils. This specific profile of fatty acids combined with the moderately intense fruitiness, green fruit aroma, slight bitterness, and mild pungency make up the unique character of PDO Kalamata olive oil.

The color of the oil ranges from green to yellowish green. The fruitiness of the olive has an average value of between 3 and 5, and the bitterness is between 2 and 3. For pungency, the average value ranges from 2 to 4. There are no flaws (Union of Agricultural Cooperatives of Messinia, 2009).

11.4.7 PDO Kalamata profile

The PDO Kalamata olive oil is processed at the olive mills using exclusively mechanical means (crushing/pressing or mostly centrifugation), with temperatures throughout all processing stages of up to 27 °C (Union of Agricultural Cooperatives of Messinia, 2009). The result is that the sensory and chemical properties of the extra virgin olive oil remain unaltered. The oil that results remains the natural juice contained in the fruit.

11.4.8 Sensory properties of Messini olive oil

During the winter 2011–2012, a study was conducted by the Technological Educational Institute (TEI) of Kalamata on the olive oils of the Municipality of Messini, the area of which is roughly equivalent to one-fifth of the Messinia region. The Municipality of Messini extends from the north-west coast of the Gulf of Messinia to the hills and mountains of central and western Messinia.

The taste study conducted on representative examples from the entire Messini Municipality area indicated that in extra virgin olive oil, the mild-intensity fruitiness (a grade of less than 3) was more prevalent (81%), while more moderate fruitiness (intensity between 3 and 6) was found in about 19%. The bitterness was mild in 100% of the samples (less than 3). In testing for pungency, 71% of the samples were mildly pungent (less than 3) and 29% were moderately pungent (between 3 and 6).

Overall, it was found to be a light and well-balanced extra virgin olive oil (Demopoulos, 2012).

11.5 Sensory characteristics of the major Greek olive varieties

Data were collected and elaborated by one of the authors, Efi Christopoulou.

11.5.1 Koroneiki variety

This is the dominant Greek variety, as over 60% of Greek olive oil is produced from this variety. The olive fruit is of small size (weight 0.5–1.4 g) and golden green in color. The oil is moderately to intensely fruity, full bodied, with a long aftertaste, and has moderate to strong bitter and spicy notes (usually more spicy than bitter). Without significant complexity of other aromas, it often carries a pleasant characteristic flavor and scent of green leaves and cut grass, which lend it intense freshness. Some oils of this variety also have subtle notes of artichokes, green apples, and green pepper.

11.5.2 Athinolia (or Tsounati, or Mastoidis)

This is cultivated mainly in the Peloponnese (central and southern) and western Crete. The olive fruit is of small size (weight 1.4–2.25 g) and golden green in color. The oil is moderately to strongly fruity, with an excellent harmony of bitter and spicy notes. The characteristics are similar to those of the olive oil produced by the Koroneiki variety, but of milder intensity. Some oils of this variety also have subtle notes of bitter almonds and citrus fruit.

11.5.3 Manaki (or Agouromanako)

This variety is cultivated mainly in the eastern Peloponnese. The olive fruit is of medium size (weight 2.2–2.9 g) and golden green in color with yellow hues. The oil is mild to moderate fruity, with very slight bitter and spicy notes, qualifying as a sweet oil. It is soft on the palate and leaves a pleasant fresh aftertaste.

11.5.4 Kolovi and Adramytiani

These are cultivated on Lesvos Island. The olive fruit is of medium size (weight 3.0–4.5 g) and yellow in color with green hues. The oil is intensely fruity with complex aromas. Its bouquet often includes notes of fresh tomato leaves and other fresh leaves. It is moderately bitter and spicy, depending on the ripeness of the olives.

11.5.5 Koutsourelia (or Ladolia or Patrini)

This is cultivated mainly in the northern Peloponnese and southern part of central Greece. The olive fruit is of small size (weight 1.0–1.8 g) and golden green in color. The oil is less bitter and spicy than other olive oils, and is moderately to intensely fruity, with predominant aromas of apples and bananas. Olive oils of this variety often have a characteristic flavor, aroma, and retronasal sensation of bitter almonds, artichokes, and oregano.

11.5.6 Lianolia of Corfu variety

This variety is cultivated mainly in the western part of central Greece. The olive fruit is of small size (weight 1.1–1.8 g). It resembles the Koroneiki variety. It is fresh and fruity, with a harmony of bitter and spicy notes.

Table 11.7 lists other olive varieties in Greece.

Table 11.7 Other olive varieties in Greece.

Variety	Regions	Size of the fruit	Purpose of cultivation
Megaritiki	Northern Peloponnese, central Greece	Medium size (4.2 g)	Dual purpose for olive oil and table olives
Maroneias	Thraki	Medium to large size (3.0–6.0 g)	Dual purpose
Kothreiki	Central Greece, Samos	Small size (3.0–5.0 g)	Dual purpose

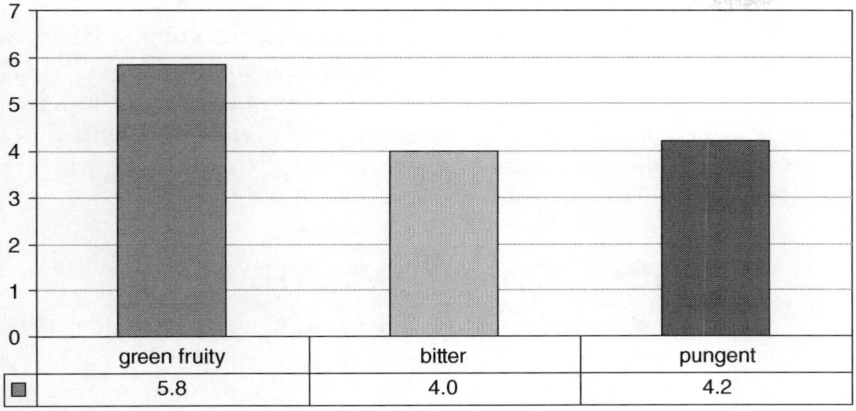

Figure 11.3 Extra virgin olive oil from PDO Kalamata: fruity, bitter, and pungent.

11.6 Three typical examples of sensory analysis

11.6.1 PDO Kalamata (Peloponnese, A1)

Figure 11.3 presents the intensity of the three main sensory attributes of extra virgin olive oil from PDO Kalamata: fruity, bitter, and pungent. This is a well-balanced sensory profile with medium intensity of bitterness and pungency and a predominance of green fruitiness.

Figure 11.4 is an analytical sensory profile of direct or retronasal aromatic olfactory, gustatory, qualitative retronasal, and tactile or kinesthetic sensations.

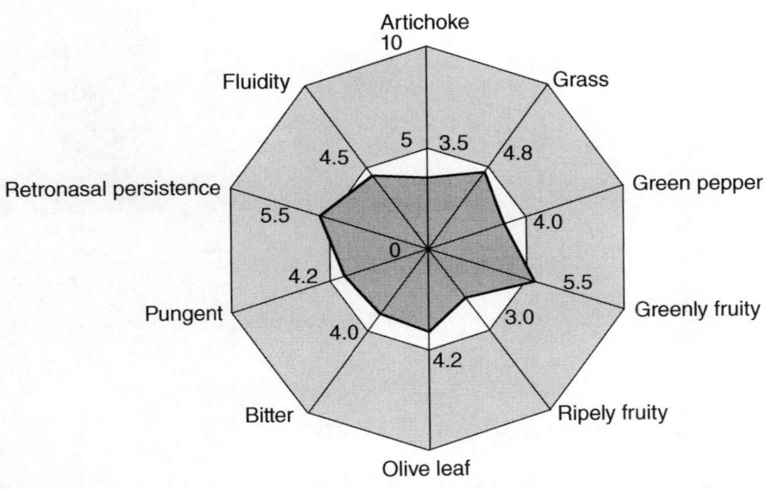

Figure 11.4 Extra virgin olive oil from PDO Kalamata: analytical sensory profile.

Odor: Very fresh and rich overall sensation; intense greenly fruity with undertones of freshly mown grass, olive leaf, and artichoke.

Taste: Fruity and harmonious, showing character, delicate in the tones of ripely fruity. Pungency is slightly more intense than bitterness. Unparalleled persistence on the palate, beginning with greenly fruity and closing with green pepper aftertaste.

11.6.2 PDO Sitia Lasithion (Crete, B19)

Figure 11.5 presents the intensity of the three main sensory attributes of extra virgin olive oil: fruity, bitter, and pungent. It is a fruity and elegant extra virgin olive oil with light to moderate notes of bitterness and pungency and a predominance of green fruitiness.

Figure 11.6 is an analytical sensory profile of direct or retronasal aromatic olfactory, gustatory, qualitative retronasal, and tactile or kinesthetic sensations.

Odor: Rich overall sensation; greenly fruity dominates in the sensory profile of this oil, followed by green apple and freshly mown grass. Weaker sensations of artichoke, olive leaf, and ripely fruity add refinement and complexity.

Taste: Fruity and harmonious; pungency is slightly more intense than bitterness. The palate is balanced with an initial greenly fruity that then turns to green apple and finishes with a taste of artichoke.

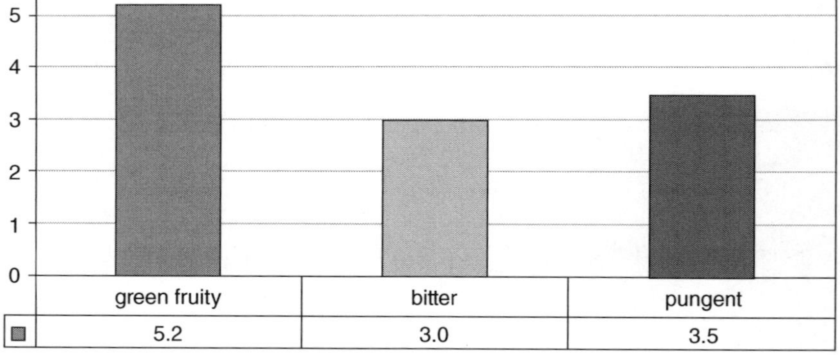

Figure 11.5 Extra virgin olive oil from PDO Sitia Lasithion: fruity, bitter, and pungent.

11.6.3 PGI Lesvos Island (Lesvos, C20)

Figure 11.7 presents the intensity of the three main sensory attributes of extra virgin olive oil: fruity, bitter and pungent. This is a well-balanced sensory

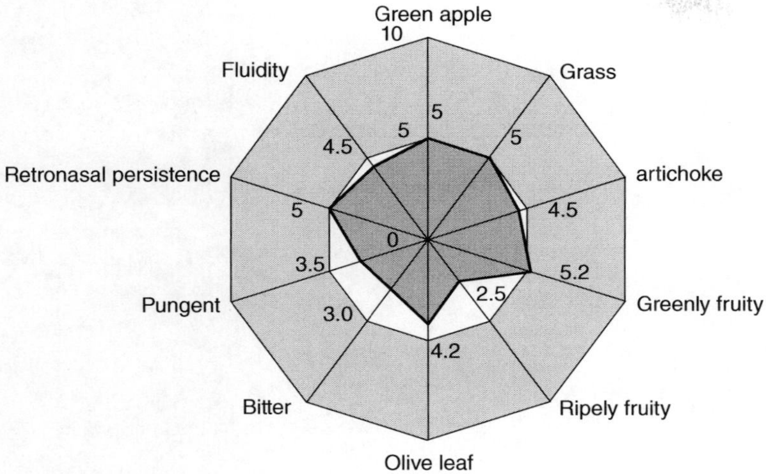

Figure 11.6 Extra virgin olive oil from PDO Sitia Lasithion: analytical sensory profile.

profile with moderate notes of bitterness and pungency and a predominance of green fruitiness.

Figure 11.8 is an analytical sensory profile of direct or retronasal aromatic olfactory, gustatory, qualitative retronasal, and tactile or kinesthetic sensations.

Odor: Excellent overall sensation; tomato leaf, vigorous and persistent, dominates, followed by a greenly fruity sensation. Weaker sensations of apple, freshly mown grass, and green and ripely fruity add refinement and complexity.

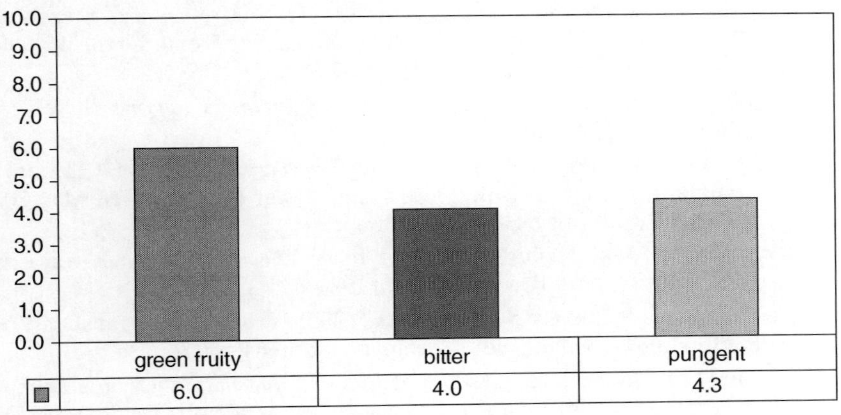

Figure 11.7 Extra virgin olive oil from PGI Lesvos Island: fruity, bitter, and pungent.

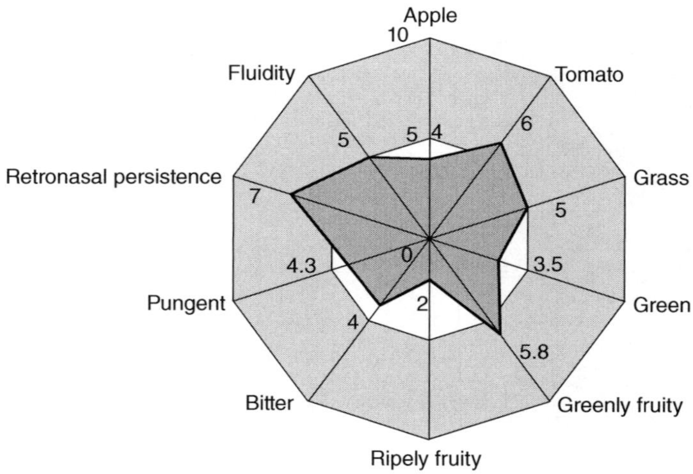

Figure 11.8 Extra virgin olive oil from PGI Lesvos Island: analytical sensory profile.

Taste: Harmonious and elegant, delicate in the tones of ripely fruity. It has a balanced bitter and pungent content closing with a tomato leaf and greenly fruity aftertaste.

References

climatemps.com (2013) *Greece*, http://www.greece.climatemps.com (last accessed February 2013).

Demopoulos, V. (2012) Organoleptic properties of the olive oil from the Municipality of Messini (title translated; original title in Greek: *Γευσιγνωστικά χαρακτηριστικά ελαιόλάδόυ Καλλικρατικόύ Δήμόυ Μεσσήνης*), presented in February 2012 at Messini Town Hall; Olive Oil Tasting Team at the Kalamata Technological Educational Institute (TEI), 28, 29, 30, 33.

ELSTAT (2008) *Provisional Data. Crop Areas by Region and Category*, Hellenic Statistical Authority, ELSTAT, Piraeus.

ELSTAT (2012) Information from virtual leaflet *Greece in Numbers* (2011) and the ELSTAT database available at http://www.statistics.gr (last accessed May 2012), Hellenic Statistical Authority, ELSTAT, Piraeus.

Elytis, O. (1985) *O Mikros Naftilos*, Ikaros, Athens; *The Little Mariner*, verse XVI, Copper Canyon Press, Port Townsend, WA, 1988.

Encyclopaedia Papyros – Larousse – Britannica (1990) *Messinia* (title translated; original title in Greek: *Μεσσηνία*), vol. 41, Papyros, Athens, pp. 360, 363.

Faklaris, P., and Stamatopoulos, V. (2003) *Minutes of International Conference, The Olive and Olive Oil from Antiquity to Present Times*, Academy of Athens, Publications of the Hellenic Folklore Research Centre No. 19, pp. 33–47.

Greek National Tourism Organization (2013) *Greek Islands*, http://www.visitgreece.gr/en/greek_islands (last accessed February 2013).

Homer (eighth century BC), *Odyssey, Rhapsodies and Verses*, K-364, Z-79, Ψ 195–198.

International Olive Council (2011) *Geographical Indications*, http://www.internationaloliveoil.org/estaticos/view/388-geographical-indications (last accessed February 2013).

Karpodini-Dimitriadi, E. (2003) *The Peloponnese – a Travel Guide to Museums, Monuments and Archaeological Sites* (title translated; original title in Greek: *Πελόπόννησός – Ταξιδιωτικός όδηγός μόυσείων, μνημείων και αρχαιόλόγικών χώρων*), Ekdotiki Athinon, Athens, pp. 162–164.

Lampropoulos, V., Panagiaris, G., Papachristodoulou, I., Klapadaki, G., Velitzelos, E., and Velitzelos, D. (2004) Conservation of fossilized olive leaves from Santorini Island – Greece, *Flora Tertiaria Mediterranea*, vol. 6, section 8, Verlag Documenta Naturae, Munich, pp. 1–27.

Maps of World (2013) *Where is Greece*, www.mapsofworld.com/greece/greece-location-map.html (last accessed February 2013).

Matsouka, P. (2007) *Messinia. Place – Time – People.* (title translated; original title in Greek: *Μεσσηνία. Τόπός – Χρόνός – Ανθρωπόι*), Militos, Athens p. 21.

Mediterrana (2013) *About Greece*, http://www.mediterrana.net/about-greece.html (last accessed February 2013).

Mpalatsouras, G. (1999). *The Olive Culture (I Elaiourgia)*, published by the author, Athens, vol. 4, pp. 57–63.

Nileas (2012) Nileas Growers Group for Integrated Olive Oil Production, historical data provided by G. Hatzi-Spiliopoulou, archaeologist with the 7th Ephorate of Prehistoric and Classical Antiquities in Olympia, available at http://www.nileasoliveoil.gr/ (last accessed May 2012).

Pontikis, K. (1981) *Olive Cultivation* (title translated; original title in Greek: *Ελαιόκόμία*), Karaberopoulos, Athens, pp. 103, 106, 109.

Union of Agricultural Cooperatives of Messinia (2009) Application to amend the specifications for the "Kalamata" Protected Designation of Origin olive oil, *Official Journal of the European Union*, 2012/C 186/10, pp. 21–25.

Vemmos, S. (2007) The super intensive olive groves. *Olive and Olive Oil (Elia and Elaiolado) Magazine*, (54), 36–43.

YPAAT (2005) Greek Ministry of Agriculture, *Official Government Journal*, Φ.ΕΚ., Issue 1552/B.

12
Olive Oils from California

Alexandra Kicenik Devarenne[1] **and Susan Langstaff**[2]
[1]*CalAthena, Petaluma, CA, USA*
[2]*Applied Sensory LLC, Fairfield, CA, USA*

12.1 Overview of olive oils from California

California produces most of the domestic extra virgin olive oil made in the United States, an amount that represents about 2% of total US olive oil consumption. The annual consumption of olive oil in the United States is almost 70 million gallons (265 million liters) or 0.24 gallons (0.9 liters) per person per year. Currently, the US imports 98% of the olive oil that it consumes and most of the imported olive oil originates from the Mediterranean basin. Ready access to the large and growing US market presents an extraordinary opportunity for the state's olive oil industry.

12.2 California climate and geography

Although the olive tree, *Olea europaea L.*, is not native to California, olive trees thrive there because of the state's climate. California has what is technically a Mediterranean, or two-season, climate in which the rainfall is concentrated in the mild winter months and the long summers are hot, dry, and sunny. The resemblance to the Mediterranean area proper is instantly apparent and it is supported by numerous analogies of temperature, rainfall, and humidity with stations in southern France, Italy, Spain, Greece, and Algeria. There is one important difference between the growing areas in California and the Mediterranean: the Mediterranean region is east–west in orientation and borders an inland sea, whereas the olive tree-growing region of California is orientated north–south separated only by the coastal ranges

Olive Oil Sensory Science, First Edition. Edited by Erminio Monteleone and Susan Langstaff.

from the Pacific ocean. This produces variants of the "Mediterranean" climate unknown in the Mediterranean itself.

Olive orchards are primarily located in seven regions of California, all of which have Mediterranean climates: North Coast, Central Coast, South Coast, Sierra Foothills, Sacramento Valley, San Joaquin Valley, and Inland Southern California (Sunset, 2012).

The coastal counties are known for their mostly smaller scale, high-quality production. Cool Pacific fogs sweep in through gaps in the coastal mountain ranges to cool the valleys – but only where they can reach. If the mountains are too dense, and if there are no river valleys to allow them passage, then the fogs stay where they are and the orchards heat up under the hot California sun. In Mendocino and Lake Counties, on California's north coast, there are areas where the fog makes the days as cool as in northern Europe. Conversely, there are other parts where the fog cannot reach, and the area is more suited to olive varieties from hot climates (Halliday, 1993).

The olive-growing regions of the coastal areas in California are dwarfed in comparison with those of the Central Valley. This immense basin – the Central Valley – lying between the coastal ranges and the Sierra Nevada mountains, is divided into the Sacramento Valley at the north and the San Joaquin Valley at the south. There is little rainfall in the San Joaquin Valley, and not many years ago it was a baking desert. The valley is still blistering hot, but under irrigation its rich soil can be enormously productive.

To the south, around Los Angeles, lies another region for growing olive trees. In Southern California, population and housing pressures have put many farmers out of business. There are many isolated trees or fragments of old groves, but the land along the Southern California coast is too expensive for olive growing. Although least important today, it was in this southern region that California's first olive trees were planted, two centuries ago.

12.3 History

California was first settled by Europeans starting in 1769 when Spanish Franciscans made their way from Baja California and established a series of missions there. They brought the first olive trees with them, primarily to supply oil for sacramental use but also for soap, food, and lighting. The trees were planted in gardens adjacent to their missions and were descendants of those brought from southern Spain to Central America in the sixteenth century. Around 1870, several small orchards with many different European varieties were planted for olive oil, all along the California coast from San Diego up to Sonoma, and in various foothill areas of the Sierra Nevada mountains. By 1885, California olive growers were producing oil from ~2000 acres (809 ha), but they could not compete well against seed oils and olive oil imports from Europe (Vossen and Devarenne, 2005).

Starting in 1900, the "California Style" firm, black olive was developed and olives were widely planted in the newly drained Central Valley to produce table olives. The table fruit industry ultimately grew to more than 35 000 acres (14 164 ha), most of which are located in the southern San Joaquin Valley. For many years, a small California olive oil industry used damaged or undersized olives discarded by the table fruit industry to produce oil (Connell, 2005; Sutter, Sibbett, and Ferguson, 2005; Vossen, 2005; USDA National Agricultural Statistics Service, California Field Office, 2006).

In the 1930s, some Italian immigrants started to make oil, largely for their personal use, from old Mission, Sevillano, and Manzanillo trees (Quest-Ritson, 2006). Starting in about 1990, new plantings of oil varieties and production of European-style oils began to occur. Olive oil came to be seen as a high-quality product, a symbol of Mediterranean civilization, and a desirable status symbol. Its health benefits were also widely praised. Several small orchards were planted first in the North Coast region and later in the South and Central Coast regions. These orchards were planted mainly to Italian varieties such as Frantoio, Leccino, Taggiasca, and Coratina at "high-density" spacing of 250–300 trees/acre (618–741 trees/ha). In contrast to California's traditional groves – widely spaced at 50–100 trees/acre (124–247 trees/ha) – high-density (HD) spacing aims to increase production rates by more than doubling the number of trees per acre. High-density plantings are expected to reach full production at about 7 years.

The first super-high-density (SHD) orchards in California were planted in 1999 (UC Davis Olive Center, 2009). These orchards have a spacing of 12–13 ft (3.6–4 m) by 4–6 ft (1.2–1.8 m), which results in 500–900 trees/acre (1236–2224 trees/ha) and are expected to reach full production at 5 years. The orchards were designed for over-the-row machine harvesting and are planted to Arbequina, Koroneiki, and Arbosana varieties. The majority of the olives in California are planted in the Central Valley: the Sacramento Valley to the north and the San Joaquin Valley to the south.

12.4 Consumption and production

During the past 15 years, there has been a surge of interest in olive oil as an ingredient to flavor gourmet food and as a healthier alternative to other forms of fats and oils in the US diet. Olive oil sales in US markets increased by 100% from 1991 to 2003 and are continuing to increase. EU estimates put the continuing growth of US olive oil consumption at 4% per annum (Datamonitor, 2010). Currently, about 8% of all fats and oils consumed in the United States is olive oil. Because of its large population, the United States ranks as third overall in world consumption of olive oil (North American Olive Oil Association, 2012). The EU estimate from 2010 puts the US consumption of olive oil at about 1 l per person per year, or a total of 66.6 million gallons (252 million liters) by 2013 (Datamonitor, 2010).

Most domestic production of olive oil is in California, where production has been increasing, from 247 500 gal (936 900 l) in 1999–2000 to 1.2 million gal (4 542 500 l) of olive oil from the 2010–2011 harvest. A crop failure in 2010–2011 resulted in flat production, but a projected 2 million gal (7 571 000 l) is expected in 2012–2013. During the period 1999–2006, Californians planted just less than 8000 acres (19 760 ha) of oil olive cultivars, bringing the total area devoted to olive oil production to 10 500 acres (4249 ha) in 2006. In 2012, there is an estimated 60 000 total acres (148 200 ha) planted to olives in California, just over half of that being oil olive cultivars (INFOSCAN Statistics, 1999; USDA Economic Research Service, 1999; North American Olive Oil Association, 2004; Senise Barrio and Carman, 2005; Vossen and Devarenne, 2005, 2006; Datamonitor, 2010; COOC, 2011; UC Davis Olive Center, 2012).

12.5 Production systems

In the past, olive orchards in California were planted at the wide spacing typical of dry farming. Such "traditional" systems with up to 50 trees/acre still exist, but mostly as relics. Modern irrigated table olive orchards in California were sometimes still planted with very wide spacing (25–35 ft or 7.6–10.6 m between trees) to permit large trees with good light exposure, but even for table olives the trend is to tighter spacing. Modern California olive oil producers are planting trees at about 175–270 trees/acre (430–670 trees/ha) in the high-density (HD, also called medium-density) system and about 560–900 trees/acre (1380–2225 trees/ha) in the SHD system. The more trees that are planted per acre, the faster the planting comes into full bearing, but when full production is reached, yields are equal (Civantos López-Villalta and Pastor, 1996; Tous *et al.*, 1999). The HD system can be planted to steeper ground and harvested with tree shakers or various hand methods. It allows for the use of any variety and is less intensively managed. The SHD system is designed to accommodate an over-the-row modified grape harvester, so the land must be flat or gently sloped. So far, only three varieties, Arbequina, Arbosana, and Koroneiki, have been planted in the system in California. Although sometimes referred to as dwarf varieties, this is not actually the case. The cultivars used in SHD are simply lower in vigor, so they are more easily managed in this production system that requires the trees to remain small enough to accommodate over-the-row harvesters. Establishment costs are higher because of higher tree and infrastructure numbers, but overall production costs are lower than for the hand-harvested HD groves, primarily because of the efficiency of mechanical harvesting (Vossen, 2002, 2007b; Vossen *et al.*, 2004).

12.6 California designations of olive oils

The two most significant designations for olive oil being processed in California are:

1. *California Olive Oil Council (COOC) Certified:* The COOC is a trade organization based in Berkeley, CA. The COOC Seal Program certifies COOC members' California-produced olive oil as conforming to the "extra virgin" standard for chemistry and sensory analysis. The COOC extra virgin standard is the same as that of the International Olive Council (IOC) except that the free fatty acid limit is 0.5% instead of 0.8%.
2. *Certified Organic:* Organic certification is based on national standards developed by nationally approved certifiers, for both olive production and olive oil processing. A number of certifying entities can be found in California. In the 2005 California Olive Oil Industry Survey, it was found that a total of 66% of the olive oil production in California was being farmed with organic methods, although only 16% was certified organic (Vossen and Devarenne, 2005). Organic continues to be an important market niche, especially for boutique producers.

12.7 Chemistry of California olive oils

In 2009–2010, a study of the chemical composition of California olive oil was undertaken by researchers at the University of California with funding from the Foreign Agricultural Service of the US Department of Agriculture (USDA) under the direction of the COOC (Wang, 2011). The primary objective of this study was to understand better how California extra virgin olive oils compared with the chemical standards of the IOC. The IOC standards are based on olive oils produced in the Mediterranean region from a particular group of cultivars. The experience of olive oil producers outside this Mediterranean region, and working with other varieties, is that certain of the IOC parameters do not accommodate the natural variation found in some of their extra virgin olive oils.

This study analyzed 60 samples from the 2010 harvest. It was found that a large percentage of the California olive oils made from Arbequina olives exceeded the IOC limit of $\leq$4.0% for the sterol campesterol. Since this represents over 50% of California production, it is a significant issue if the IOC limit is applied in international trade. The 2010 US Standard for Olive Oil and Olive-Pomace Oil has a higher limit for campesterol of $\leq$4.5%. Only one of the oils in the study, a Koroneiki, failed that USDA limit, but there were several that fell between 4.0 and 4.5%.

The other IOC parameter that was exceeded by a significant number of oils was the limit of $\leq$1.0% for linolenic acid. The USDA standard has a higher limit for linolenic acid of ≤1.5%. In this study, none of the samples failed the USDA limit for linolenic acid.

Nine of the samples in the study exceeded the limit of ≤0.3% for heptadecenoic acid that is used by both the USDA and IOC standards. These were oils made from the varieties Ascolana, Kalamon, Manzanillo, and Sevillano. These results suggest that a level of 0.4% of heptadecenoic acid might be more appropriate for California olive oil.

12.8 Olive varieties in California

Although there are over 150 varieties of olives in California, less than two dozen are being grown with any frequency. The California olive oil industry functioned for many years as a salvage operation of the table olive industry, and the traditional table olive cultivars still occupy an important niche in the state's olive oil production. Owing to large planting starting in the early 2000s, the varieties used in SHD planting, mostly Arbequina, followed by Arbosana and Koroneiki, have become the most common oil olive varieties by far. Other fairly common varieties are Ascolano, Coratina, Frantoio, Leccino, Manzanillo, Maurino, Mission, Moraiolo, Pendolino, Picholine, Sevillano, and Taggiasca. Plantings of Aglandau, Farga, Itrana, Kalamon, Lucques, Nocellara del Belice, Picual, and other cultivars can be found, but acreage is still small. Some innovative California olive growers are exploring different varieties for their flavor characteristics and horticultural suitability in various climatic zones.

12.8.1 Super-high-density varieties

12.8.1.1 Arbequina This cultivar from northern Spain is the most abundant in the SHD olive acreage in California. The patented IRTA i-18 selection of Arbequina developed by Institut de Recerca i Tecnologia Agroalimentaries (IRTA), a research institute in Catalunya, Spain, was the most planted olive variety in the first wave of SHD development in California. This clone was selected for both sensory and horticultural characteristics, with early productivity a high priority. Arbequina in later SHD plantings drew from several sources. Arbequina in California is harvested starting in mid to late October. The fruit is usually fairly green at harvest; harvest commences early in order to assure that all the fruit is picked before frost hits the northern part of the state.

The classic flavor profile of California Arbequina is very fruity, with green notes in the nose and a mostly soft, ripe impression on the palate. Herbaceous, almond, tropical, and wood/hay/straw flavors are characteristic. There can be some pungency on the finish, but bitterness is rare except in cases of extreme deficit irrigation and very early harvest. The style of artisan

Arbequina production extends from very green to very ripe, creating a variety of flavor profiles from this cultivar.

12.8.1.2 Arbosana Another cultivar of northern Spanish origin, Arbosana, is also used for SHD production. Although used sparingly in the early plantings, Arbosana quickly attracted attention for both its appealing flavor profile and its horticultural qualities. It is regarded by many SHD growers as the most easily managed of the three SHD cultivars in the orchard. Like Arbequina, Arbosana in California is harvested when fairly green. It is the latest maturing of the main three SHD cultivars.

Arbosana has more intense grassy and leafy green notes and a little more bitterness and pungency than Arbequina. Its ripe aromas are floral and apple and other ripe fruit. It is often blended with Arbequina, but some California producers produce a monocultivar Arbosana oil.

12.8.1.3 Koroneiki Koroneiki is the primary oil olive variety in Greece. The fruit is small and strongly attached to the tree, making it difficult to harvest, especially when green. Mechanical harvest makes this variety much more practical; in California it is planted almost entirely in SHD systems. It is more vigorous than Arbequina and Arbosana, so it requires careful management in the orchard, but it also can produce large yields of high-quality oil. It is the most cold sensitive of the SHD cultivars.

Koroneiki is the highest in polyphenols of the varieties used in California SHD. When harvested green, it can have pronounced bitterness and pungency in addition to high fruitiness. The green notes most characteristic of California Koroneiki are herbaceous and green banana. As Koroneiki becomes riper, tropical and ripe banana notes become dominant. Koroneiki is often found in a blend with Arbequina, but it has done also well in competition and the marketplace as a monocultivar oil.

12.8.2 Classic California table varieties

12.8.2.1 Manzanillo The premier table olive cultivar in California, Manzanillo is dominant in Tulare, Kings, and Fresno counties in the southern San Joaquin Valley. Manzanillo is also abundant in Tehama and Butte counties in Northern California, but can be problematic because it is somewhat sensitive to cold winter temperatures. An additional problem with growing Manzanillo in cold winter environments is that freeze damage exacerbates olive knot, a bacterial disease to which Manzanillo is very susceptible (Sutter, Sibbett, and Ferguson, 2005).

For table olive production, Manzanillo fruit is harvested when very green, usually by mid-October. For oil, however, November is a more common time for harvest. Manzanillo oil is notoriously difficult to extract, particularly when the water content of the fruit is high – a condition that is very common when the trees have been generously irrigated to produce larger-sized fruit for curing. The oil obtained from earlier harvest California Manzanillo is usually regarded as superior to that from riper fruit.

12.8.2.2 Sevillano (Gordal) Planted for table olive production, Sevillano is concentrated mostly in the Corning area of Tehama county in Northern California, but small plantings have been established in Butte and Tulare counties and elsewhere. It is somewhat resistant to cold damage, and is sometimes planted with Manzanillo as a pollenizer.

The fruit matures relatively late and is harvested in October for table processing and somewhat later for oil. The oil content of Sevillano is extremely low, but it is sought after in California for olive oil because of its unique varietal flavor. Growers prefer to sell Sevillano for table olive processing because of the high price it commands, but undersized fruit or extra production can be used for olive oil production. The oil is prized for its intense aroma and balanced flavor. Grassy green notes combined with nutty and stone fruit ripe flavors characterize the best Sevillanos.

12.8.2.3 Mission Mission is the only cultivar that is considered uniquely Californian. The variety traces back to the days of the Franciscan missions – hence the name – although there is no agreement as to whether the trees were imported as cuttings, small trees, or seeds. It is truly dual purpose – table and oil – and is the most cold resistant of the traditional cultivars in California. Its cold hardiness made it popular for table olive planting in the northern part of the state.

Mission is a late-maturing variety. Harvest for most table olive styles starts in late October and continues through November. When used for oil, this high-yielding variety is harvested from late November all the way until the end of February or even later. Very ripe Mission can be used to create a classic California style of oil usually called Late Harvest Mission. This popular oil has a buttery and pineapple-fruity quality. Mission olive oil has aggressive bitterness when green, so it is usually harvested for oil when moderately mature. The oil has green notes of pine, mint, and straw, and a ripe character that is distinctly buttery.

12.8.2.4 Ascolana (Ascolano) Most plantings of Ascolana are in the San Joaquin Valley, particularly in Kings and Tulare counties. Originally from Italy, Ascolana is resistant to cold injury and grows well in all olive-producing areas of California. Once a popular table variety, Ascolana fell out of favor with the packers because it is delicate and bruises easily. As a consequence, many of the trees were removed or grafted over to other varieties in table olive-producing areas. The rise of olive oil production has given the cultivar new value.

Ascolana produces a medium yield of very fruity, high-quality olive oil. The varietal flavor of Ascolana is reminiscent of stone fruit (apricots or peaches) with herbaceous green notes and moderate bitterness and pungency. When riper, the oil takes on a tropical character. The fruit matures early and is picked for table olive production from mid-September to early October and for oil mostly in November.

12.8.3 Major Italian varieties in California

12.8.3.1 Frantoio Frantoio is a popular variety with small- and medium-sized California producers, especially in the coastal and foothill areas of the state. Although purported to be cold sensitive, it has proven to be fairly adaptable in California and is found in all the olive-growing areas of the state. Usually planted as part of a "Tuscan blend," Frantoio is the most popular of the group in California for single varietal oils because of its high oil yield and excellent sensory quality.

This variety ripens unevenly; it is not uncommon to see fruit ranging from totally green to black on a single tree. Although it is a late-maturing cultivar, most California Frantoio is harvested at around the 50:50 point, when half the fruit has some color and half is still green. The green flavors of California Frantoio are usually dominated by fresh-cut grass and artichoke, enhanced by ripe floral and nutty notes.

12.8.3.2 Leccino The other mainstay of the California "Tuscan blends," Leccino is a more cold-tolerant variety and also ripens early. This resistance to low temperatures and early ripening have made it a popular olive in areas of the state where a short growing season or cold winters limit the choices. It the first of the "Tuscan blend" varieties to mature, and is harvested first when the grower has that option. It is often the "olio nuovo" ("new oil") offering because it is the first through the mill. Its modest polyphenol content makes it well suited to this; even when green and straight off the press, the oil is not too intense.

12.8.3.3 Taggiasca Originally from Liguria in northern Italy, Taggiasca is used in California both for monocultivar oils and for blends. Traditionally it is made into ripe style delicate oils in Liguria, but in California it is harvested early. California Taggiascas are usually medium to robust in style, with bright green flavors of grass, green almond, artichoke, and some ripe floral notes. The bitterness and pungency are pronounced, but harmonious.

12.8.3.4 Coratina Coratina is the only southern Italian variety to be planted with any frequency in California. In the early days, in many cases it was planted along with the "Tuscan blend," but there have been more monocultivar Coratina oils appearing in recent years. When produced as a single varietal oil harvested in December, California Coratina has intriguing ripe floral and black cherry notes in addition to grassy and herbaceous green flavors. It has bitter and pungent characteristics, but these are in harmony when the fruit is harvested late in the year. Very green Coratina in California, like very green Coratina everywhere, is extremely bitter, making it a difficult choice for short season locations.

12.9 Olive oil regions in California

Olive orchards are primarily located in seven regions of California (Figure 12.1), all of which have Mediterranean climates. The majority of olive

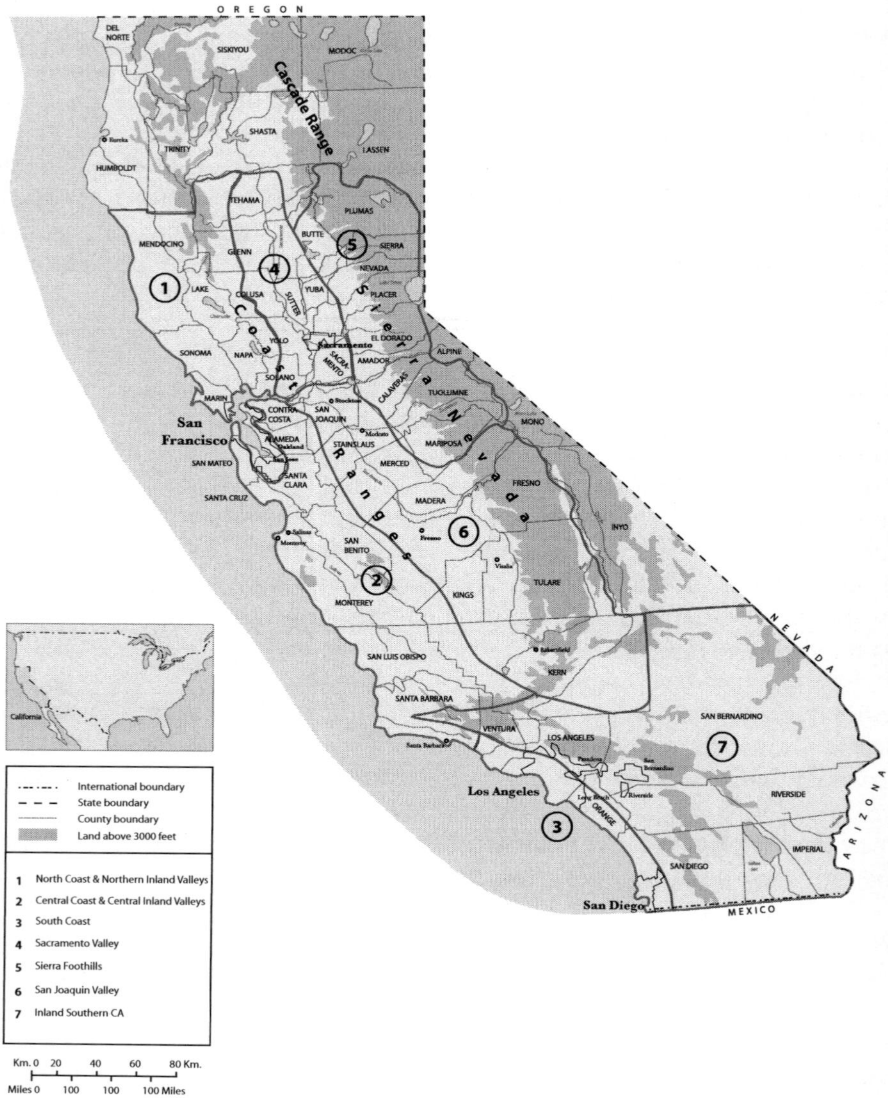

Figure 12.1 Olive-growing regions in California. For a color version of this figure, see the color plates section.

oil production is in the ~40–60 miles (~60–100 km) wide Central Valley, divided into the Sacramento Valley to the north and the San Joaquin Valley in the south.

1. North Coast region (Alameda, Contra Costa, Lake, Marin, Mendocino, Napa, Santa Clara, San Mateo, and Sonoma counties).
2. Central Coast region (Monterey, San Benito, San Luis Obispo, Santa Cruz, and Santa Barbara counties).
3. South Coast region (Los Angeles, Orange, San Diego, and Ventura counties).
4. Sacramento Valley region (Butte, Colusa, Glenn, Sacramento, Shasta, Solano, Sutter, Tehama, Yolo, and Yuba counties).
5. Sierra Foothills region (Amador, Calaveras, El Dorado, Mariposa, Nevada, Placer, Plumas, Sierra, and Tuolumne counties).
6. San Joaquin Valley (Fresno, Kern, Kings, Madera, Merced, San Joaquin, Stanislaus, and Tulare counties).
7. Inland Southern California (San Bernardino, Riverside, and Imperial counties).

Table 12.1 summarizes the main olive cultivars in California and common sensory properties of the oil produced in the various regions.

12.9.1 Region: North Coast

The North Coast region includes the counties of Alameda, Contra Costa, Lake, Marin, Mendocino, Napa, Santa Clara, San Mateo, and Sonoma.

12.9.1.1 Climates and growing season The North Coast region is a tranquil scene of picturesque landscapes with mountain ranges, rolling hills, and deep valleys, all helping to produce some of the best olives for olive oil in the state. Influenced by the cool coastal air and fog from the Pacific Ocean, the region enjoys a temperate climate with a long growing season of sunny, warm days followed by cool evenings – the ideal climate for growing olive trees (National Gardening Association, 2012a).

Luther Burbank called Sonoma County "the chosen spot of all the Earth as far as Nature is concerned." A vastly diverse range of topography, including numerous small valleys with distinct microclimates, the Russian River, and the Pacific Ocean, all characterize the region. A moderate climate with a cooling maritime influence, Sonoma County embodies ideal and diverse olive-growing weather: from valley to hillside, moist ocean coast to dry inland, and cool southern regions that complement the warmer, more northern areas (Wine Institute, 2012a).

Defined by mountain ranges and a proximity to the Pacific Ocean, the Napa Valley enjoys a temperate climate with a long growing season of sunny, warm days followed by cool evenings. Within the Napa Valley there are specific

Table 12.1 Main olive cultivars in California and typical sensory properties of the oil produced in the various regions.

Region	Main styles	Cultivars	Typical sensory properties
North Coast	Tuscan blends	Frantoio, Leccino, Moraiolo, Maurino, Pendolino	Green fruity, high bitterness, high pungency, fresh-cut grass, artichoke, green almond, herbaceous
Central Coast	Arbequina and Arbequina blends	Arbequina, Arbosana, Koroneiki	Varies from soft ripe fruity to green and pungent
	Tuscan blends	Frantoio, Leccino, Moraiolo, Maurino, Pendolino	Green fruity, high bitterness, high pungency, fresh-cut grass, artichoke, green almond, herbaceous
	Table varieties and other blends	Ascolano, Mission, Coratina, Aglandau, Tanche, Picholine, Picual, Lucca Taggiasca	Diverse but most are medium robust to robust
South Coast	Blends	Frantoio, Leccino, Moraiolo, Maurino, Pendolino, Mission, Manzanillo, Ascolano, Nocellara del Belice, Coaratina, Picual, Taggiasca, Picholine, Aglandau	Diverse

Sacramento Valley	Arbequina	Arbequina	Very fruity with both green grass/herbaceous aromas and ripe floral/tropical notes; soft and nutty on the palate with a slightly peppery finish
	Arbequina blends	Arbequina, Arbosana, Koroneiki	Similar to above description but usually with more pungency and bitterness and a greater complexity
	Table varieties	Ascolana	Stone fruit and herbaceous aroma; if harvested riper, strong tropical aromas with nutty ripe flavors
		Sevillano (Gordal)	Grassy, herbaceous aroma with resinous/eucalyptus note; low bitterness and pungency with intense fruitiness and ripe nutty flavors
		Mission	Later harvest styles are buttery, mild with pineapple–tropical notes and light to moderate bitterness; earlier harvest styles have distinct bitterness with pine and herbaceous green notes
Sierra foothills	Monocultivars and blends	Mission, Frantoio, Leccino, Moraiolo, Maurino, Pendolino, Ascolana, Coratina, Picholine, Barouni	Diverse but mostly robust style
San Joaquin Valley	Arbequina and Arbequina blends	Arbequina, Arbosana, Koroneiki	Mild and fruity with added bitterness, pungency, and complexity in blends
	Table varieties	Manzanillo	Herbaceous, grassy, green banana
Inland Southern California	Blends and monocultivars	Frantoio, Leccino, Moraiolo, Maurino, Pendolino, Mission	Diverse

areas with distinct microclimates and terrains formed by a varied topographical configuration of hills, exposures, and elevations.

Lake County surrounds Clear Lake, the largest natural lake in California. The olive orchards are planted throughout the county, from the agriculturally rich valley to the rocky red volcanic soil around Mount Konocti, a dormant volcano in the Pacific Rim chain. These elevations provide cooler winter conditions and a later start to the growing season. Summer growing conditions are suitably warm to ripen the olives and the elevation allows rapid cooling in the evening.

Marin County is bordered on three sides by the cooling influence of the Pacific Ocean and San Francisco Bay and the area grows mostly early maturing varieties.

Very cold summer water along the west coast of the United States has a direct effect on the making of high-quality olive oil in coastal California. Cold ocean current running down the Pacific coast from Alaska wells up off north California, partly as a result of the strong coastal winds turning the water over. The colder water below comes to the surface and meets the warmer air, causing it to cool and condense into massive fog banks. Meanwhile, on land, the baking sun heats up the interior valleys, where the hot air rises and pulls the cooler air and the fogs from the coast inland, through any gaps in the coastal ranges, to fill its space. The temperature falls, fogs roll in throughout the night, only to disperse again the next day with the midday heat (Clarke, 2002).

12.9.1.2 Principal olive oil styles The cooling influence of the ocean in the North Coast region makes it suitable for producing olive oils of greener styles because it is difficult to get fruit to ripen fully ahead of the winter frosts. The olive oils of the North Coast are known for complexity and high quality. Because of high land costs and lower average yields, almost all of the production in this area is boutique in scale and often linked to the high-quality wine production for which the area is famous (California Vineyards, 2012). The North Coast was the epicenter of the California olive oil renaissance in the 1990s, the movement that spurred the development of the current olive oil industry in the state.

12.9.1.3 North Coast Tuscan blends In the 1990s, several small-scale coastal California producers imported olive varieties from Italy (Frantoio, Leccino, Moraiolo, Maurino, and Pendolino) to produce a "California Tuscan blend" olive oil. The results have been excellent from the perspective of oil quality, producing oils reminiscent of the blends made with the same varieties that are produced in various parts of Italy. The olives are typically harvested when approximately 50:50 green and colored, and the blends in general are green fruity in character, with high bitterness and pungency. Fresh-cut grass, artichoke, green almond, and herbaceous notes dominate (Vossen, 2003, 2005; Devarenne *et al.*, unpublished work). As the

"California Tuscan" style has evolved, some producers are allowing the fruit to ripen more and are developing more complex green/ripe flavor profiles with these blends. In recent years there has also been a rise in monocultivar production of these varieties, especially Frantoio (Vossen, 2007a).

12.9.2 Region: Central Coast

The Central Coast region includes the counties of Monterey, San Benito, San Luis Obispo, Santa Cruz, and Santa Barbara.

12.9.2.1 Climates and growing season This region has a huge diversity of climates, due primarily to the influence of the Pacific Ocean fog (Wine Institute, 2012b).

The climate of Monterey County reflects the cooling influence of the Monterey Bay and lack of abundant rainfall. There are enough warm days to ripen olives, but the marine influence predominates. Cool growing conditions allow for a long season and slow fruit maturation. The steep slopes and rolling hills provide good drainage, and Monterey soil temperatures are cooler than other parts of the state, limiting crop size.

The city of Paso Robles, situated 20 miles inland from the Pacific Ocean, is in San Luis Obispo County, half way between San Francisco and Los Angeles. The area is characterized by warm, clear days, generally unencumbered by clouds, fog, or severe winds. Nighttime temperatures drop by ~40 °F (~22 °C), cooled by a marine layer that moves over the region after sunset. Proximity to the ocean, orientation of the numerous canyons and valleys, and varying elevations produce diverse macroclimates, allowing the production of both cool- and warm-loving olive varieties (Halliday, 1993).

In Santa Barbara County, the north–south coastal range of mountains abruptly turns to run almost east–west for 50 miles, framing the valleys in a unique transit to the Pacific Ocean. This is the only stretch of land from Alaska to Cape Horn constituting an east–west traverse. The unique topography allows the flow of fog and ocean breezes to shape distinct microclimates and makes the region one of the coolest olive-growing areas in California. However, warmer daytime temperatures in the inland areas allow a wide variety of olives to be grown. Terrain and climates vary widely, from steep, wind-swept hillsides to rolling inland valley orchards where summer temperatures often reach ~100 °F (~40 °C).

Although this is an area of microclimates, like the rest of the California coast, the Central Coast generally has more heat than the North Coast. This permits greater latitude in harvest maturity because of the possibility of more ripening before the threat of frost forces harvest.

12.9.2.2 Principal olive oil styles Similarly to its neighbor to the north, the Central Coast produces high-quality oils on a mostly boutique scale. Since planting in the area started slightly later than in the North Coast region, there are more SHD plantings of Arbequina, Arbosana, and Koroneiki to be found in the area. The SHD groves of the Central Coast are mostly small, and there is a significant amount of boutique SHD production. The small- and medium-sized growers of the Central Coast have embraced a broad selection of cultivars, ranging from the Italian varieties of the "California Tuscan blend" to the classic California table cultivars. The oils of the Central Coast are known for their complexity and intense flavor.

12.9.2.3 Arbequina and Arbequina blends The small and medium SHD producers of the Central Coast have created a niche in distinctive artisan production of Arbequina, Arbosana, and Koroneiki. Although some of these producers are using over-the-row mechanical harvesters, many of the olive groves are too small or on unsuitable terrain and they harvest their olives by hand instead. The flavor profiles of these small-batch Arbequina and Arbequina blend oils vary from soft ripe fruity to green and pungent.

12.9.2.4 Central Coast Tuscan blends The Tuscan blends of the Central Coast comprise the classic cultivars of the blend: Frantoio, Leccino, Moraiolo, Maurino, and Pendolino. The quality of these Central Coast blends is on a par with those of the North Coast and the sensory profiles are similar.

12.9.2.5 California table cultivars and other blends Central Coast producers use a diversity of varieties in both monocultivar oils and blends. There are a number of boutique producers using Ascolana, Mission, and Coratina, sometimes in combination with the Tuscan blend cultivars, to create signature oils. Some plantings of French varieties such as Aglandau, Tanche, and Picholine are also found, in addition to Picual, Lucca, and Taggiasca. These highly individualized oils are diverse, but most of them are medium robust to robust in character.

12.9.3 Region: South Coast

The South Coast region includes the counties of Ventura, Los Angeles, Orange, and San Diego.

12.9.3.1 Climates and growing season The climate of the South Coast olive-growing region has long been recognized as ideal for growing olives for oil. In fact, some of the earliest known olive trees in America were planted here. Stretched along Southern California's beaches, this region is almost completely dominated by the ocean. It has a mild marine climate (milder than Northern California's maritime climate) because south of Point Conception, the Pacific is comparatively warm. The winters are mild, the summers cool, and the air seldom really dry. On many days in spring and early summer, the sun does not break through the high overcast until afternoon. Record heat usually comes in early October, carried to the coast by Santa Ana winds. The wind's power and dryness usually cause more problems than the heat itself. Morning mists, sunny days, cool summer nights, and Pacific Ocean breezes characterize the region (National Gardening Association, 2012b).

12.9.3.2 Principal olive oil styles Large swaths of the coast of Southern California, once important agricultural land, are now built up. Ventura County still has significant amounts of farmland, as does San Diego County, but olives are a minor crop in the region. As water becomes scarcer, some citrus and avocado growers are looking to olives as a less thirsty alternative.

12.9.3.3 South Coast blends The production of the South Coast is small and eclectic. The principal varieties planted in recent years are the Tuscans – Frantoio, Leccino, Pendolino, Maurino, Moraiolo – but there are significant numbers of older trees to be found. These are mostly Mission, Manzanillo, and Ascolana. There are also other oil varieties to be found, such as Nocellara del Belice, Coratina, Picual, Taggiasca, Picholine, and Aglandau, but most of them end up in blends rather than monocultivar oils.

12.9.4 Region: Sacramento Valley

The Sacramento Valley region includes the counties of Butte, Glenn, Sacramento, San Joaquin, Shasta, Solano, Sutter, Tehama, Yolo, and Yuba.

12.9.4.1 Climates and growing season The Sacramento Valley in the northern part of the Central Valley is one of the traditional table olive-producing regions of California. The Sacramento Valley runs for ~120 miles (~190 km) from Red Bluff in the northern end of the valley to the city of Sacramento. Bordered by the Sierra Nevada to the east and the Coast Ranges to the west, this level, sun-drenched, agriculturally rich area is unaffected by ocean influences. The Sacramento and San Joaquin Valleys meet at the river delta ~100 miles (~160 km) east of San Francisco, roughly encompassing portions of Solano, Yolo, Sacramento, and San Joaquin counties. A small portion of Solano County receives the cool maritime influence with ocean breezes flowing through the San Francisco Bay and the Delta (Wine Institute, 2012c).

12.9.4.2 Principal olive oil styles Olive production is concentrated in the area around Orland and Corning. Home to one of the state's two remaining large table olive processors, the Sacramento Valley has orchards of Manzanillo, Sevillano, Mission, and Ascolana. These table olive varieties are made into olive oil by several small- and medium-sized mills in the Sacramento Valley and the North Coast. The Sacramento Valley was also home to the first SHD plantings in California and the state's largest olive oil producer is located there. There is more acreage of oil olives in this region than anywhere else in the state, most of it planted in SHD.

12.9.4.3 Arbequina, Arbosana, and Koroneiki This is the region that produces most of California's olive oil. Since the majority of the SHD acreage is planted in Arbequina, this variety is the one that most defines the style. Classically, California Arbequina is a very fruity oil, with both green grass/herbaceous aromas and ripe floral/tropical notes. The oil is nutty and soft on the palate, with a slightly peppery finish. The addition of Arbosana and Koroneiki in a blend is common, and the addition of these varieties can add more pungency and bitterness and greater complexity. Monocultivar Koroneiki and Arbosana varietal oils are becoming more common.

There are a number of boutique producers in the Sacramento Valley working also with Arbequina, Arbosana, and Koroneiki. Although the SHD system is designed to allow over-the-row mechanical harvesting, because of their small scale some of these orchards are hand harvested. These boutique producers are making a range of oils using these cultivars. By manipulating irrigation and/or harvesting the fruit very green, they create surprisingly robust oils even with the usually mild Arbequina variety.

12.9.4.4 California table varieties The table olive orchards of the Sacramento Valley are used to produce a number of monocultivar and blend olive oils. The Ascolana variety, although out of favor with table olive processors, has become a popular olive with oil producers. It gives a moderate yield of extremely distinctive olive oil, characterized by stone fruit and herbaceous aromas. If harvested riper, strong tropical notes dominate the aroma and it has nutty ripe flavors on the palate. This variety has balanced bitterness and pungency when it harvested moderately green.

Sevillano (Gordal) is used to make a popular and distinctive monocultivar oil. The oil yield is low, but the flavor is excellent. The aroma is grassy and herbaceous, with a resinous/eucalyptus note. There is low bitterness and pungency, and intense fruitiness. Ripe nutty flavors predominate on the palate.

The Sacramento Valley also produces Mission. This olive is used to make olive oil with a range of styles from green to ripe depending on the producer and the intended market. The traditional classic style for California Mission was very ripe, usually harvested after New Year. The resulting oil is buttery and mild, developing pineapple–tropical notes the riper it gets. Many producers are making medium-ripe style Mission oils, with mostly ripe fruit notes and light to moderate bitterness. The challenge with the Mission olive is to capture the green fruit notes of pine and green olive without overwhelming, harsh bitterness.

Ripe Mission is also a popular choice for making citrus oils in California. The best of these oils are made in the mill, not by infusing a finished olive oil with flavorings. The olives are crushed with fresh citrus fruit – lemons, oranges, mandarins, limes, yuzu, and so on – and then malaxed and separated as usual. If made well, with good olives and good citrus in the right proportions, the resulting flavored olive oils have a balance of good olive oil and fresh, fragrant citrus flavor. Although controversial with some purists, they are extremely popular products in the market.

12.9.5 Region: Sierra Foothills

The Sierra Foothill region includes the counties of Amador, Calaveras, El Dorado, Mariposa, Nevada, Placer, Plumas, Sierra, and Tuolumne.

12.9.5.1 Climates and growing season The Sierra Foothills stretch along the western portion of the Sierra Nevada, with Amador, El Dorado, and Calaveras counties in the center. Olive trees are nestled throughout the nooks and crannies of the foothills, with orchards generally located at high enough elevations to create a four-season climate. The shallow, mountainside soils create moderate stress on the trees, producing low to moderate yields and high quality.

12.9.5.2 Principal olive oil styles The olive oil producers of the Sierra Foothills range from small–medium to very small. This boutique production has attracted considerable attention, however, for the high quality and complexity of the oils.

12.9.5.3 Foothill blends and monocultivars The oils of the foothills are a diverse group of both monocultivars and blends. The temperature extremes and early freezes of the Sierra Foothills have contributed to the development of oils mostly in a robust style. The cold hardiness of the olive variety is also a factor; Mission, the classic Tuscans, Ascolana, Coratina, and Picholine all appear in oils from the foothills. There is also an unusual monocultivar oil produced there from Barouni, a Tunisian variety that was planted in the early part of the twentieth century primarily as a fresh market product for home canning. This cultivar produces a low yield of distinctive, high-quality oil.

12.9.6 Region: San Joaquin Valley

The San Joaquin Valley region includes the counties of Fresno, Kern, Kings, Madera, Merced, San Joaquin, Stanislaus, and Tulare.

12.9.6.1 Climates and growing season In the San Joaquin Valley there are 5 million acres of irrigated farmlands planted to cotton, grains, vegetables, fruits, and nuts. One of the richest agricultural areas in the world, the San Joaquin Valley measures ~220 miles (~350 km) in length and ~40–60 miles

(~60–100 km) in width, extending from around Stockton south to Bakersfield. The Sierra Nevada mountains form the eastern border of this grand expanse of land, and the lower, more irregular Coast Ranges define it to the west.

Irrigation water for this land with limited rainfall comes from two huge reservoir and canal systems that bring water from the length of the Sierras to the valley farmers. The soils, from north to south, are consistently fertile. Because it is a valley protected on three sides, the weather is extremely predicable, with most precipitation occurring during the cool winter months long after the completion of most harvests.

12.9.6.2 Principal olive oil styles The olive oils of the San Joaquin Valley are similar to those of the Sacramento Valley. There are sizable SHD plantings, the majority from the northern end down to around Fresno. The table olive orchards in the San Joaquin Valley are concentrated towards the southern part of the valley around Visalia. The northern part of the San Joaquin Valley is home to California's oldest olive oil company. They specialize in the production of monocultivar oils from classic California table varieties.

12.9.6.3 Arbequina, Arbosana, and Koroneiki The majority of the olive oil produced in the San Joaquin Valley is from recent SHD orchards. These oils are generally mild and fruity, with similar varietal character to the SHD production of the Sacramento Valley.

12.9.6.4 Classic California table varieties Manzanillo is the most common table variety in the San Joaquin Valley. Fruit that is not used for canning can be made into oil instead. When harvested at the right degree of maturity, Manzanillo produces a good quality oil. The best California Manzanillos are usually early harvest, contributing herbaceous, grassy, and green banana notes. Because most of the Manzanillo in the state is heavily irrigated to produce large fruit for canning, it is a notoriously difficult variety to process into oil.

12.9.7 Region: Inland Southern California

The Inland Southern California area comprises San Bernardino, Riverside, and Imperial counties.

12.9.7.1 Climates and growing season The Southern California interior valleys are hilly and affected by thermal belts. Hilltops and valleys are colder in winter (with the possibility of frost) and are warmer in summer than the slopes and hillsides from which cold air drains. This climate is little influenced by the ocean. The days are fairly sunny with most of the rain falling in the winter. The temperature swing over the year is more extreme, with hotter summers and colder winters than the coastal climates to its west. Drying winds such as the Santa Ana's can desiccate plants in some areas.

12.9.7.2 Principal olive oil styles The area around Temecula in Riverside County has become a boutique olive oil-producing area. As in coastal Southern California, water scarcity is a concern here. The avocado growers of areas in northeast San Diego County, just west of Temecula, are considering olives as an alternative.

12.9.7.3 Inland Southern California blends and monocultivars The oils of this region are diverse. Similarly to other boutique producers in the state, the Tuscan mainstays – Frantoio, Leccino, Pendolino, Moraiolo, and Maurino – were early favorites. Also, typical of Southern California, there is a well-established population of old trees, mostly Mission, which are being used to make both extra virgin and flavored olive oils. There is also an assortment of other varieties, including some recently planted SHD in the Imperial Valley.

12.10 Conclusion

Olive oil production in California is in a period of growth and recognition. In a recent consumer magazine tasting of olive oils available in the United States, two California oils were top ranked; one of them was a "best buy" SHD Arbequina and the other a boutique "California Tuscan" blend (Consumer Reports, 2012). This result is a microcosm of the California industry. The success of the SHD system at producing good extra virgin olive oil at an affordable price has been critically important as it has introduced large numbers of Americans to fresh California olive oil. Simultaneously, the boutique part of the industry has received recognition in international competitions and the media, attracting attention to the high quality of the oils coming out of California. Sitting as it does on the edge of one of the world's largest markets for olive oil, the California industry is well positioned to become a significant supplier to the US market as its production increases.

References

California Vineyards (2012) *California Wine Regions and Maps*, http://california-vineyards.com/wine-regions/ (last accessed 3 June 2013).

Civantos López-Villalta, L., and Pastor, M. (1996) International Olive Oil Council (IOOC) Production Techniques, in *World Olive Encyclopedia*, IOOC, Madrid, pp. 147–194.

Clarke, O. (2002) *Oz Clarke's New Wine Atlas*, Harcourt, Orlando, FL.

Connell, J.H. (2005) History and scope of the olive industry, in *Olive Production Manual*, 2nd edn (eds G.S. Sibbett and L. Ferguson), Publication 3353, University of California Agricultural and Natural Resources, Oakland, CA, pp. 1–10.

Consumer Reports (2012) How to find the best extra-virgin olive oil, *Consumer Reports Magazine,* September, http://consumerreports.org/cro/magazine/2012/09/how-to-find-the-best-extra-virgin-olive-oil/index.htm (last accessed 3 June 2013).

COOC (2011) *About the COOC*, California Olive Oil Council, Berkeley, CA, http://www.cooc.com/About%20COOC_10.4.11.pdf (last accessed 3 June 2013).

Datamonitor (2010) *Study on the Promotion of Consumption of Olive Oil and Olives in the USA and Canada*, IOC Document Pro.6/CO/9/09, Report CYCM0372, Datamonitor, London.

Halliday, J. (1993) *Wine Atlas of California*, Viking, New York.

INFOSCAN Statistics (1999) *Market Sales of Olive Oil by Year: 1993–98*, Information Resources, Chicago.

National Gardening Association (2012a) *Region Description: Northern California Coastal and Inland Valleys*, National Gardening Association, Williston, VT, http://www.garden.org/regional/report/description/full/3 (last accessed 3 June 2013).

National Gardening Association (2012b) *Region Description: Southern California Coastal and Inland Valleys*, National Gardening Association, Williston, VT, http://www.garden.org/regional/report/description/full/4 (last accessed 3 June 2013).

North American Olive Oil Association (2004) *US Imports of Olive Oil 1982–Present*, North American Olive Oil Association, Neptune, NJ.

North American Olive Oil Association (2012) *About Olive Oil: Consumption*, North American Olive Oil Association, Neptune, NJ, http://www.aboutoliveoil.org/consumption.html (last asccessed 3 June 2013).

Quest-Riston, C. (2006) *Olive Oil*, Eyewitness Companions, DK Publishing, New York.

Senise Barrio, O., and Carman, H. (2005) Olive oil: a "rediscovered" California crop, University of California Giannini Foundation, *Agricultural and Resource Economics Update*, **8** (5), May/June, 1–4.

Sunset (2012) *Sunset Climate Zones: California/Nevada*, http://www.sunset.com/garden/climate-zones/sunset-climate-zones-california-nevada-00400000036331/ (last accessed 3 June 2013).

Sutter, E.G., Sibbett, G.S., and Ferguson, L. (2005) Olive cultivars and propagation, in *Olive Production Manual*, 2nd edn (eds G.S. Sibbett and L. Ferguson), Publication 3353, University of California Agricultural and Natural Resources, Oakland, CA, pp. 19–25.

Tous, J.M., Romero, A., Plana, J., and Baiges, F. (1999) Planting density trial with 'Arbequina' olive cultivar in Catalonia (Spain), Proceedings of the 3rd ISHS Symposium on Olive Growing, *Acta Horticulturae*, **474**, 177–180.

UC Davis Olive Center (2009) *Survey: Super-High-Density Olive Production in California*, http://olivecenter.ucdavis.edu/files/survey%20111509%20revised.pdf (last accessed 3 June 2013).

UC Davis Olive Center (2012) Unpublished statistics.

USDA Economic Research Service (1999) *Imports and Exports of Olive Oil by Country 1994–98*, USDA Economic Research Service, Washington, DC.

USDA National Agricultural Statistics Service, California Field Office (2006) *California Fruit and Nut Review*, USPS 598-290, vol. 26, no. 8, http://www.nass.usda.gov/ca (last accessed 3 June 2013).

Vossen, P.M. (2002) Super-high-density olive oil production, *OLINT Magazine [special English edition]*, **1**, 17–23.

Vossen, P.M. (2003) California 'Arbequina' and 'Arbosana' olive oils get a very high rating from the local tasting panel, *OLINT Magazine [special English edition]*, **2**, 15.

Vossen, P.M. (2005) Producing olive oil, in *Olive Production Manual*, 2nd edn (eds G.S. Sibbett and L. Ferguson), Publication 3353, University of California Agricultural and Natural Resources, Oakland, CA, pp. 157–173.

Vossen, PM. (2007a) Olive oil: history, production and characteristics of the world's classic oils. *HortScience*, **42** (5), 1093–1100.

Vossen, P.M. (ed.) (2007b) *Organic Olive Production Manual*, Publication 3505, University of California Agricultural and Natural Resources, Oakland, CA.

Vossen, P.M., and Devarenne, A. (2005) *California Olive Oil Industry Survey Statistics 2004*, UC Cooperative Extension Sonoma County Report, http://cesonoma.ucanr.edu/files/27314.pdf (last accessed 3 June 2013).

Vossen, P.M., and Devarenne, A. (2006) California olive oil industry statistics update, *UC Cooperative Extension Sonoma First Press Newsletter*, vol. 1, no. 4.

Vossen, P.M., Connell, J.H., Klonsky, K., and Livingston, P. (2004) *Samples Costs to Establish a Super-High-Density Olive Orchard and Produce Oil – Sacramento Valley*, University of California Cooperative Extension Division of Agriculture and Natural Resources, Oakland, CA.

Wang, S. (2011) *TASC 2011 Project Report: Composite Chemical Picture of US Olive Oil: Removal of Potential Trade Barriers – Year 2*, US Department of Agriculture Foreign Agricultural Service, Washington, DC.

Wine Institute (2012a) *The Appellations of California Wine – North Coast*, Wine Institute, San Francisco, http://www.wineinstitute.org/resources/consumerfeaturedstories/article338 (last accessed 3 June 2013).

Wine Institute (2012b) *The Appellations of California Wine – Central Coast and Southern*, Wine Institute, San Francisco, http://www.wineinstitute.org/resources/consumerfeaturedstories/article339 (last accessed 3 June 2013).

Wine Institute (2012c) *The Appellations of California Wine – Sacramento and San Joaquin Valleys, Delta, and Sierra Foothills*, Wine Institute, San Francisco, http://www.wineinstitute.org/resources/consumerfeaturedstories/article340 (last accessed 3 June 2013).

13
Olive Oils from Australia and New Zealand

Leandro Ravetti[1] and Margaret Edwards[2]
[1]*Modern Olives/Boundary Bend Ltd, Lara, Victoria, Australia*
[2]*Matiatia Grove/The Waiheke Olive Oil Company Ltd, Oneroa, Waiheke Island, New Zealand*

13.1 Overview of olive oil industry

13.1.1 Australia

Olive-growing history in Australia is fairly recent with the first orchard established around Sydney in the early 1800s. By 1844, olives were introduced in South Australia near Adelaide and by 1890 Australia had the very first olive-growing attempts along the Murray River in Victoria (Hartmann, 1962). The evolution of the olive-growing area was fairly slow until the first half of the twentieth century, when it jumped from a few hundred hectares to 1500 ha, reaching 3000 ha by 1960. Following the 1960 peak, olives remained a cottage industry with a declining area under cultivation until the resurgence of the industry towards the end of the century.

The olive oil industry began to gather momentum in the early 1990s, with several factors acting as catalysts for this growth: increasing popularity of Mediterranean food and growing market demand; a perceived opportunity to replace imported olive oil; a perceived opportunity to receive price premiums for Australian oils based on their quality and for the ability to supply fresh oil during the European summer; the establishment of large-scale investment plantations driven by the introduction of the Australian Tax Office Product Ruling system in 1998, which allowed investors to receive an upfront tax deduction for some or all of the initial and ongoing cost of their investment; and the establishment of large numbers of boutique or life-style groves. Nationally, olive oil production has climbed from less than 1000 t in 2002 to

Olive Oil Sensory Science, First Edition. Edited by Erminio Monteleone and Susan Langstaff.

17 000 t in 2011 (Miller, 2011). It is estimated that 12 000 t of this production have been consumed in Australia, representing 30% of the total domestic market. The remaining 5000 t have been exported, with the United States, European Union, China, and several South East Asian countries as the main destinations.

When considering the new Australian olive industry, we have to think about a modern olive-growing model pursuing maximum profit with sustainable production methods. In order to achieve this objective, it has been necessary to pursue high fruit and oil yields, the highest price for the final product, and minimum production costs. Olive oils from Australia have quickly gained a reputation as a consistent high-quality product. Best horticultural practices in combination with timely harvesting and purpose-built processing facilities using state-of-the-art technology are the main reasons that support the production of fresher and higher quality oils (Figure 13.1).

Industry statistics estimate that Australia has ~30 000 ha of olive groves and more than 2000 producers. Several surveys conducted since 2005 found that 90% of growers had less than 10 ha and another 9% had between 10 and 100 ha. The remaining 1% of Australian growers, with more than 100 ha, have been producing ~90% of the country's olive oil crop. The Australian olive oil industry is highly vertically integrated, with most medium and large groves having their own processing plants and commercial brands. In

Figure 13.1 Mechanical harvesting of olives in Australia. For a color version of this figure, see the color plates section.

Australia, there are ~100 olive mills at an estimated average of one mill for every 300 ha. Most of those mills utilize modern continuous two-phase extraction systems (Ravetti, Miller, and Mailer, 2008).

Australian growers are represented by the Australian Olive Association (AOA), which has been the national peak body for the Australian olive industry since 1995. The AOA has a Board of Directors who are elected by their State Associations and the Large Enterprise Growers to discuss current issues and future strategies to ensure the industry's continued prosperity, growth, and sustainability.

The AOA implements an ongoing consumer awareness program (CAE) to promote the benefits of Australian extra virgin olive oil (EVOO) to consumers. Each October, the AOA holds the National Olive Industry Conference and Trade Exhibition, which brings together national and international experts to discuss the latest in technology and research. The industry also gathers at this time to celebrate the season's success at their Gala Awards Presentation Dinner, when the best oils and olives of the season are recognized.

The AOA is responsible for setting and maintaining quality standards for Australian products, which is achieved by growers joining the Australian Olive Industry's Code of Practice (COP). The AOA Code of Practice guarantees the authenticity and quality of certified products and distinguishes them from imported and uncertified domestic products. To be certified, products must be Australian, have undergone strict organoleptic and chemical testing, and comply with the Australian Standard for Olive Oils and Olive-Pomace Oils (AS 5264-2011).

Until July 2011, Australia did not have an existing Standard for olive oils or olive-pomace oils. Following a rigorous development process taking almost 8 years, involving multiple industry stakeholders and almost 800 public comments, Standards Australia approved a new olive oil standard (AS 5264-2011), bolstering consumer protection.

The objective of the Australian Standard was to establish a scientifically based and consumer-oriented standard for olive oils and olive-pomace oils traded in Australia. While the Australian Standard and other international standards have a number of areas in common, particularly associated to worldwide recognized analytical methodologies and critical limits, AS 5264-2011 differs significantly in a number of respects. Some of those points of difference include a simpler and clearer commercial denomination of the different categories of olive oils and olive-pomace oils in order to avoid the current misleading and confusing terms such as "extra light" and "pure." Additionally, a review and some modification of the range limits for a number of chemical parameters were undertaken in order to avoid genuine olive oil, particularly Australian, being excluded for its natural variation in composition. The introduction of recently developed analytical methods, which are capable of detecting modern refining techniques not currently detectable by the older methodologies included in the above standards, was also an integral part of the standard.

13.1.2 New Zealand

The New Zealand olive oil industry fits a boutique classification. It is young and extremely small with only a few orchards having 10 000 trees or more. In spite of the country being a leader in horticultural research and innovation, it took over 150 years from the first planting of olive trees for a small commercial olive oil industry to be established. Although 25 years later this industry is still in its infancy, New Zealand has already been recognized for the high quality of its olive oils, a number of which consistently win awards at both national and international competitions. Until relatively recently, New Zealanders did not include olive oil in their diets, but a growing interest in the Mediterranean-style diet and its apparent health benefits has seen a steady increase in consumption. Given the lack of a well-developed domestic industry, most olive oil has been imported, but as the local product has become available in the market, consumers have become increasingly aware of the freshness, flavor, and health benefits of New Zealand extra virgin olive oils. Fresh, high-quality New Zealand olive oils have the desirable attributes of flavor, presence of healthy antioxidants, and freedom from adulteration, but limited availability of land suitable for olive orchards and high labor costs make olive oil production expensive.

The New Zealand food culture has its roots in Polynesia and Britain, so it is surprising to many people, New Zealanders included, to learn that there is evidence of olive trees being planted as early as 1830. Early records from Charles Darwin's visit show that olives existed in Northland before 1835 and documents from 1843 indicate that olives were growing successfully in the Auckland region. Between 1860 and 1880, two prominent early settlers, Logan Campbell and Sir George Grey, independently attempted to establish an olive oil industry. The ventures lasted only a few years before being abandoned. In the following 100 years, at least three governmental studies were undertaken to research the feasibility of establishing a New Zealand olive industry, but the subsequent reports were largely ignored. It was not until 1986 when the first commercial olive orchard and nursery were established in Marlborough that a modern industry began. Importation of the Barnea cultivar and acquisition of stock from the International Olive Council's tree bank in Spain formed the basis of the nursery and subsequent plantings around the country (Noser, 1997). Following the development of a fledgling industry, the mid-1990s saw the industry experience a relative boom. However, a number of plantings failed, largely due to a lack of knowledge, and also inappropriate site selection and/or cultivars.

Because no national data have been collected, it is very difficult to obtain accurate information on the olive industry in New Zealand. Based on data obtained from Olives New Zealand (ONZ) members, who responded to orchard surveys carried out in 2008 and again in 2012 (Sheridan, 2012), coupled with anecdotal information, it is thought that there are around 600 000 olive trees planted in New Zealand. Production figures are also extremely difficult to obtain but in 2006 ~190 000 l of extra virgin olive oil were produced. This increased to around 230 000 l in 2011.

Although New Zealand has no indigenous olive trees, there are many examples of "local" trees that have been propagated from olives brought into the country in the nineteenth century. These trees still thrive but, with one exception, J5, they are rarely used for the production of high-quality olive oil. During the early 1990s, as the New Zealand olive industry developed, Barnea was promoted as the olive tree of choice for any commercial olive grove, anywhere in the country. However, in many regions, particularly in northern New Zealand, Barnea did not thrive and has been replaced with more suitable varieties. Although many orchards have a mix of varieties, it is the traditional Mediterranean cultivars that have proved to be the most reliable nationwide and are now the mainstay of the modern New Zealand industry. Unlike Australia, with its wide choice of olive varietals, commercial olive orchards in New Zealand are planted with a relatively narrow range of cultivars, the most widely planted being Frantoio, Leccino, Barnea, Picual, Koroneiki, Manzanilla, and J5 (the "local" variety) along with small plantings of Pendolino, Picholene, Moraiolo, Ascolano, Chemlali, Verdale, and Nocellara. Strict quarantine requirements have limited the importation of additional cultivars.

In the early 1990s, ONZ was established as the industry body to undertake research and to provide networking opportunities for its members. For its first 10 years, the industry concentrated largely on horticultural aspects, but in 2004 ONZ instituted a certification program for New Zealand olive oils to help raise the quality and profile of locally produced oils. To qualify for certification, New Zealand oils must meet the ONZ chemical analysis requirements of $<0.5\%$ free fatty acids and peroxide value <15 meq. O_2/kg. They must also undergo sensory analysis by an International Olive Council recognized sensory panel. Producers whose oils pass the tests are licensed to display a certification seal on their bottles. This seal distinguishes certified products in the marketplace and provides assurance for consumers that the oil they purchase meets the ONZ requirements for extra virgin olive oil. ONZ also runs an annual national competition and a conference.

13.2 Main chemical characteristics of olive oils

13.2.1 Australia

Owing to the typically early harvest, the use of modern harvesting and processing technology, and the vertical integration of its production, Australian olive oils are typically characterized by low levels of free fatty acids and peroxide values with a significantly high proportion of this production (>90%) being extra virgin (Mailer, 2007).

The ample range of combinations between different environmental conditions and genetics that characterizes the Australian olive industry produces oils with diverse chemical characteristics. Some of the most typical examples of these variations could be observed in the fatty acid profiles of Australian oils or in their sterol fraction. Based on a comprehensive study initiated by the Australian Oil Research Laboratory (AORL) (Mailer and Ayton, 2007) and complemented by the work of Modern Olives Laboratory Services (MOLS) (Guillaume *et al.*, 2011) with the support of the AOA and Rural Industries Research and Development Corporation, the Table 13.1 and Table 13.2 summarize the results of more than 2000 analyses encompassing all Australian environments and varieties.

13.2.2 New Zealand

Harvesting the fruit at the optimum time, prompt milling using modern two-phase processing equipment, and appropriate storage ensure that the majority of New Zealand olive oils presented for certification complied with the ONZ standards. Data gained from analysis of almost 1000 of these oils over the past 8 years showed that 95% of them met the ONZ requirement of <0.5% free fatty acids and peroxide value <15 meq. O_2/kg in addition to meeting the International Olive Council's sensory standard for extra virgin olive oil.

Owing to climate variations, New Zealand oils tend to show differences in fatty acid composition and the sterol fraction between the warmer northern regions and colder areas in the south. These variations are demonstrated in Table 13.3 and Table 13.4, which give the results of analyses of 132 samples for fatty acid composition and 120 samples for sterol composition by The New Zealand Institute for Plant and Food Research.

In 2007, ONZ obtained a Sustainable Farming Fund research grant (Ministry of Primary Industries, New Zealand) to test the applicability of overseas data, specifically olive maturity indices, and to gain knowledge to develop new, locally appropriate measures to assess optimal maturity where total phenolics and oil yield are maximized. This 3 year study, entitled Effects of Harvest Maturity on Extra Virgin Olive Oil, was undertaken by The New Zealand Institute for Plant and Food Research Ltd in collaboration with Massey University, Albany. Instrumental and chemical measures, and also sensory analysis, were carried out and a strong correlation between harvest time, phenolic content, and sensory attributes was established (Requejo-Jackman, Farrell, and Olsson, 2011). This study provided extremely valuable information for the industry.

Table 13.1 Summary of records for fatty acid composition of Australian olive oils (% w/w).

	C16:0	C16:1	C17:0	C17:1	C18:0	C18:1	C18:2	C18:3	C20:0	C20:1	C22:0	C24:0
Average	11.87	0.98	0.09	0.11	2.31	73.81	9.25	0.73	0.39	0.31	0.10	0.09
Median	11.70	0.90	0.10	0.10	2.30	74.30	9.40	0.70	0.40	0.30	0.10	0.10
Standard deviation	1.57	0.33	0.03	0.05	0.44	3.86	2.75	0.10	0.05	0.05	0.02	0.03
Maximum	20.26	3.56	0.50	0.50	5.40	84.15	23.79	1.71	0.70	0.60	0.20	0.20
Minimum	6.70	0.29	0.00	0.00	0.20	53.87	2.21	0.30	0.16	0.03	0.04	0.00
No. of samples	1859	1859	1859	1859	1859	1859	1859	1859	1859	1859	1859	1859

Table 13.2 Summary of records for sterol composition of Australian olive oils (% w/w).

	Cholesterol	Brassicasterol	Campesterol	Stigmasterol	Δ^7-Stigmastenol	β-Sitosterol	Total sterols	Erythrodiol + uvaol
Average	0.17	0.02	3.76	0.72	0.25	94.15	1705.45	1.18
Median	0.13	0.03	3.70	0.69	0.20	94.11	1723.24	1.00
Standard deviation	0.13	0.02	0.68	0.29	0.11	0.76	333.95	0.70
Maximum	0.80	0.24	5.00	2.30	1.19	96.70	2862.00	5.53
Minimum	0.00	0.00	1.88	0.00	0.00	92.00	789.23	0.20
No. of samples	756	756	756	756	756	756	756	687

Table 13.3 Summary of records for fatty acid composition of New Zealand olive oils (% w/w methyl ester).

	C16:0	C16:1	C17:0	C17:1	C18:0	C18:1	C18:2	C18:3	C20:0	C20:1
Average	8.92	0.62	0.03	0.06	1.79	78.50	8.62	0.92	0.28	0.25
Median	8.92	0.61	0.02	0.06	1.78	78.77	8.64	0.96	0.28	0.24
Standard deviation	1.28	0.23	0.01	0.01	0.38	2.09	1.35	0.21	0.05	0.04
Maximum	11.98	1.20	0.09	0.10	2.86	83.66	12.36	1.66	0.43	0.39
Minimum	7.00	0.02	0.01	0.04	0.89	72.10	5.68	0.43	0.14	0.16
No. of samples	132	132	132	132	132	132	132	132	132	132

Source: data collected and analyzed by The New Zealand Institute for Plant and Food Research.

13.3 Principal olive varieties in Australia and New Zealand

13.3.1 Barnea

Australia: Barnea produces medium-fruity oils with a character that combines notes of fresh-cut grass and sweet fruits ranging from green apple to ripe banana depending on the maturity of the olives. The intensity of the aroma in Barnea oils tend to decrease fairly sharply once the fruit is completely colored. Its delicate flavor and mild almond aftertaste usually combine with levels of bitterness and pungency ranging from very mild to fairly robust depending on the rainfall levels during the growing season, the irrigation regime, and the harvest time.

New Zealand, North and South Islands: The oil has a complex aroma of green olives and fresh-cut grass with spicy chili pepper notes. The spicy, green olive fruitiness carries through to the palate along with bitter herbs, green banana, green apple, and green leafy notes. The oil generally has a high level of bitterness and chili pepper-like pungency. Because Barnea is grown in frost-prone regions in New Zealand, it is usually harvested early and is robust in style.

13.3.2 Picual

Australia: Picual delivers extremely fruity oils with dominant sharp characters and good lengthy flavor. Early-harvest Picual oils from Australia are characterized by a pleasant combination of medium bitterness and pungency together with an aroma that reminds of fig leaves, arugula (rocket) salad, and green tomatoes or tomato bush. As the fruit ripens, the levels of bitterness and pungency drop but the intensity of the aroma remains firm. The typical green notes of the early harvest give way to riper characteristics dominated by ripe tomatoes and ripe tropical fruits.

Table 13.4 Summary of records for sterol composition of New Zealand olive oils (% w/w).

	Campesterol	Stigmasterol	Clerosterol	β-Sitosterol	Sitostanol	5-Avenasterol	5,24-Stigmastedienol	7-Stigmastenol	7-Avenasterol	Apparent β-sitosterol
Average	3.92	1.15	0.56	83.84	2.48	7.06	0.50	0.09	0.58	94.93
Median	3.92	0.99	0.57	84.24	1.935	7.125	0.37	0.03	0.36	95.02
Standard deviation	1.05	0.49	0.14	2.60	1.42	3.10	0.73	0.13	0.52	1.28
Maximum	6.32	2.6	0.79	88.42	5.11	13.39	0.74	0.33	2.37	97.27
Minimum	1.99	0.38	0.18	78.47	0.6	1.64	0.03	0.02	0.16	92.3
No. of samples	120	120	120	120	120	120	120	120	120	120

Source: data collected and analyzed by the New Zealand Institute for Plant and Food Research.

New Zealand, North Island: The aroma of Picual oil from olives grown in this region is predominantly of tomato leaves and stalks, fresh green herbs, citrus, and green bananas, with aromatic fresh culinary herbs, citrus, and tomatoes on the palate. Typically there are moderate to moderately high levels of bitterness and pungency in oils from early-harvested fruit, whereas oils from mid-harvest Picual have sweeter, ripe tomato coulis characteristics.

13.3.3 Frantoio

Australia: Frantoio produces mild to robust oils with intensity of aroma and palate strongly linked to the area of cultivation. The typical sensory profile of this variety in Australia is characterized by different grassy notes, artichoke, and olive fruit aromas. The palate normally displays nutty flavors ranging from green almonds to walnuts depending on the ripeness of the olives.

New Zealand, North Island: Early-harvested Frantoio produces robust oils with strong fruity aromas and flavors of fresh green bean, bitter salad greens, artichoke, and fresh walnuts, usually with high levels of bitterness and pungency. Oils from later harvested fruit are typically medium in style with ripe olive fruitiness and vanilla toffee and walnut flavors. Frantoio, either as a single varietal or more commonly in a blend with Leccino or Koroneiki, tends to dominate the market in the upper North Island.

New Zealand, South Island: In the South Island, this variety produces fresh, lively oils with complex, intense aromas. Green olive, fresh-cut grass, and green apple notes dominate the flavor. The levels of bitterness and chili-like pungency are usually moderate to high, in keeping with oils bordering on the robust style.

13.3.4 Coratina

Australia: Coratina oils in Australia typically display medium-intensity aromas combining notes of green almonds, fresh-cut grass, and citrus zest. These characters are well supported by a lengthy palate and medium to robust mouthfeel. High levels of bitterness and pungency are characteristic in this variety.

13.3.5 Arbequina

Australia: Arbequina is well known for its sweet and delicate oil with low levels of bitterness and a subtle peppery finish. Despite this mild mouth feel, Arbequina oils can be fairly fruity with a clear dominance of notes linked to fresh-cut grass and green apples when processed with a low maturity index. Vanilla or custard characters become more dominant in riper fruit.

13.3.6 Koroneiki

Australia: Koroneiki is a highly regarded variety that consistently delivers fruity oils of medium to robust styles. Its complex fragrance is characterized by a varying combination of citrus, pine resin, fresh-cut grass, and green banana skin notes. Bitterness and pungency together with length of flavor are usually moderately high and in balance with the intensity of the aroma.

New Zealand, North Island: Koroneiki produces some of the finest oils from this region. Typically, the oil is very complex and has high levels of fruitiness. When harvested early, the oil is usually robust and has strong fresh herb and fresh-cut grass aromas. On the palate, green notes of arugula (rocket), bitter salad, green banana, citrus, and black pepper dominate. As the fruit ripens, the levels of bitterness and pungency drop and along with fresh herb and green grass, citrus, ripe banana, and floral notes become evident. This oil is classified as medium.

New Zealand, South Island: As the fruit is harvested early in this region, this variety has complex, strong, green fruity, cut grass, and herbaceous aromas with green banana, fresh herbs, and citrus notes on the palate. The oil typically has moderate to high levels of bitterness and pungency, and is usually robust.

13.3.7 Manzanillo

Australia: This dual-purpose variety normally produces interesting medium oils with very fragrant characters. Its rather particular flavor combines typical olive characteristics with delicate confectionary notes and fresh strawberries or strawberry jam. The levels of bitterness and pungency seem to vary significantly depending on the rainfall or irrigation regime and the time of harvest.

New Zealand, North and South Islands: Much of the Manzanillo crop is used for table olives. However, when crushed, the resulting oil is very different from the typical New Zealand oils. Manzanillo oil has interesting "warm spicy" aromas and flavors reminiscent of banana cake, tropical fruit, and caramel confectionary along with ripe olive fruitiness. It is generally medium in style with moderate to low levels of bitterness and moderate pungency.

13.3.8 Hojiblanca

Australia: This variety is not widely planted in Australia but over the past few years has been producing superbly complex and fruity oils. Hojiblanca oils tend to have an almost creamy texture with mild to medium levels of bitterness and pungency. Its aroma is characterized by a balanced combination of green and sweet notes where green tomatoes, unripe tropical fruits, floral fragrances, and arugula are well represented.

13.3.9 Leccino

New Zealand, North Island: Leccino in the North Island region generally ripens early and produces oils with a ripe olive fruit and ripe banana aroma. These oils typically have flavors of soft salad leaves and nuts along with sweet caramel toffee notes. With moderate to moderately low levels of bitterness and pungency, they are almost always medium in style.

New Zealand, South Island: The Leccino oils from this region usually have fresh green fruity aromas with fresh nutty notes. Flavors of green almond and green tea with a hint of caramel sweetness are common. Generally there are moderate to high levels of bitterness and pungency. Much of the Leccino oil produced is used in Tuscan-style blends.

13.3.10 J5

New Zealand, North Island: This is a New Zealand variety. Unlike most olives, J5 trees grow well in soils that are not well drained and tolerate relatively humid conditions well. This variety has mild olive fruity aromas and more complex flavors of ripe olives, red apples, and soft salad leaves. It has a fresh nutty aftertaste. With moderate levels of bitterness and pungency, it is almost always medium in style.

13.4 Overview of olive growing regions and principal olive oil styles

13.4.1 Australia

Olive oil in Australia is produced in almost every state, with more than 40 informally established olive-growing regions, many of those mirroring some of the country's most recognized wine regions. Australia's olive oil regions are mainly in the southern, cooler parts of the country, with most olive groves located in South Australia, New South Wales, Victoria, and Western Australia. The main varieties cultivated include Arbequina, Barnea, Coratina, Corregiola, Frantoio, Hojiblanca, Koroneiki, Leccino, Manzanillo, Nevadillo Blanco, Pendolino, Picual, Picholine, and South Australian Verdale.

In order to simplify the analysis of Australia's growing areas, they have been grouped into four main regions, Northern, Central Northern, Central Southern and Southern Regions, based on their distinctive climatic conditions, varieties performance. and oil styles (Figure 13.2).

Table 13.5 summarizes the main olive cultivars in Australia and typical sensory properties of the oil produced in the four main regions.

13.4.1.1 Northern Region The Northern Region is the northernmost olive-growing region in Australia. It covers totally or partially the areas of Darling

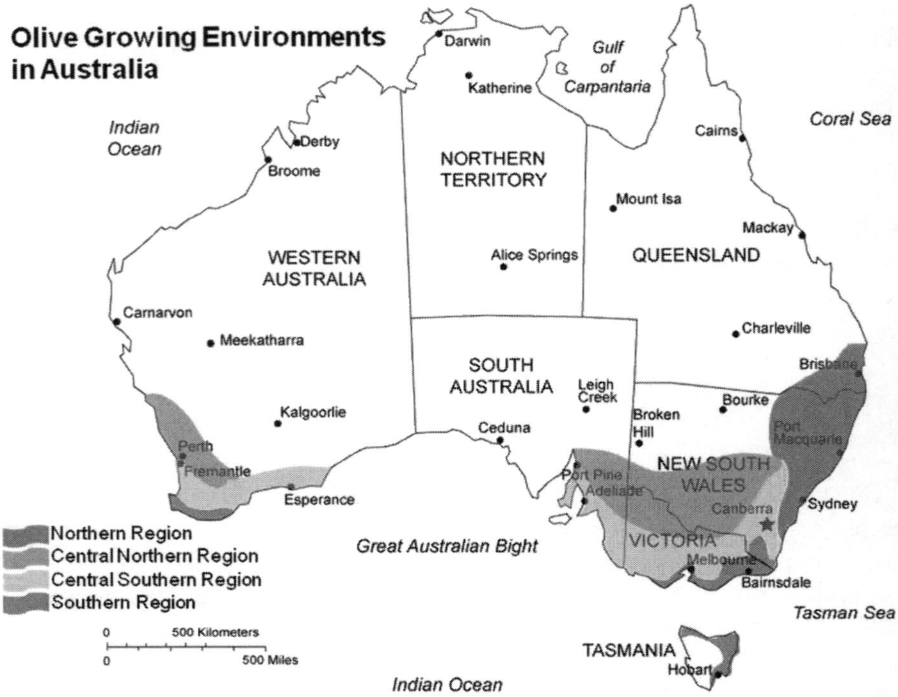

Figure 13.2 Olive-growing environments in Australia. For a color version of this figure, see the color plates section.

Downs, Brisbane, West Moreton, and Wide Bay-Burnett in Queensland and also South, Central, and North Coast, Hunter, Northern Inland, and Orana in New South Wales. This region has long warm to hot summers, moderate and variable rainfall and a winter of cool clear days. Cold, frosty nights predominate, particularly inland, with often a rapid transition from summer to winter. Average maximum temperatures of the six hottest months of the year range from 24 to 30 °C. Unlike the other growing regions in Australia, the Northern Region does not present a typical Mediterranean rainfall pattern, with summer usually providing the highest falls and the remaining months providing generally even falls. Summer rain tends to be more variable than winter rain due to the incidence of thunderstorms. Winter frosts are fairly common and represent a limiting factor for those groves located inland and in low-lying areas. Less than 8% of Australia's production comes from this region.

Groves in the Northern Region were some of the first groves to be planted during the resurgence of the Australian olive industry. Consequently, they have utilized mostly those varieties that were available in the early stages of the industry, such as Manzanillo, Frantoio, Nevadillo Blanco, and Corregiola. The typical early flowering as a consequence of the warm spring and a fruit

Table 13.5 Main olive cultivars in Australia and common typical properties of the oil produced in the various regions.

Region	Main cultivars	Common sensory properties
Northern	Manzanillo, Frantoio, Nevadillo Blanco, Corregiola	Fresh and sweet fruit aromas with a gentle mouthfeel displaying low levels of bitterness and pungency
Central Northern	Picual, Coratina, Picholine	Complex and intense aromas dominated by characters such as ripe tomato, banana, red apple, and tropical fruits; mild to medium in style. Moderate levels of bitterness and pungency
Central Southern	Barnea, Arbequina	Milder and earlier ripening varieties produce well-balanced oils
	Picual, Koroneiki, Hojiblanca	Dominant green grassy notes with unripe fruit characters; more robust than oils from the Central Northern region
	Hardy's Mammoth, Sevillano, Manzanillo	Great complexity and prominent fruit characters
Southern	Central and Northern Italian varieties such as Frantoio, Pendolino and Leccino	Very distinctive green notes of unripe olives matched by fresh-cut grass combined with a nutty, unripe aftertaste. Moderate to high levels of bitterness and pungency

ripening period coinciding with hot summer temperatures and common rain events have led to a characteristic sweet and mild nature of the oils produced in this region. These oils usually share lively fresh and sweet fruit aromas combined with a gentle mouthfeel displaying very low levels of bitterness and pungency.

13.4.1.2 Central Northern Region The Central Northern Region is the warmest of the Australian olive-growing regions, displaying a typical Mediterranean rainfall pattern. It encompasses the areas of Central West, Riverina, and Murray in New South Wales, Mallee in Victoria, Murray Lands, Riverland, and Clare Valley in South Australia, and the Wheat Belt and Moore River areas in Western Australia. This region has long hot and dry summers and cool winter days. Average maximum temperatures of the six hottest months of the year range between 27 and 30 °C. In the Eastern states, the annual rainfall is usually scarce and it is fairly evenly distributed through the year but is more concentrated in the winter and spring months. Meanwhile, Western Australia shows a more significant concentration of winter rainfall, with up to 10 times more rain during this period than during the summer months. Approximately half of Australia's production is grown in this region.

Oils from the Central Northern Region tend to have complex and intense aromas dominated by fruity characters such as ripe tomato, banana, red apple, and tropical fruits. These oils typically show mild to medium styles. Over the years, late-ripening and more robust varieties from this region, such as Picual, Coratina, and Picholine, became well-known oils around the world for their vibrant fruitiness and well-balanced medium character with moderate levels of bitterness and pungency. Their popularity has turned them into some of the leading oils for both the domestic and export markets.

13.4.1.3 Central Southern Region The Central Southern Region shares various similarities with the Central Northern Region but it typically displays slightly higher rainfall levels and slightly cooler temperatures during the growing season. It covers totally or partially the areas of Southern Inland in New South Wales, Loddon, Hume, Grampians/Wimmera, Central Highlands, and Western Districts in Victoria, Limestone Coast, Yorke Peninsula, Fleurieu Peninsula, Adelaide Hills, and Barossa Valley in South Australia, and Great Southern in Western Australia. This region experiences a more oceanic climate with typically dry and warm summers and mild, cool winter days. Average maximum temperatures of the six hottest months of the year range between 22 and 27 °C. Similarly to the Central Northern Region, the rainfall pattern in Western Australia is more sharply concentrated during the winter period compared with the Eastern States. On average, the region may experience 20–50% more rainfall than the Central Northern Region. More than 40% of Australian olive oil is a product of this region.

Oils from the Central Southern Region tend to have dominant green, grassy notes with unripe fruit characters. These oils are normally more robust than the oils from the Central Northern Region, showing more intense palate and mouth sensations. Milder and early-ripening varieties such as Barnea and Arbequina produce well-balanced oils that are becoming the backbone of many Australian commercial blends. Other varieties such as Picual, Koroneiki, and Hojiblanca from this region have produced outstanding complex oils that have been nationally and internationally awarded in several occasions (Gawel, 2010). Occasionally, some oils from table olive varieties, such as the local Hardy's Mammoth, Sevillano, and Manzanillo, have produced very interesting and rather unique results amongst boutique producers, with oils displaying great complexity and prominent tropical fruit characters.

13.4.1.4 Southern Region The Southern Region covers all growing areas in Tasmania; Gippsland, Mornington Peninsula, Ballerine Peninsula, Alpine, and South West Coast regions in Victoria and also the South West Coast in Western Australia. All areas in this region have fairly temperate and stable weather due to the strong influence of the Southern Ocean. Average maximum temperatures of the six hottest months are fairly low and relatively marginal for olives, ranging between 18 and 22 °C. Rainfall patterns in the region are also Mediterranean but with even higher overall levels than the Central Southern Region. This region contributes only 1% to Australia's olive oil production.

Oils from the Southern Region share many similarities with those from New Zealand. Central and Northern Italian varieties such as Frantoio, Pendolino, and Leccino tend to dominate the production of this region, although some new varieties are gradually coming into production. These oils are characterized by very distinctive green notes of unripe olives matched by fresh-cut grass combined with a nutty, unripe aftertaste. The levels of bitterness and pungency are usually moderate to high, supporting the robust style of these oils.

13.4.2 New Zealand

13.4.2.1 Climate and growing season The islands that make up New Zealand have a land mass slightly smaller than Italy. The two main islands, the North and South Islands, lie from latitude 35° south to 47° south. The central mountain ranges in both islands have a significant influence on the climate that varies considerably, from subtropical in the north to cool temperate in the far south. The mountains also provide a barrier for the prevailing westerly winds, making the areas to the east of the mountains in both islands the driest and most suitable for olive growing. However, it is the large mass of surrounding ocean that moderates extremes in temperatures. New Zealand is noted for its wind; an average wind speed of 14 km/h is common.

Olive trees have been planted from the north of the North Island almost to the southernmost region of the South Island. Apart from the very high-rainfall western coastal areas, no region is particularly unsuitable for olives, but there are specific regional issues. For example, in the warmer, more humid areas of the north, fungal disease is generally worse, whereas frost is a problem in the south of the country. Olive fly and olive knot are absent but birds are a problem everywhere (Lang, 2004).

Compared with the more southern olive-growing regions, spring arrives earlier and is warmer in the northern and Hawke's Bay regions of the North Island. Summer is warm to very warm and cooler autumn temperatures arrive later than in the south of the country. Historically, these northern regions also have a much greater number of growing degree days (GGDs) in spring and autumn than the southern regions. For example, the Auckland region in the North Island would usually have 858 GGDs for spring and autumn (September to November and March to May), compared with 445 GGDs for the same periods in the South Island Nelson region. All of the olive-growing regions generally have a high number of sunshine hours, ranging from around 1900 to 2400 h per year.

These differences in climate, coupled with the differences in orchard location, mean that olive fruit maturity is achieved at different times during the season. Normally this will be earlier in the North Island regions and progressively later further south. For example, the commercial harvest usually begins in Northland in late March or early April and ends in early June, whereas in the South Island, the commercial harvest is only commencing in June and weather permitting can be extended until the beginning of July (Requejo-Jackman, Farrell, and Olsson, 2011).

As in most olive-growing countries of the world, New Zealand's industry is regionalized. The main growing regions (Figure 13.3) in the North Island are the Upper North Island, which includes Northland; Auckland, Waiheke Island, and Bay of Plenty; Central/Southern North Island comprising Hawke's Bay and Wairarapa and the small Lower North Island region

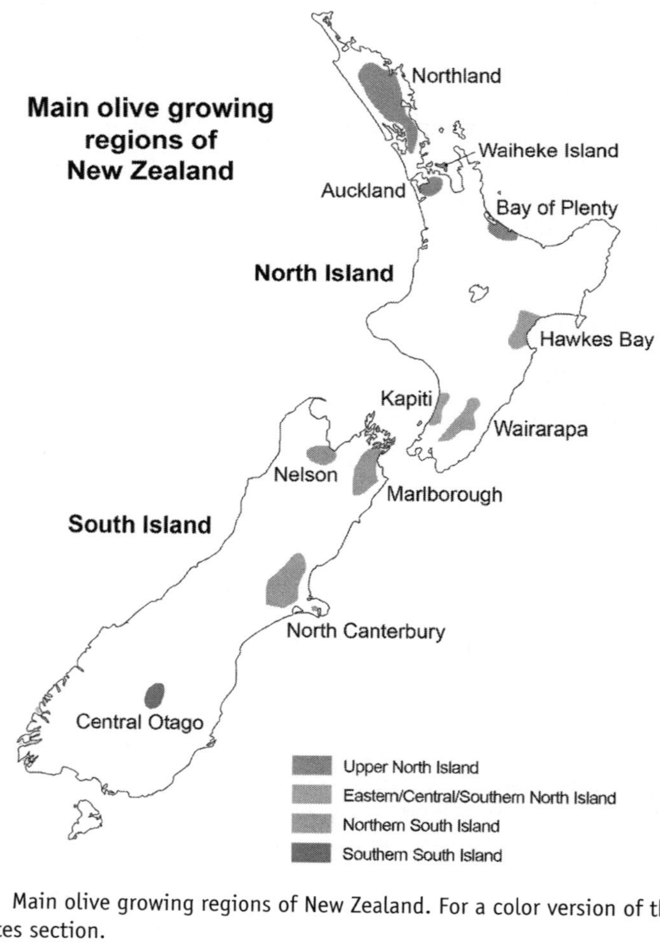

Figure 13.3 Main olive growing regions of New Zealand. For a color version of this figure, see the color plates section.

of Kapiti. The South Island is divided into the Upper South Island region of Nelson/Marlborough and North Canterbury and the central, southern region of Central Otago.

In general, olive oils produced in New Zealand are noted for their lively, complex green fruity aromas and flavors. As the industry has developed, regional styles of oils have begun to emerge, but the climatic conditions coupled with the comparatively small number of olive varieties grown mean that there is much less variation in style than in Australia or the northern hemisphere. The majority of high-quality oils produced in New Zealand are classified as medium (moderate levels of bitterness and pungency) or robust (high levels of bitterness and pungency). While the olive variety will largely determine the characteristics of New Zealand oils, style is more a reflection of climate, orchard site, and harvesting time. In general, the oils from early-harvested fruit have significantly higher levels of bitterness and pungency than oils produced from later harvested olives (Edwards, 2007). Table 13.6 summarizes the main olive cultivars in New Zealand and typical sensory properties of the oil produced in the various regions.

13.4.2.2 New Zealand olive-growing regions – North Island

13.4.2.2.1 Upper North Island (Northland, Waiheke Island, Auckland, Bay of Plenty) The climate in this region is subtropical with warm, humid summers and mild winters. In summer, the maximum daytime air temperatures will range from 22 to 26 °C whereas winter maximum air temperatures are from 12 to 17 °C. Rainfall differs amongst these areas, with the Auckland and Northland regions having the highest rainfall ($\sim$1300 mm per year). Although most rainfall normally occurs in the winter, summer and autumn storms of tropical origin can bring high winds and heavy rainfall, and it is these that have the greatest impact on the olive industry. The growing season is potentially the longest in the country.

About 100 000 trees (ONZ members' surveys) are thought to be planted. The most common cultivars are Frantoio, Leccino, Koroneiki and the New Zealand variety J5, along with smaller plantings of Picual. The "mother" tree used for the propagation of J5 still thrives on the edge of marshy land in Northland. Whereas J5 is one of the main varieties grown in the north of this region, it is not common in Hawke's Bay or further south. Although large numbers of Barnea trees were planted in the 1990s, these failed to thrive and have been replaced with more suitable varieties.

Oils from the Upper North Island are diverse, depending upon cultivar and harvest time. Frantoio and early-harvested Koroneiki and Picual are usually robust in style, characterized by green notes and high levels of bitterness and pungency. Oils from late-harvest Koroneiki and Picual have more ripe fruit attributes and are sweeter, with lower levels of bitterness and pungency. Leccino and J5 have ripe olive and nutty characteristics; both are usually made

Table 13.6 Main olive cultivars in New Zealand and typical sensory properties of the oils produced in the various regions.

Region	Sub-region	Main cultivars	Common sensory properties
North Island	Upper North Island	Frantoio	Robust with strong fruity aromas and flavors of fresh green beans, bitter salad greens, artichoke, and fresh walnuts; high levels of bitterness and pungency. Oils from later harvested fruit have ripe olive fruitiness, vanilla toffee, and walnut flavors
		Leccino	Aromas of ripe olive fruit and ripe banana. Flavors of soft salad leaves, nuts, and sweet caramel toffee. Moderate to moderately low levels of bitterness and pungency; medium style
		Koroneiki	Very complex, high levels of fruitiness. When harvested early, robust with strong fresh herb and fresh cut grass aromas. Green flavors of arugula (rocket), bitter salad, green banana, citrus, and black pepper. Oil from riper fruit has citrus, ripe banana, and floral notes; medium style
		J5	Mild olive fruity aromas; more complex flavors of ripe olives, red apples, and soft salad leaves with fresh nutty aftertaste. Moderate levels of bitterness and pungency; medium style
		Picual	Aromas of tomato leaves and stalks, fresh green herbs, citrus, and green bananas, with fresh culinary herbs, citrus, and tomatoes on the palate. Moderate to moderately high levels of bitterness and pungency in oils from early-harvested fruit; oils from mid-harvest have sweeter, ripe tomato coulis characteristics
	Eastern/ Central/ Southern North Island	Frantoio	Similar to Frantoio from Upper North Island
		Barnea	Complex aroma of green olives and fresh-cut grass with spicy chili notes. Spicy, green olive fruitiness carries through to the palate along with bitter herbs, green banana, green apple, and green leafy notes. High level of bitterness and chili pepper-like pungency
		Leccino	Similar to Leccino from Upper North Island
		Manzanillo	Interesting "warm spicy" aromas and flavors reminiscent of banana cake, tropical fruit, and caramel confectionary along with ripe olive fruitiness. Medium in style with moderate to low levels of bitterness and moderate pungency

Table 13.6 (*Continued*)

Region	Sub-region	Main cultivars	Common sensory properties
South Island	Northern South Island	Frantoio	Fresh, lively oils with complex, intense aromas. Green olive, fresh-cut grass, and green apple notes dominate the flavor. Bitterness and chili-like pungency moderate to high; oils border on robust style
		Leccino	Fresh green fruity aromas with fresh nutty notes. Flavors of green almond and green tea with a caramel sweetness. Moderate to high levels of bitterness and pungency. Much of the Leccino oil produced is used in Tuscan-style blends
		Barnea	Similar to Barnea from Eastern/Central/Southern North Island
		Koroneiki	Complex, strong green fruit, cut grass and herbaceous aromas with green banana, fresh herbs, and citrus notes on the palate. Moderate to high levels of bitterness and pungency; robust style
		Manzanillo	Similar to Manzanillo from Eastern/Central/Southern North Island
	Southern South Island	Frantoio	Similar to Frantoio from Northern South Island
		Leccino	Similar to Leccino from Northern South Island

in a medium style with moderate to moderately low levels of bitterness and pungency.

13.4.2.2.2 Eastern/Central/Southern North Island (Hawke's Bay, Wairarapa, Kapiti) The Hawke's Bay and Wairarapa regions share a similar dry, sunny climate with warm, settled weather in summer. Maximum daytime air temperatures range from 20 to 28 °C. Frosts are frequent in winter. Rainfall of around 1000 mm occurs mainly in the cooler months.

The Kapiti region is often windy because of its exposure to disturbed weather systems from the west but has few climate extremes. The most settled weather occurs during summer and early autumn. Summers in Kapiti are warm with typical summer daytime maximum air temperatures ranging from 19 to 24 °C.

With a conservative estimate of 151 000 trees (ONZ members' surveys), this region is the most important for olive oil production. Several of the larger commercial groves with plantings in excess of 10 000 trees are now well established in Hawke's Bay. Frantoio, closely followed by Barnea, Leccino, and Manzanillo, are the most common varietals in this region. There are also significant plantings of Picual and lesser plantings of Koroneiki.

Frantoio and Leccino oils from the Eastern/Central/Southern North Island are similar to their counterparts from the Upper North Island. Barnea oils display green characteristics and have a high level of bitterness and chili pepper-like pungency. Manzanillo oils are medium in style with "warm spicy" aromas and low levels of bitterness and pungency.

13.4.2.3 New Zealand olive-growing regions – South Island

13.4.2.3.1 Northern South Island (Nelson, Marlborough, North Canterbury) Much of this region is sheltered by mountains and has at least 2300 sunshine hours annually, more than any other New Zealand region. The summer is warm, dry, and settled with typical daytime temperatures ranging from 20 to 26 °C. The most unsettled weather comes in late winter and early spring. Frosts are common in winter and can have an impact on the olive harvest.

This region has a combined total of around 68 500 trees (ONZ members' surveys). Although Marlborough was the first area to have substantial commercial olive plantings in the 1990s, much of the land in recent times has been acquired for wine grape growing. Frantoio and Leccino along with Barnea are the most important varieties. Small areas of land are planted with Koroneiki and in areas with specific microclimates there are very small numbers of Manzanillo.

Oils from the Northern South Island are characterized by very distinctive notes of green fruits (olive, apple, almond, banana) matched by fresh-cut grass and herbs. The levels of bitterness and pungency are usually moderate to high, supporting the robust style of these oils.

13.4.2.3.2 Southern South Island (Central Otago) Although it has the lowest rainfall and high sunshine hours, Central Otago has the lowest number of GGDs of all the olive-growing regions in New Zealand. Because of the early onset of frost and possibility of snowfalls, it is difficult to achieve high production figures.

This is the smallest olive-growing region in New Zealand with ~3550 trees (ONZ members' surveys). Frantoio and Leccino are the most commonly grown varietals. Typically, their flavor characteristics and levels of bitterness and pungency are very similar to those of Frantoio and Leccino oils from the northern South Island region.

13.5 Conclusion

13.5.1 Australia

With an expected peak production by 2017 of ~30 000 t of olive oil, the Australian olive industry volumes will continue being small in world terms. Nonetheless, the leading role of Australian orchards in grove modernization and mechanization, the uncompromising view of Australian producers on EVOO quality, and the public recognition of its Standard have revolutionized this traditional industry.

There is a generalized consensus that mechanization of the most critical orchard practices is essential for the long-term sustainability of the global olive industry. Nonetheless, it is even more critical for the future success of the olive industry and also for consumer satisfaction that the current international and domestic analytical methods and chemical limits for olive oil are reviewed. This review has to consider the natural variability of olive oils, the need to be able readily to identify and control new adulteration techniques, and the inclusion of a measurement of the freshness of olive oil as an important parameter for oil classification.

13.5.2 New Zealand

As the volume of oil increases, the future for New Zealand olive oils lies in the ability of producers and Olives New Zealand to educate New Zealand consumers to recognize both the superior taste and nutritional qualities of fresh, locally produced certified extra virgin olive oils and to encourage their purchase and liberal use as part of healthy, balanced diets. Given the small domestic market, exporting New Zealand olive oils to niche markets around the world will also be essential. However, if the industry is to develop successfully in this world of trade globalization, New Zealand producers must strive to make extra virgin olive oils that are significantly better than the current world standards for both quality and safety, and through skilful promotion be recognized as such in overseas markets (Edwards, 2007).

Acknowledgments

Leandro Ravetti would like to express his appreciation to Dr Rodney Mailer, Research Fellow and Adjunct Professor at Australian Oils Research Laboratory, Paul Miller, President of the Australian Olive Association, Lisa Rowntree, CEO of the Australian Olive Association, Pablo Canamasas, oil production manager at Boundary Bend Olives, Claudia Guillaume, laboratory manager at Modern Olives, and Tim Smith, marketing director at Cobram Estate, for their great assistance reviewing this chapter.

Margaret Edwards would like to express her thanks and appreciation to the following people for their assistance in reviewing this chapter: Dr Marie Wong, Associate Professor, Institute of Food, Nutrition and Human Health, Massey University, Albany, Cecilia Requejo-Jackman, Scientist, The New Zealand Institute for Plant and Food Research Ltd, and Dr Roger Harker, Science Group Leader, The New Zealand Institute for Plant and Food Research Ltd.

References

Edwards, M. (2007) The development of the New Zealand extra virgin olive oil industry, in *Handbook of Australasian Edible Oils* (ed. C. O'Connor), Oils and Fats Specialist Group of New Zealand Institute of Chemistry, Auckland, pp. 38–60.

Gawel, R. (2010) Analysis of award winning Australian oils. Presented at theAustralian Olive Association National Conference 2010, Adelaide.

Guillaume, C., Ravetti, L., Ray, D.L., and Johnson, J. (2011) Technological factors affecting sterols in Australian olive oils. *Journal of the American Oil Chemists' Society*, **89** (1), 29–40.

Hartmann, H.T. (1962) Olive growing in Australia in 1962. *Economic Botany*, **16** (1), 31–44.

Lang, S. (2004) *Olives New Zealand Best Management Practice/Integrated Olive Production Electronic Manual*, from Olives New Zealand Members' web site, http://www.olivesnz.org.nz (last accessed 2 April 2012).

Mailer, R.J. (2007) *The Natural Chemistry of Australian Extra Virgin Olive Oil*, New South Wales Department of Primary Industries and Rural Industries Research and Development Corporation, Kingston, ACT.

Mailer, R.J., and Ayton, J. (2007) *A Survey of Australian Olive Cultivars to Determine Compliance with International Standards*, New South Wales Department of Primary Industries and Rural Industries Research and Development Corporation, Kingston, ACT.

Miller, P. (2011) *Australian Olive Association Annual President's Report.* Presented at the Australian Olive Association National Conference 2011, Wangaratta, VIC.

Noser, G. (1997) *Olives: the New Passion*, Penguin Books (NZ), Auckland.

Ravetti, L., Miller, P., and Mailer, R. (2008) The modern Australian olive industry. Presented at the VIth International Olive Symposium, Evora, Portugal.

Requejo-Jackman, C., Farrell, M., Olsson, S., Beresford, M., Harker, R., Wohlers, M., McGhie, T., Comeskey, D., Wang, Y., Woolf, A., and Wong, M. (2011) *Effect of Harvest Maturity on Extra Virgin Olive Oil. Year 3. A Report to Olives New Zealand Inc.*, from Olives New Zealand Member's web site, http://www.olivesnz.org.nz (last accessed 2 April, 2012).

Sheridan, G. (2012). *Olive Grove Census. A Report to Olives New Zealand Inc.*, from Olives New Zealand Member's web site, http://www.olivesnz.org.nz (last accessed 20 May 2012).

14 Olive Oils from South America

Adriana Turcato and Susana Mattar
CRESA: Laboratorio de Análisis Sensorial de Aceite de Oliva y Alimentos, Facultad de Ciencias de la Alimentación, Universidad Católica de Cuyo, San Juan, Argentina

14.1 The origins of olive growing in South America

The history of olives in America began around 1492 with the arrival in America of Christopher Columbus, who brought with him hundreds of different varieties and species of olives which then mixed with the native flora and fauna of the new lands. The first olive trees were brought from Seville, Spain, to the West Indies and later they spread to the rest of South America. In 1510, there were olive trees in Mexico and these were taken to Peru, Argentina, and Chile.

Olive growing was introduced in Argentina during colonial times through the first religious orders in the region. Olives were first planted in Argentina around 1550. It is not clear which was the actual introduction path or the first site where olives were planted, but it is well documented that olive cultivation closely followed the first population settlements.

There are two different versions of olive introduction: one of the stories indicates that members of a Spanish Embassy, Hernan Mirabal, Rodrigo de Quintero, Bartolomé de Mansilla, Nicolás de Garnica, and Pedro de Cáceres, were the first to bring the olive tree to Argentina. None of them made specific reference to olive trees in their memoirs but they quoted the term of "fruiting trees."

Other authors have suggested that the first olive trees were introduced in Santiago del Estero, the first city founded in 1554 by explorers arriving from

Olive Oil Sensory Science, First Edition. Edited by Erminio Monteleone and Susan Langstaff.

Peru. Later, other Spanish troops arriving from Chile brought olive cuttings with them and those cuttings were planted in the current region of Arauco in La Rioja province.

Owing to the favorable development of the crop in America, the Spanish King Charles III and his Peruvian deputy Pedro Fernández de Castro, Conde de Lemos (1667–1672), ordered all olive trees from Peru to the River Plate to be cut down, fearing that the olive industry in the new continent could overshadow Spain as world leader. Legend tells that only one tree was spared and that it was located in the township of Aimogasta in La Rioja province in a grove belonging to the Avila family, direct descendants of the La Rioja founders.

Following a number of selections, crossings, and propagations, olive growing spread again to other regions such as Cordoba, Mendoza, and San Juan. Cuttings were taken also to Chile and Peru, giving rise to a typical Argentinean variety, Arauco. This variety is most likely the result of some crossbreeding of southern Spanish varieties.

While it is true that the industry has been present in South America for more than four centuries, domestic consumption of olive products only started to increase towards the end of the nineteenth century with the first significant wave of Mediterranean immigrants. With a growing demand and an almost nonexistent domestic production, the olive trade became highly dependent on Spanish products until, at the beginning of the twentieth century, economic and political tensions with Spain led to a relative restriction to access to olives and olive oil.

14.2 Olive growing in Argentina

Following the events mentioned above, the Argentinean government decided to promote the olive industry in different regions across the country. In 1932, Law No. 11,643, also called "of olive growing promotion," was passed, and in 1954, during the National Olive Growing Conference in La Rioja province, the "Be a patriot, plant an olive" motto was established. This strong promotion led to the development of new growing regions.

This historical period left excellent research publications that show the coordinated scientific efforts of the Argentinean olive industry during that era. Some of those publications include relevant topics such as the influence of olive varieties on the chemical composition of olive oil (Karman de Sutton *et al.*, 1963) as good examples of the long olive-growing tradition in the country. Most of the scientific work was carried out at the Estación Experimental Agropecuaria de Catamarca (INTA), Facultad de Ciencias Exactas y Naturales, Universidad Nacional de Buenos Aires, and Consejo Nacional de Investigaciones Científicas y Técnicas (CONICET), carrying out fatty acid composition analysis at the Instituto de la Grasa de Sevilla.

The olive industry in Argentina has experienced several fluctuating periods as a consequence of diverse historical or political reasons. During the 1960s, olive growing was severely affected by strong domestic seed oil competition. From that decade onwards, the olive oil industry experienced a period of poor development until the 1990s, when the olive industry and its distribution received the support of several promotional policies such as the "Tax Deferral Law." This law benefited the provinces of Catamarca, La Rioja, and San Juan, where most of the new developments took place. From that moment on, the expansion of the olive-growing area has been significant and sustained over a period of time, jumping from 30,000 to ~90 100 ha in 2007 according to data from the Instituto Nacional de Tecnología Agropecuaria (INTA) (INTA, 2007). La Rioja is the leading region with 29% of the planted area, followed by Catamarca with 22%, San Juan with 20%, Mendoza with 20%, Cordoba with 5%, and Buenos Aires with 3% (INDEC-CNA, 2002). According to its main purpose, 67% of the area is dedicated to olive oil production and the remaining 33% is mostly directed to table olive production.

At the beginning of the 1990s, the olive-growing world entered a period of deep changes, introducing a dynamic that was not typical of a traditionally static and undeveloped industry. Within this context, Argentina became one of those "new players" in the world of olive growing as the largest producer of olive oil and table olives outside the Mediterranean basin.

According to the National Food Agency (Dirección Nacional de Alimentos), the greatest challenge that Argentina faces is the need to break into a highly competitive global market which is currently dominated by quality, history, and brand awareness.

A quick snapshot of the most important aspects related to the growth of the industry in the past decade shows:

- The widespread use of technology, particularly in relation to harvest mechanization and the use of higher capacity equipment for high-quality olive oil production.
- The appearance of new players in the global olive oil market, including Argentina.
- The ongoing growth in global olive oil consumption leading to a potential excess demand.

In the context of Argentinean olive oil production and according to estimates by INTA, the typical traditional grower manages a grove ranging from 5 to 100 ha, with trees planted, on average, 30 years ago at a density of 80–100 trees/ha and yielding 3 t/ha, which are harvested by hand. On the other hand, the new producer is characterized by groves ranging from 100 to 2000 ha, planted with more suitable varieties and higher densities of 300–800 trees/ha. Yields in those groves are higher at 10–12 t/ha, which are harvested mechanically. Those smaller farms with less than 100 ha represent just 14% of the cultivated area, while the remaining 86% is mostly covered by groves averaging 100–500 ha.

Following this olive-growing expansion, research activities around olive growing became more relevant for a number of institutions such as INTA, universities in the growing regions of Catamarca, La Rioja, San Juan, Mendoza, and Córdoba, and scientific study centers such as Plapiqui in Bahía Blanca and CRILAR in Anillaco, La Rioja. All these activities had the ongoing support of the different regional governments through the Consejo Federal de Inversiones (CFI), where the 2020 Olive Strategic Plan was developed.

Nowadays, Argentina has two International Oil Council (IOC)-recognized taste panels: one at the Universidad Católica de Cuyo in San Juan and the other at INTA, Catamarca. The latter panel is composed entirely of trained blind staff. Furthermore, Uruguay, Chile, Peru and Argentina are currently training additional panels.

14.3 Other olive-growing countries in South America

The olive distribution in South America is shown in Figure 14.1.

14.3.1 Chile

In Chile, the olive tree was introduced mainly through the central valley. Plantations spread to the Limary valley to the north and the Bio-Bio River to the south. Chile is bounded on one side by the Pacific Ocean and on the other side by the Andes mountain range. This geography produces different micro- and mesoclimates which affect the adaptation of different varieties of olive trees in the country. Olives grow abundantly between the Atacama and Maule regions.

The modern history of olive growing in Chile starts in 1946, when Italian immigrants planted some commercial olive groves just south of Santiago, producing, by 1953, some of the first Chilean olive oils. In the 1990s, the main olive varieties used worldwide were introduced, as were the latest agricultural technologies.

The Chilean olive industry is currently showing a significant development, with a total planted area of 24 000 ha. The groves are located mostly in the central and northern parts of the country, covering the semi-desert region of Coquimbo with a typical sunny and dry climate, the southern region of Maule with characteristic cold and wet winters, and the regions of Concepción

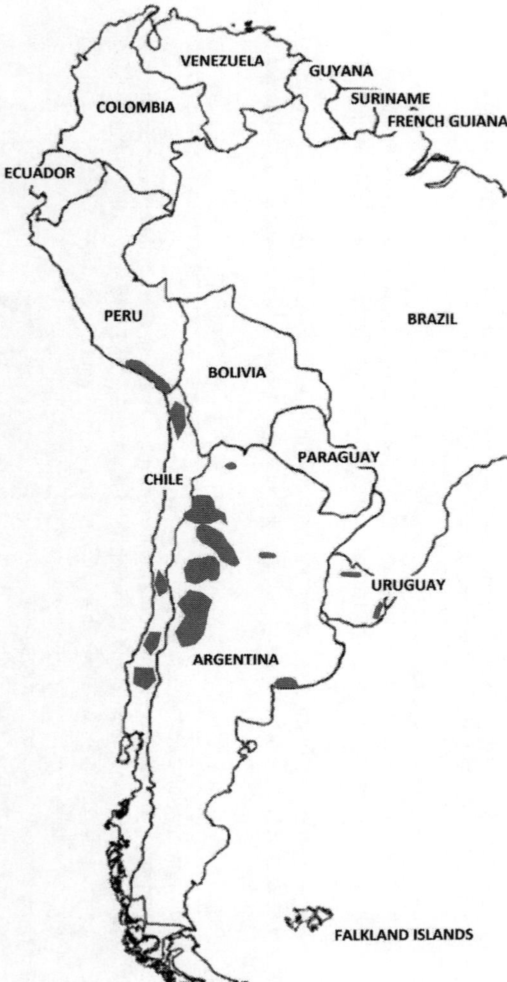

Figure 14.1 Olive distribution in South America. For a color version of this figure, see the color plates section.

and Arica further north. Despite being a relatively young olive-producing country, Chile has managed to reach important milestones and it is currently a highly mechanized industry.

The main olive varieties in Chile are Arbequina, Frantoio, Arbosana, Picual, Leccino, Koroneiki, and Coratina, with smaller quantities of Biancolilla, Noccelara del Belice, Grappolo, Barnea, and the indigenous variety, Racimo. Table 14.1 summarizes the main olive cultivars in Chile and typical sensory properties of the oils produced.

Table 14.1 Main olive cultivars in Chile, Peru, and Uruguay and typical sensory properties of the oil produced.

Country	Regions	Main cultivars	Typical sensory properties
Chile	Limary valley to the north and the Bio-Bio River to the south	Arbequina Frantoio Coratina Leccino Picual Barnea Arbosana	Virgin olive oils: from lightly fruity to medium fruity. Arbequina, Frantoio, and Empeltre are noted for their intense fresh fruit aromas, sweet, sour, and spicy but with persistence. Virgin olives oils rich in fruit; hints of ripe tomato, apple, and banana and aromatic herbs. Taste: complex with vegetable flavor. Bitterness and pungency are distinctive and balanced. There are excellent extra virgin olive oil blends: bi- and trivarietals. Sensory analysis has shown that olive oils obtained from Frantoio are characterized by bitter and pungent tastes whereas Arbequina virgin olive oils are sweeter and with a slightly more pronounced aroma (Estay, Servili, and Selvaggini, 2012)
Peru	In the center and in the extreme south-west of the country. between the Cordillera of Andes and the Pacific Ocean	Coratina Picual Frantoio Pendolino Leccino Empeltre	*Extra virgin olive oil of cv. Coratina* From medium to intense fruity perceptions. Pungent and bitter with notes of apple, ripe tomato, and green herbs. Taste: green, herbaceous notes of artichoke. Taste is complex and strong *Extra virgin olive oil of cv. Picual* Average intensity of fruity perceptions. Pungent and bitter with notes of green herbs. Vegetable notes of apple and artichoke and notes of aromatic herbs and fruit, especially fresh walnut and almond. There are excellent extra virgin olive oil blends
Uruguay	Cologne on the south-west banks of the Rio de la Plata and Maldonado in the south and east	Arbequina Picual Manzanilla Coratina Frantoio Barnea	*Extra virgin olive oil of cv. Arbequina* Medium-sweet fruity perception. Aroma: tomatoes, nuts, and herbs with notes of citrus fruits. Flavor perceptions: tomato and herbs with slight bitterness and spiciness with greater persistence. Oils are sweet; very harmonious, and of medium complexity. There are excellent extra virgin olive oil blends: *Extra virgin olive oil of cv. Arbequina–Coratina* Medium fruity. Aroma: fresh herbs, olive leaves, tomato, banana notes, and flowers. Flavor perceptions: almonds, tomato, and herb notes. Low-intensity bitterness and spiciness. Oils are sweet, balanced, of great harmony and complexity *Extra virgin olive oil of cv. Barnea–Coratina–Picual* Medium fruity. Aroma is rich in vegetable notes, fresh herbs, and olive leaves with tomato and fig notes. Taste: fresh herbs, fruity perceptions. It is more complex with a medium persistence of

14.3.2 Peru

The Peruvian olive history goes as far back as 1559, when the Spanish noble Antonio de Ribera arrived from Seville with selected cultivars.

Olives are mostly produced in the central and south-western parts of the country where the narrow band of land between the Andes and the Pacific Ocean determines a favorable microclimate to grow olives. Peru's olive industry is small even in comparison with those countries considered to be low producers. The Peruvian olive-growing area covers ~11 000 ha planted in the regions of Chilca, Huaral, in the Ica, Chincha, and Pisco valleys and in the areas of Arequipa and Tacna. Peruvian producers are continually working to better the quality by reducing the degree of free fatty acidity in the oil.

Domestic production is ~900 t, which is mostly destined for local consumption as a complement to Peruvian cuisine. A small percentage is exported to Ecuador, Bolivia, and Colombia. However, demand from Canada, Japan, and Australia has contributed to a rise in production and exports. There are 40 processing plants.

The most dominant olive varieties are Frantoio, Coratina, Pendolino, Leccino, and Empeltre. Furthermore, Peru has a local variety similar to the Argentinean Arauco called Zappa. Table 14.1 summarizes the main olive cultivars in Peru and typical sensory properties of the oils produced.

14.3.3 Uruguay

Olives reached Uruguay from Argentina. The planted area has increased dramatically over the past few years, reaching ~9000 ha. The groves are concentrated in the eastern coastal region of Maldonado and also in the northern part of Salto and the eastern side of Rivera.

The most common varieties are Arbequina, Leccino, Frantoio, and Barnea. Oils from these groves are of very good quality. The Premium Quality Seal is the first quality indicator for Uruguayan olive oils. This seal recognizes the authenticity and excellence of Uruguayan extra virgin olive oil and certifies compliance with quality standards established by the Uruguayan Olive Oil Association and monitored by the School of Chemistry of the University of the Republic.

The seal is granted every year by the Uruguayan Olive Oil Association to brands that pass checks carried out at the chemistry and sensory evaluation laboratories of the School of Chemistry. Uruguay has university experts in olive oil tasting trained in Spain.

Local production for 2012 reached 550 000 l; however, Uruguayans consume just 0.4 l per year per capita. Table 14.1 summarizes the main olive cultivars in Uruguay and typical sensory properties of the oils produced.

14.4 Brief geographic description of Argentina

Argentina is located in the south-eastern part of South America. It covers a total area of 2 780 400 km^2 and is the second largest country in Latin America. It shares borders with Bolivia and Paraguay towards the north, with Brazil in the north-east, with Uruguay and the Atlantic Ocean in the east, and Chile in the west.

Argentina's name comes from the Latin word *argentum*, which means "silver," and that was how the first European conquerors named it.

From the geographical point of view, three large zones can be distinguished: plains in the central part of the country, table lands in the south, and mountains in the west. Due to its diverse landscape and great length, Argentina displays a wide range of climates. Temperate climates are dominant but they have different characteristics according to the region: warm and humid summers with mild dry winters in the north, cool winters and rainy summers in the central part of the country, warm summers and cold winters with snow in the mountains, and intense polar-like winters in the south.

Rivers and underground aquifers abound throughout the country. Lakes and glaciers, such as the Perito Moreno glacier, are common in the Patagonia region. There is a significant development of hydro energy plants with numerous dams on the country's most important rivers.

14.4.1 Olive-growing regions

Olive growing in Argentina has been more developed in lower latitudes than in the traditional European growing regions. The dominant climate in most olive-growing areas in the country is hilly arid. A number of circumstances experienced by the country in the past few years, such as frosts and drying winds at flowering time, contributed to significant fluctuations in olive oil production levels.

In La Rioja Province, the main growing regions are Aimogasta, Chilecito, and Capital; in Catamarca Province Pomán, Central Valley, and Tinogasta; in San Juan, Valle del Tullum, Ullum, Zonda, Pocito, Jáchal, and Sarmiento (Cañada Honda and San Carlos). In Córdoba Province, Tras la Sierra is the only relevant region, as is Bahía Blanca in Buenos Aires.

Catamarca is a province with a significant olive-growing area located in the north-west of Argentina. Catamarca's Central Valley is framed by the Ancasti mountains towards the east and the Ambato mountains towards the west. Catamarca is part of the Argentine semi-arid region. There are two main types of climate: a warm climate in the north-eastern part of the province with good levels of summer rainfall and high temperatures, and also mild, humid winters, and the hilly, arid climate covering most of the region with dominant dry winds and low rainfall levels. The region counts on a number of rivers and creeks that are fed by the snow melting from the mountains.

In La Rioja Province, the towns of Aimogasta and Pomán are found in the Pipanaco Valley. These are very productive areas in the province in addition to La Rioja Capital, on the foothills of Sierra de Velasco, and the Chilecito Valley, between these mountains and Sierra de Famatina to the west.

Not all of these valleys are located at the same altitude. Whereas the Central Valley of Catamarca and La Rioja Capital Valley are between 400 and 450 m above sea level, the Pomán and Aimogasta Valleys are at 800 m, the Chilecito Valley at 950 m, and the region of Tinogasta in Catamarca at 1100 m. These variations lead to significant climatic differences despite being located at similar latitudes, which in turn determine different characteristics in their olive oils. In these two regions, their oils share similar sensory characteristics.

Coratina is one of the dominant varieties with sensory attributes such as herbal fruitiness with green grassy chlorophyll notes. The oils display important levels of bitterness and pungency. Changlot olive oils in the region offer medium green fruitiness with notes of olive fruit, intense bitterness and pungency, and low sweetness.

The Province of Mendoza is located in the western side of the country against the Andes Mountains. It has a dry and arid climate with a large temperature amplitude and little rainfall. Summer is warm and humid with average temperatures of 25 °C. Winter is cold and dry with average temperatures below 8 °C and occasional frosty nights.

The main rivers in the province start from the snow melting from the Andes and feed an irrigation system similar to the San Juan channel irrigation system. Mendoza did not have the benefit of the Tax Deferral Law and this is the main reason why most of its groves are of the traditional type. The Arauco variety is dominant in the region, with oils displaying high levels of bitterness and pungency and also deep green notes of grass and herbs.

San Juan Province is located in the western part of the country. It borders La Rioja to the north, La Rioja and San Luis to the east, and Mendoza to the south, while Chile shares the western border. It has a total area of 89 651 km^2, of which 80% is covered by steep hills and high mountains. The climatic conditions determined that irrigation brought life to the Province's landscape. Through a system of channels, dams, and furrows, the region developed an irrigation and drainage that covers ~2000 km.

Annual rainfall in the region is always below 200 mm and it is dominated by a semi-arid climate. The typical warm and dry wind of the region is called Zonda. It is generated by a high-pressure system in the Pacific Ocean that discharges its moisture on the Chilean side of the Andes and reaches Argentina as a fast-moving, dry, and warm wind that can reach up to 100 km/h. The maximum temperature is 43.8 °C, minimum –8.0 °C, and average 17.2 °C. With access to irrigation, land is suitable for a wide range of crops. Nonetheless, the region specialized in grapes and, most recently, olives.

The main variety among traditional producers is Arauco, which is mostly utilized for table olive production even though its oil is also highly regarded. With the Tax Deferral Law explained above, a new wave of modern olive groves was established in San Juan. These investments are characterized by large productive areas, high density of planting, and higher yields. San Juan is the third-largest producing region in Argentina. Nonetheless, the ongoing investments are projecting this province as the potential productive leader in the country. The most important areas from the productive point of view are 25 de Mayo, Sarmiento, Pocito, Rivadavia, and Rawson, which comprise 84% of the total olive-growing area in the region (INTA, 2007, based on date provided by the Departamento de Hidráulica de San Juan, Ministerio de la Producción).

Approximately 20% of the olive production in the region is used for table olive production while the remaining 80% goes for olive oil production. San Juan olive groves are characterized by the highest yields per hectare in the country, almost doubling the production in some cases. There is no doubt that the soil and climatic conditions of San Juan are responsible for these differences and also for the particular aroma and flavor of its oils.

Table 14.2 summarizes the main olive cultivars in Argentina and typical sensory properties of the oils produced in the five main regions.

14.5 Characterization of San Juan's olive oils

In a research project that was carried out on olive oils from San Juan, different varietal characteristics were studied over several seasons covering almost the entire past decade. Some 250 samples of olive oils from the varieties Arbequina, Arauco, Frantoio, Coratina, Koroneiki, Hojiblanca, Barnea, Changlot, and Picual were evaluated (Turcato *et al.*, 2010).

The method used was COI/T.20/Doc. No. 22AP1 and COI/T.20/Doc. No. 15/Rev. 1. The sensory assessment was carried out by the Universidad Católica de Cuyo panel, which has been accredited by the IOC since 2006.

Statistical analyses were carried out to determine the coefficient of variation for the different descriptors that characterized the olive oil varietals over the time of the project and the median values of their intensities. Olive oils with intense fruitiness of 7 on the scale and other oils with firm pungency of 6 were detected.

Through the analysis of the main components, the correlations between the different descriptors and the chemical parameters were determined (Mattar and Turcato, 2006). It was possible to determine the difference between the different studied varieties ($P \leq 0.05$) with significant variations between the intensities of the different attributes.

Table 14.2 Main olive cultivars in Argentina and typical sensory properties of the oil produced in the various regions.

Region	Main cultivars	Typical sensory properties
Catamarca	Arbequina Picual Coratina Koroneiki Frantoio Manzanilla Barnea Noccelara del Bellice	*Extra virgin olive oil of cv. Coratina* Aroma: fig leaves, fresh herbs with notes of tomato and artichoke. Flavor: notes of fresh herbs and greens. Bitter and pungent, high intensity and persistent *Extra virgin olive oil of cv. Frantoio* Fruity. Medium bitter and pungent. Aroma: fruity notes of banana, ripe tomato, and apple. Taste: sweet, complex with green vegetables and banana. More pungent than bitter. Very harmonious *Extra virgin olive oil of cv. Arbequina* Aroma: fruity. Taste: fresh herbs, tomato, apple, and banana with appreciable sweetness and pungency. Harmonious, complex, and persistent. The oils from cv. Arbequina are highlighted by their sweetness *Extra virgin olive oil of cv. Arauco* Green fruity sour/spicy. Aroma descriptions: tomato, fig, and green banana notes. Flavor perceptions: herbs, greens, and tomato notes. Medium-high intensity with persistence of bitter and spicy. Medium complexity and harmony *Extra virgin olive oil cv. Changlot* Features of fruitiness, herbal perception, green leaves, artichoke, and green tomatoes with intense notes of bitter and spicy *Extra virgin olive oil of Arbequina* Fruity and ripe sweet. Aroma: herbs, banana, tomato, and leaf notes. Flavor: nuts, especially almonds, and notes of tomato. Medium persistency of bitter and spicy. This oil is sweet, complex, and very harmonious. There are excellent extra virgin olive oil blends: *Bivarietals are common: Arbequina–Coratina and Arbequina–Picual* Virgin olive oils, medium fruity. Vegetable notes of artichoke and aromatic herbs, high intensity of fruit, especially apple. Also fresh walnut and almond. Taste is complex and strong. *Extra virgin olive oils bivarietal: Arbequina–Arauco* A typical Argentinean bivarietal composed of Arauco, the native varietal. Medium green fruity, bitter, and pungent. Aroma descriptors: herbs, tomato, and olive leaf notes. On the palate: tomato, soft and sweet herb notes. More bitter than pungent. The differences found between the production areas were significant for Arbequina cultivars but not for Coratina, which showed less variability
Cordoba	Arbequina Arauco Manzanilla	
La Rioja	Arbequina Arauco Picual Coratina Koroneiki Frantoio Manzanilla	
Mendoza	Arbequina Arauco Picual Coratina Koroneiki Frantoio Manzanilla	
San Juan	Arbequina Arauco Changlot Picual Coratina Manzanilla	
Others: Rio Negro Bahia Blanca Coronel Dorrego	Arbequina Frantoio Manzanilla Empeltre Farga	

Table 14.3 Median, standard deviation (SD), and coefficient of variation (CV) of the varieties grown in San Juan Province.

	Arbequina		Picual		Frantoio		Arauco	
Attribute	Median ± SD	CV (%)	Median ± SD	CV (%)	Median ± SD	CV (%)	Median ± SD	CV (%)
Fruitiness	5.4 ± 0.3	11.20	5.4 ± 0.5	13.43	4.6 ± 0.4	20.87	6 ± 0.3	14.55
Apple	2.4 ± 0.2	15.03	1 ± 0.3	10.39	1 ± 0.4	16.98	1.2 ± 0.4	12.76
OMF[a]	1.6 ± 0.4	22.09	1.4 ± 0.4	17.98	0.8 ± 0.3	15.08	1 ± 0.5	19.87
Green	1.4 ± 0.5	10.20	1.8 ± 0.7	10.98	0.6 ± 0.4	13.06	4.9 ± 0.2	10.87
Bitterness	2 ± 0.2	20.5	4.8 ± 0.3	7.98	2.8 ± 0.7	11.65	5.9 ± 0.2	14.67
Pungency	3.8 ± 0.4	19.09	4.6 ± 0.2	10.09	4.8 ± 0.5	7.54	6 ± 0.2	12.04
Sweetness	2.4 ± 0.2	7.89	1 ± 0.4	13.00	1.2 ± 0.2	20.84	0.6 ± 0.2	19.98

[a]OMF: other mature fruits.

Source: Reproduced from Turcato, A.R., Paez Rivera, H., Briones, V., and Mattar, S. (2010) with kind permission from Grasas Y Aceites.

Table 14.3, Table 14.4, Table 14.5, and Table 14.6 indicate the most frequent descriptors listed by the tasters in their sensory profile sheets. In some varietal oils, it was possible to observe consistent attributes whereas in other varietals a wide range of attributes dominated.

Pungency levels in the Arauco variety are clearly higher than those of the rest of the varieties. Intensities of around 6 ± 0.2 were found, leading to unbalanced oils that despite being of extra virgin quality left a persistent spicy feeling in the throat. It is important to highlight that this variety is mostly utilized for table olives and, consequently, is normally harvested with a very low maturity index. The typical high levels of pungency in this variety determine that its oils are excellent for use in blends with oils of milder characteristics rather than being utilized as a single varietal by the Argentinean consumer.

Table 14.4 Median, standard deviation (SD), and coefficient of variation (CV) of the varieties grown in San Juan Province.

	Coratina		Koroneiki		Hojiblanca		Barnea		Changlot	
Attribute	Median ± SD	CV (%)	Median ± SD	CV (%)	Median ± SD	CV (%)	Median ± SD	CV (%)	Median ± SD	CV (%)
Fruitiness	5 ± 0.3	11.20	6.5 ± 0.5	13.43	5 ± 0.3	14.55	5.7 ± 0.4	20.87	7 ± 0.3	14.55
Apple	3 ± 0.2	15.03	2 ± 0.3	10.39	4 ± 0.4	20.76	4 ± 0.4	16.98	4 ± 0.6	22.76
OMF[a]	2 ± 0.4	21.09	3 ± 0.4	17.98	4 ± 0.5	19.87	5 ± 03	15.08	3 ± 0.5	19.87
Green	4 ± 0.5	10.20	5 ± 0.7	10.98	3 ± 0.2	10.87	3 ± 0.4	13.06	6 ± 0.8	10.87
Bitterness	4.9 ± 0.2	20.5	5.5 ± 0.3	7.98	2 ± 0.2	14.67	2 ± 0.7	11.65	6 ± 0.2	11.67
Pungency	4.6 ± 0.4	16.09	6 ± 0.2	10.09	2 ± 0.2	12.04	3 ± 0.5	7.54	5 ± 0.9	15.04
Sweetness	1.7 ± 0.2	7.89	1.5 ± 0.4	23.00	4 ± 0.2	19.98	5 ± 0.2	18.84	2 ± 0.2	7.98

[a]OMF: other mature fruits.

Source: Reproduced from Turcato, A.R., Paez Rivera, H., Briones, V., and Mattar, S. (2010) with kind permission from Grasas Y Aceites.

Table 14.5 Median and standard deviation (SD) for the different sensory descriptors found by the panel for the different varieties.

	Median ± SD			
Attributes	Arbequina	Frantoio	Arauco	Picual
Green grass	3 ± 0.8	4 ± 0.7	5 ± 0.4	5 ± 0.6
Green apple	5 ± 0.5	0.00	0.00	0.00
Citrus	0.00	2 ± 0.8	0.00	4 ± 0.9
Grapefruit	0.00	0.00	0.00	5 ± 0.8
Artichoke	0.00	1 ± 0.3	0.00	2 ± 0.7
Herbs	2 ± 0.4	3 ± 0.2	4 ± 0.2	3 ± 0.9
Spices	0.00	0.00	0.00	0.00
Olive leaf	3 ± 0.9	2 ± 0.8	5 ± 0.9	4 ± 0.6
Lettuce	1 ± 0.5	0.00	0.00	0.00
Ripe banana	0.00	0.00	0.00	0.00
Almond/walnut	2 ± 0.6	0.00	0.00	0.00
Tomato/capsicum	1 ± 0.6	3 ± 0.1	3 ± 0.5	2 ± 0.8
Hydrocarbon	0.00	0.00	0.00	0.00
Fig leaf	0.00	0.00	0.00	2 ± 0.1

Source: Reproduced from Turcato, A.R., Paez Rivera, H., Briones, V., and Mattar, S. (2010) with kind permission from Grasas Y Aceites.

Table 14.6 Median and standard deviation (SD) for the different sensory descriptors found by the panel for the different varieties.

	Median ± SD				
Attribute	Coratina	Koroneiki	Hojiblanca	Barnea	Changlot
Green grass	3 ± 0.8	5 ± 0.7	2 ± 0.3	0.00	5 ± 0.3
Green apple	0.00	0.00	3 ± 0.4	3 ± 0.4	0.00
Citrus	2 ± 1	3 ± 0.8	0.00	0.00	0.00
Grapefruit	1 ± 0.5	0.00	0.00	0.00	0.00
Artichoke	0.00	2 ± 0.3	2 ± 0.2	0.00	0.00
Herbs	5 ± 0.4	4 ± 0.5	2 ± 0.2	1 ± 0.9	3 ± 0.8
Spices	3 ± 0.1	2 ± 0.1	3 ± 0.5	0.00	0.00
Olive leaf	5 ± 0.9	4 ± 0.8	2 ± 0.9	0.00	4 ± 0.7
Lettuce	1 ± 0.9	0.00	0.00	0.00	0.00
Ripe banana	0.00	0.00	0.00	5 ± 0.5	0.00
Almond/walnut	0.00	0.00	5 ± 1.0	3 ± 0.8	0.00
Tomato/capsicum	0.00	0.00	3 ± 0.5	0.00	3 ± 0.8
Hydrocarbon	0.00	0.00	0.00	0.00	2 ± 0.9
Fig leaf	0.00	0.00	0.00	1 ± 0.1	0.00

Source: Reproduced from Turcato, A.R., Paez Rivera, H., Briones, V., and Mattar, S. (2010) with kind permission from Grasas Y Aceites.

Frantoio and Picual oils have shown also relatively high values, with pungency levels of 4.8 ± 0.5 and 4.6 ± 0.2, respectively, but without becoming overly aggressive.

On the other hand, oils from the Arbequina variety have an average pungency level of 3.8 ± 0.4, which is typical of this variety that contributes to the harmony of these oils.

Among sweet oils, Arbequina is usually the most dominant variety (in the analysis of the first four as Hojiblanca and Barnea have important sweet notes) with an average value of 2.4 ± 0.2. The varieties Picual, Frantoio, and Arauco show average intensities of sweet notes which are clearly lower with average values of 1 ± 0.4, 1.2 ± 0.2, and 0.6 ± 0.2, respectively.

Bitterness in the Arauco variety oils reaches medium to high levels of 5.9 ± 0.2. With Picual, bitterness is as important as pungency but it does not reach levels that require their reporting. Arbequina and Frantoio oils show average bitterness levels of 2 ± 0.2 and 2.8 ± 0.7, respectively, which are pleasant intensities for the palate of the typical Argentinean consumer.

All varietal oils have relevant levels of fruitiness, which is of great importance for extra virgin olive oils. Arbequina, and also Picual and Frantoio, shows a medium-high fruitiness with an average score of 5.4 ± 0.3. Oils from the Arauco variety have even higher scores with an average of 7.0 ± 0.3, confirming their character as highly fruity oils.

14.6 Sensory profiles

Arbequina olive oils are characterized by medium-sweet fruitiness (Turcato *et al.*, 2010). Compared with other oils, their level of bitterness is low and pungency is low to moderate. It is possible to highlight apple and other ripe fruit notes such as walnut in addition to notes of olive leaf, lettuce, and tomato. These oils have good harmony without a dominant attribute (Figure 14.2). According to unofficial data from local consultants, Arbequina is the dominant variety in the region with ~50% of the total planted area.

A similar case is found with Frantoio variety. This cultivar showcases oils with medium fruitiness but its levels of sweetness are lower and bitterness and pungency are slightly higher. The most important descriptors are fresh cut grass, citrus, artichoke, herbs, and capsicum (Figure 14.3). Frantoio has shown some problems from the productive point of view with significant levels of alternate bearing that, in some cases, led to its replacement or top grafting with more productive varieties. Its cultivation covers ~15% of the olive growing area in the region.

Arauco oils display high levels of pungency that can be excessive at times, and also medium to high bitterness levels with low intensity of sweet characters. Fruitiness is usually high in this varietal oil. The most dominant notes are herbs, olive leaves, fresh-cut grass, and, occasionally, capsicum and tomato (Figure 14.4). These are oils that are not easily accepted by the average

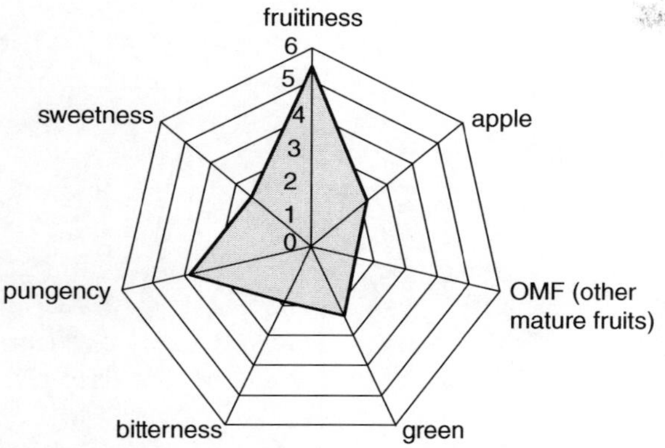

Figure 14.2 Sensory profile of variety Arbequina. *Source:* Reproduced from Turcato, A.R., Paez Rivera, H., Briones, V., and Mattar, S. (2010) with kind permission from Grasas Y Aceites.

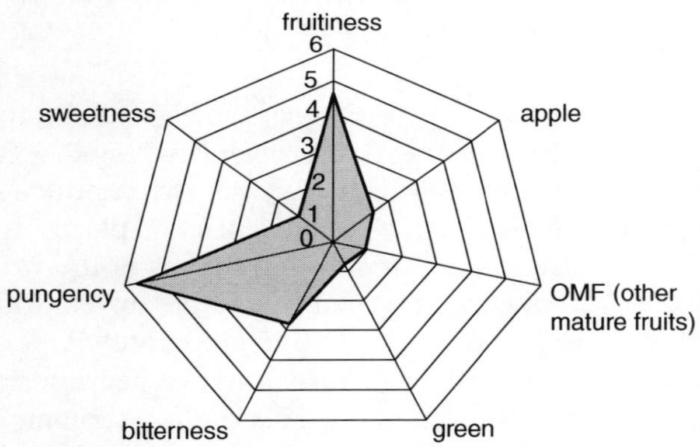

Figure 14.3 Sensory profile of variety Frantoio. *Source:* Reproduced from Turcato, A.R., Paez Rivera, H., Briones, V., and Mattar, S. (2010) with kind permission from Grasas Y Aceites.

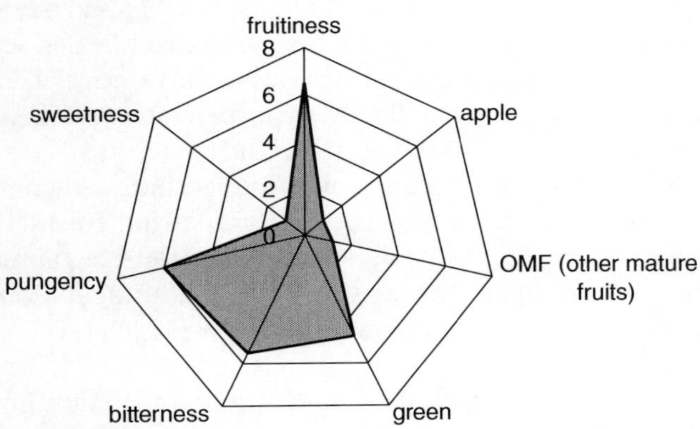

Figure 14.4 Sensory profile of variety Arauco. *Source:* Reproduced from Turcato, A.R., Paez Rivera, H., Briones, V., and Mattar, S. (2010) with kind permission from Grasas Y Aceites.

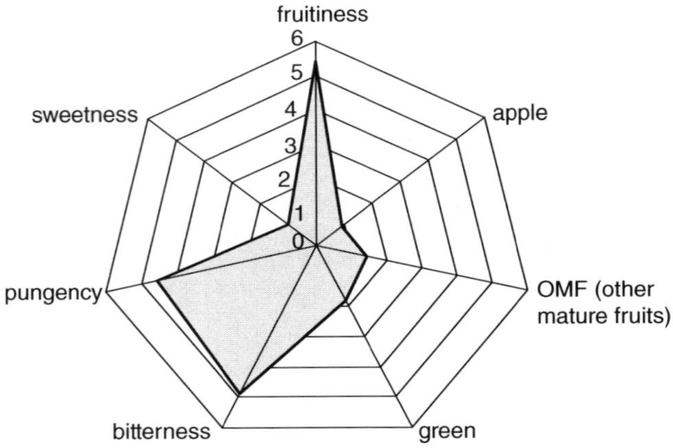

Figure 14.5 Sensory profile of variety Picual. *Source:* Reproduced from Turcato, A.R., Paez Rivera, H., Briones, V., and Mattar, S. (2010) with kind permission from Grasas Y Aceites.

Argentinean consumer but they are very useful to blend with other oils. Approximately 17% of the olive-growing area in the region is planted with Arauco. This variety is also known as "criolla" in some places. It is cultivated only in Argentina and it is not planted in other areas of the world. It is typically utilized for the production of green style table olives, with up to 80% of its crop used for this purpose. The tree is moderately vigorous depending on the region. It does not show an early onset of bearing and it is better adapted to temperate to warm climates. It is quite alternant, but this tendency is reduced when harvested for green table olive production. It is rather sensitive to low temperatures, particularly to winter and spring frosts.

Picual produces oils with high levels of pungency and bitterness, depending on the maturity index at the moment of harvest. This variety shows some interesting complex fruit notes such as citrus, grapefruit, artichoke, and grass (Figure 14.5). If the malaxing temperature was above normal, it is possible to detect a characteristic fig-leaf flavor. Approximately 11% of the olive-growing area in San Juan is planted with Picual.

Coratina oils show medium fruitiness with intense bitterness and pungency. They have positive attributes reminiscent of grapefruit, fresh-cut grass, dry grass, citrus, herbs, olive leaves, and, in some cases, lettuce, amongst others. These attributes are particularly persistent in the mouthfeel. Some Coratina oils may be classified as unbalanced owing to the very high levels of bitterness and pungency (Figure 14.6).

Koroneiki oils have a pleasant medium fruitiness with dominance of ripe fruits and red fruits. They have also green and herbal notes. They are generally balanced with medium bitterness levels and more intense and persistent pungent notes (Figure 14.7).

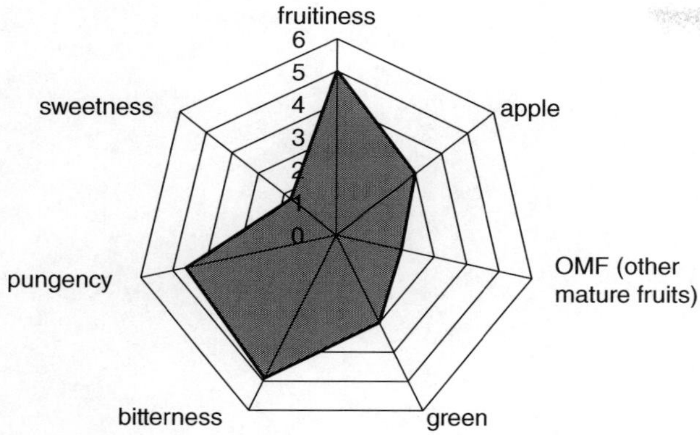

Figure 14.6 Sensory profile of variety Coratina. *Source:* Reproduced from Turcato, A.R., Paez Rivera, H., Briones, V., and Mattar, S. (2010) with kind permission from Grasas Y Aceites.

Hojiblanca is characterized by medium fruitiness with typical notes of apple, artichoke, peppercorn, tomato, and walnut. They are normally sweet oils with pleasant bitterness and pungency levels which are not aggressive in the mouth. Green almond is a common descriptor (Figure 14.8).

Barnea oils typically show a paler color and lighter mouthfeel than other varietals. Their fruity notes are characterized by banana aromas and flavors and fig-leaf and almonds are present to a lesser extent. They are generally mild with low levels of bitterness and pungency (Figure 14.9).

Changlot variety produces an oil characterized by dry grass notes. It also displays capsicum, green leaf, herbal, and olive flavors. Pungency and bitterness are clearly noticed (Figure 14.10).

For the above-mentioned varieties, such as Hojiblanca, Coratina, Koroneiki, Changlot, and Barnea, there are just a small number of hectares

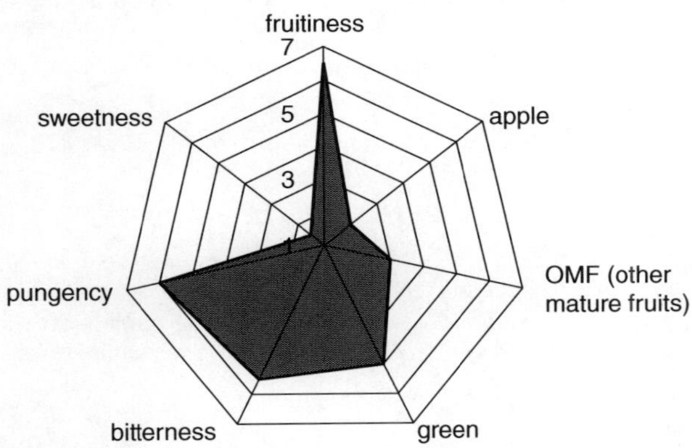

Figure 14.7 Sensory profile of variety Koroneiki. *Source:* Reproduced from Turcato, A.R., Paez Rivera, H., Briones, V., and Mattar, S. (2010) with kind permission from Grasas Y Aceites.

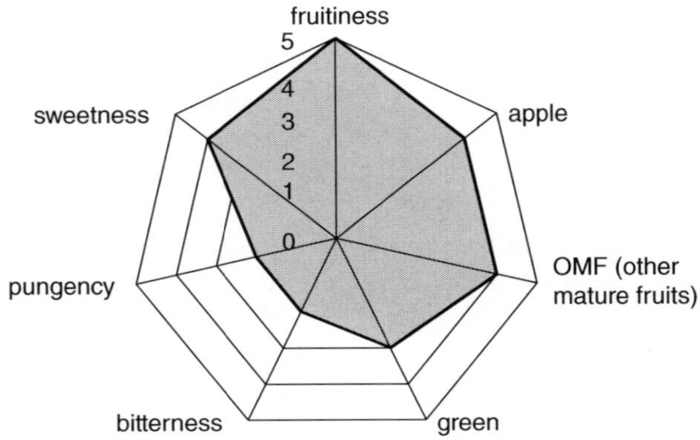

Figure 14.8 Sensory profile of variety Hojiblanca. *Source:* Reproduced from Turcato, A.R., Paez Rivera, H., Briones, V., and Mattar, S. (2010) with kind permission from Grasas Y Aceites.

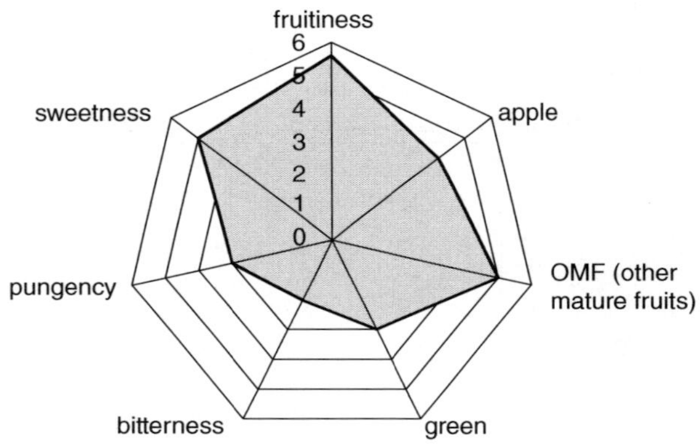

Figure 14.9 Sensory profile of variety Barnea. *Source:* Reproduced from Turcato, A.R., Paez Rivera, H., Briones, V., and Mattar, S. (2010) with kind permission from Grasas Y Aceites.

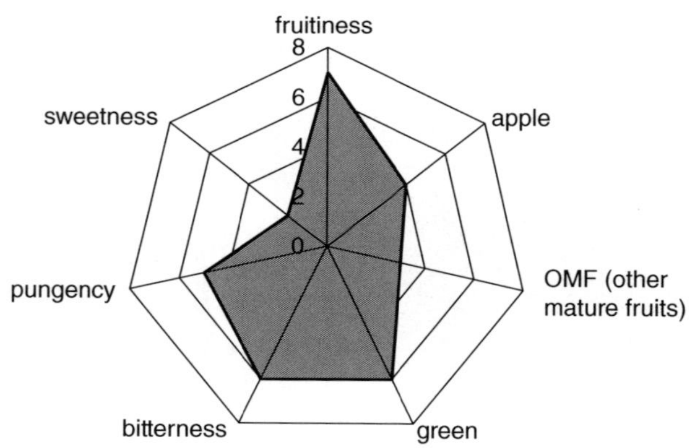

Figure 14.10 Sensory profile of variety Changlot. *Source:* Reproduced from Turcato, A.R., Paez Rivera, H., Briones, V., and Mattar, S. (2010) with kind permission from Grasas Y Aceites.

planted in the region with ~1% of the total area for Hojiblanca, Barnea, and Koroneiki and 2% for Coratina and Changlot.

14.7 Correlations between sensory and chemical parameters

All those characteristics detailed in the sensory profiles are correlated with chemical parameters. Through multivariate analysis we compared percentage of fatty acids, mono- and polyunsaturated fatty acid ratio, shelf life, polyphenol content, maturity index, and sensory attributes of bitterness, fruitiness, and pungency. A correlation matrix of those parameters was created (Mattar and Turcato, 2006).

Applying principal components analysis (PCA), a graphic representation of the correlations was obtained, namely the space of the variables defined by components 1 and 2, which are the most variable. In PCA, that space is a flat projection of an *n*-dimensional space that matches the number of descriptors.

This analysis of the data matrix produced the graphic representation in Figure 14.11, showing the direct or indirect correlations between the different parameters. Among others, it is possible to observe a negative correlation between maturity index and polyphenol content and also a direct correlation between the polyphenols and attributes of fruitiness, pungency, and bitterness. All these data differ according to the specific characteristics of each variety. Correlations between chemical and sensory variables support the sensory

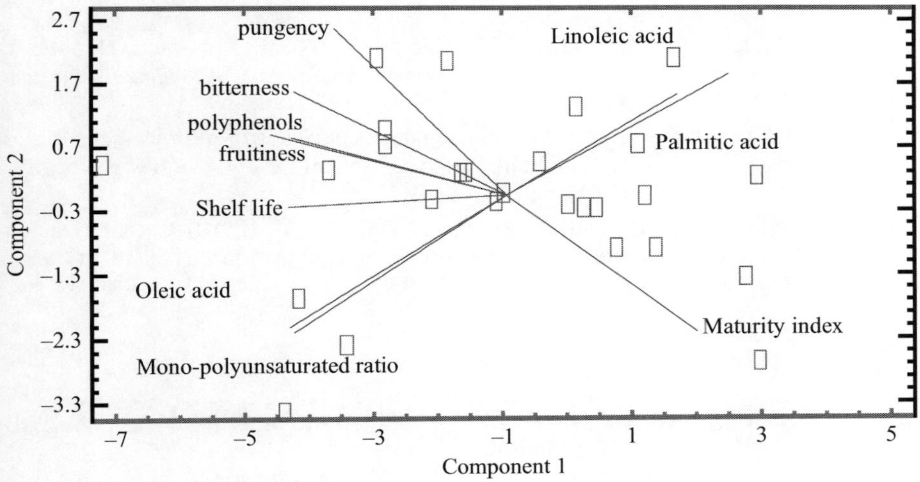

Figure 14.11 Principal component analysis of the data matrix for extra virgin olives oils from Argentina. *Source:* Reproduced from Turcato, A.R., and Mattar, S. (2006) with kind permission from Grasas Y Aceites.

evaluation carried out by the trained panel test and validated as an important tool for quality control of the oils.

14.8 Conclusion

The olive oil industry in South America is experiencing important transformations in terms of investments for the expansion of olive cultivation areas, technologies incorporated into the production and industrialization process, and the opening up of new markets for export. Sensory studies of olive oils can promote consumption by highlighting the specific oil attributes and relating these to each variety and production region.

Acknowledgments

Special thanks are due to the CRESA olive oil panel test, to the food and olive oil sensory laboratory of the Facultad de Ciencias de la Alimentación, Bioquímicas, y Farmacéuticas of the Universidad Católica de Cuyo, and to the companies that provided and continue to provide samples for analysis and study.

References

Estay, K., Servili, M. and Selvaggini, R. (2012) Characterization of extra-virgin olive oils from cvs. Arbequina and Frantoio produced in a latitudinal gradient in Chile, presented at the VII International Symposium on Olive Growing, ISHS, San Juan, Argentina.

Karman de Sutton, G., Bertoni, M., Cattaneo, P., Abitbol, J., and Denett, J. (1963) Aceites de oliva de producción nacional, *Anales de la Asociación Quimica Argentina*, **53**, 87–105. 1963.

Mattar, S., and Turcato, A. (2006) Correlación de los parámetros químico-sensoriales de los aceites de oliva de San Juan. Bs. As. Noviembre 2005. *Revista Grasas y Aceites*, No. 64, Tomo XIV, Vol. 3.

Turcato, A., Páez Rivera, H.; Briones, V., and Mattar, S. (2010) Caracterización Sensorial de los Aceites de Oliva de la provincia de San Juan. *Aceites y Grasas*, ASAGA No. 79, Tomo XX, Vol. 2, pp. 332–338.

Further reading

Alba, J., Izquierdo, J.R., and Gutiérrez, F. (1997) *Aceite de Oliva Virgen. Análisis Sensorial*, Agrícola Española, Madrid.

Boskou, D. (1998) *Química y Tecnología del Aceite de Oliva*, 1st edn, AMV Ediciones, Madrid.

Cáceres, R., Novello, R., and Robert, M. (2009) *Análisis de la Cadena del Olivo*, Estudio Socioeconómico de los Sistemas Agroalimentarios y Agroindustriales, No. 2, INTA, Publicación Nacional, ISSN 1852-4605.

Caporale, G., Policastro, S., and Monteleone, E. (2004) Bitterness enhancement induced by cut grass odorant (*cis*-3-hexen-1-ol) in a model olive oil. *Food Quality and Preference*, **15**, 219–227.

CEE (1991) *Reglamento Comunitario (CEE) No. 2568/91, Panel-Test*, European Commission, Brussels.

Civantos, L. (1999) *Obtención del Aceite de Oliva*, 2nd edn, Editorial Agrícola Española, Madrid.

Dobarganes, M.C. (1999) *Evaluación de la Calidad del Aceite de Oliva Virgen*, Instituto de la Grasa y sus Derivados, Seville.

Hartzallah, H., and Kiritsakis, A. (1999) Efecto antioxidante de los extractos fenólicos de la hoja y del fruto del olivo. *OLIVAE*, No. 27, pp. 47–49.

INDEC (2008) Censo Nacional Agropecuario 2002, www.indec.gov.ar (last accessed 17 June 2013).

Instituto de Desarrollo Rural (2008), www.idr.org.ar (last accessed 17 June 2013).

IOOC (1987) *Análisis Sensorial: Vocabulario General Básico*, COI/T.20/Doc. No. 4, International Olive Oil Council, Madrid.

IOOC (1996) *Guía para la Selección, el Entrenamiento y el Control de los Catadores Cualificados de Aceite de Oliva Virgen*, COI/T.20/Doc. No. 14/Rev.1, International Olive Oil Council, Madrid.

IOOC (2005) *Método para la Valoración Organoléptica del Aceite de Oliva Virgen Extra que Opta a una Denominación de Origen*, COI/T.20/Doc. No. 22, Resolución Res-2/93-IV/05, International Olive Oil Council, Madrid.

IRTA (1998) *Cata de Aceites en Cataluña. Dossier Oleo*, Institut de Recerca i Tecnología Agroalimentarias, Barcelona, p. 41.

Mattar, S., Carelli, A., and Ceci, L. (2012a). Relación entre análisis químicos y sensoriales en aceites de oliva de la Provincia de San Juan. *Revista la Alimentación Latinoamericana*, Agosto.

Mattar, S. and Turcato, A (2006) Correlación de los parámetros químico-sensoriales de los aceites de oliva de San Juan. *Grasas y Aceites*, No. 64, Tomo XIV.

Mattar, S., Carelli, A., Ceci, L., *et al.* (2012b) Chemical and sensory characterization of monovarietal virgin olive oils from San Juan Province, Argentina. Presented at the VIIth International Symposium on Olive Growing, ISHS, Argentina.

Turcato, A. (2010) Caracterización sensorial de los aceites de oliva de la provincia de San Juan, Argentina. Tesis de Magister, Universidad de La Serena, Chile.

Turcato, A., and Mattar, S. (2006) Perfiles Sensoriales de aceites de oliva extra virgen de la provincia de San Juan. Campaña 2001, 2002, 2003 y 2004. *Aceites y Grasas*, ASAGA No. 65, Tomo XIV, Vol. 4, pp. 664–669.

Turcato, A. R., Páez Rivera, H., Briones, V., and Mattar, S. (2010) Caracterización sensorial de los aceites de oliva de la provincia de San Juan, Argentina. *Aceites y Grasas*, ASAGA No. 79, Tomo XX, Vol. 2, pp. 332–338.

Index

Olive Oil Sensory Science, First Edition. Edited by Erminio Monteleone and Susan Langstaff.

Olive Oil Sensory Science